Limnology and Remote Sensing

A Contemporary Approach

Springer
London
Berlin
Heidelberg
New York
Barcelona
Hong Kong
Milan
Paris
Santa Clara
Singapore
Tokyo

K.Ya. Kondratyev and N.N. Filatov

Limnology and Remote Sensing

A Contemporary Approach

Springer

Published in association with
Praxis Publishing
Chichester, UK

Editors:
Professor K.Ya. Kondratyev
Counsellor of the Russian Academy of Sciences
Research Centre for Ecological Safety
St. Petersburg, Russia

Professor Nikolai Filatov
Director, Northern Water Problems Institute
Petrozavodsk
Republic of Karelia
Russia

SPRINGER–PRAXIS SERIES IN REMOTE SENSING
SERIES EDITOR: Dr. Philippe Blondel, Senior Scientist, Southampton Oceanography Centre, UK
CONSULTANT EDITOR: Kirill Ya. Kondratyev

ISBN 1-85233-112-7 Springer-Verlag Berlin Heidelberg New York

British Library Cataloguing in Publication Data
   Limnology and remote sensing : a contemporary approach. –
   (Springer-Praxis series in remote sensing)
   1. Limnology – Remote sensing
   I. Kondratyev, K. IA. (Kirill IAkovlevich)   II. Filatov, Nikolai
   551.4$'$8

Library of Congress Cataloging-in-Publication Data
   Limnology and remote sensing : a contemporary approach / [edited by]
   K.Ya. Kondratyev and N.N. Filatov.
   p. cm. – (Springer-Praxis series in remote sensing)
   Includes bibliographical references.
   ISBN 1-85233-112-7 (alk. paper)
   1. Limnology – Remote sensing.   I. Kondrat'ev, K. IA. (Kirill IAkovlevich)
   II. Filatov, N.N.   III. Series.
   GB1601.72.R42L56   1999   98-52345 CIP
   551.48 – dc21

© Praxis Publishing Ltd, Chichester, UK, 1999
Printed by MPG Books Ltd, Bodmin, Cornwall, UK

Cover design: Jim Wilkie
Typesetting: Originator, Gt. Yarmouth, Norfolk, UK

Printed on acid-free paper supplied by Precision Publishing Papers Ltd, UK

Preface

The problem of water resources on the globe has become a severe limitation for the future development of civilization. In many countries (especially in a number of developing ones with rapidly growing populations) the resources of drinking water are close to being completely exhausted. Therefore it is urgent to study and assess the various components of global water resources. Freshwater lakes are one of the most important components. That is why limnological study is a subject of growing significance. Regrettably, this has not been adequately recognized so far, in spite of numerous efforts devoted to further development of environmental sciences and various outstanding events, such as the Second UN Conference on Environment and Development (UNCED, 1992) and a subsequent UN General Assembly Special Session (UNGASS, 1997).

The aim of this book is to discuss contemporary limnological problems on a local, regional and global scale with special emphasis on the application of remote-sensing techniques to monitor lake dynamics, thermodynamics, biodynamics, and water quality. An interactive approach has been used to consider various processes from viewpoints of both numerical modelling and observations. In the latter case, a combined use of *in situ* and remote-sensing data has been recommended, making use of (for illustrative purposes) information relevant to the lakes of north-western Russia and the American Great Lakes (a specific comparative analysis has been carried out in this context). The role of geographic information systems (GIS) has been discussed and emphasized.

The book is intended for broad circles of readers, including students of different levels who are involved in studying environmental subjects.

The Editors would like to express their gratitude to Mr. M.A. Poshekhonov for his contribution to the technical preparation of the manuscript of the book.

Contents

Contributing authors: K.Ya. Kondratyev
N.N. Filatov
O.M. Johannessen
V.V. Melentyev
D.V. Pozdnyakov
S.V. Ryanzhin
E.V. Shalina
A.I. Tikhomirov

Edited by: K.Ya. Kondratyev
N.N. Filatov

1

Contemporary limnological problems: local, regional and global aspects

1.1 DIRECT ANTHROPOGENIC IMPACTS LEADING TO LAKE EUTROPHICATION AND TOXIC CONTAMINATION

Since the early 1920s fresh water lakes and reservoirs have decayed and disappeared all over the world at an ever increasing rate (*Guidelines* ..., 1989). Human activities were found to be at the origin of this devastating process (Petrova, 1990). Immense amounts of chemical compounds of an extremely wide nomenclature continuously get into inland waters with waste water, runoff, river discharge, and atmospheric precipitation/fall-out. Being almost invariably accompanied by anthropogenically driven enhancement of soil erosion and water thermal contamination, the admission of chemicals results in two major water-environment plagues: eutrophication and toxic contamination of inland waters (Henderson-Sellers, 1984; Henderson-Sellers, Markland, 1987; *Ecological Problems* ..., 1997).

Eutrophication is the term which expresses "ageing" of a lake. Basically, lake ageing occurs in a natural fashion: very much like every other landscape constituent, lakes are subject to several natural transformations lasting over geological time spans and ultimately resulting in a new environment which in its proper turn is liable to further metamorphoses. Several stages in natural "ageing" of lakes can be identified (Kondratyev *et al.*, 1987).

At the *first* stage, a lake basin gets filled with surface waters whose chemical composition is inherent in the chemism of drainage area soils. A young lake is generally *oligotrophic*, i.e. deficient in plant nutrients (the chemical compounds that could be used by lake biota for subsistence and reproduction), and usually having abundant dissolved oxygen. The initial content of nutrients in the newly formed lake determines the level of primary production photosynthesized by phytoplankton. The growth of phytoplankton as an assemblage of aquatic autotrophic vegetable organisms relies on the availability of at least 19 elements, although most of these are only required in trace amounts. Together with three principal elements (carbon, hydrogen, oxygen), a planktonic cell, for its growth and division, also requires in relatively large amounts such elements as phosphorus, nitrogen, silicon, sodium, calcium, sulfur, magnesium. The other elements (copper, iron, zinc, chlorine, boron, molybdenum, cobalt, vanadium) are required in lesser quantities. The nutrients residing in natural waters are found in two principal forms: inorganic biogenic substances existing separately from carbonaceous compounds (the so-called mineralized form), and inorganic biogenic compounds integrated into organic complexes.

Insufficiency of any of the above nutrients leads to cell growth being limited by that element. In most inland water bodies, the strictly limiting element proves to be either phosphorus or, less frequently, nitrogen, although carbon and light-limited systems do occur (Petrova, 1990). In natural conditions, the input of phosphorus to a lake is exclusively associated with tributary waters (rivers, underground flows, runoff, precipitation) while nitrogen and carbon deficits in the lake water are usually reduced or even disappear through both gas exchange at the atmosphere–water interface, and nitrogen fixation by aquatic organisms.

The concentration of phosphorus in the lake, first having reached a balance with that coming down from the drainage area, remains invariable to a certain extent, being controlled by two major parameters—the phosphorus input (loading) rate divided by the lake surface and the water residence time. The primary production of the young lake, while being subject to some cyclic oscillations driven by climatic and/or weather phases, remains, on the whole, at the same level.

Bacterial decomposition of organic matter is dictated by the same factors as phytoplankton primary production, i.e. by water temperature and availability of both light and nutrients (Kapoustina, 1996). Thus, in the initially well-balanced ecological system of the lake, the rates of production and decomposition nearly equal each other varying with available phosphorus concentration. These well-balanced competing processes ultimately determine relatively stable rates of mineralized organic matter accumulation and phosphorus storage in lake sediments.

With a slow accumulation of bottom sediments, the lake basin gradually becomes more shallow which favors a considerably stronger heating of the lake bulk water. Further, persistently warmer waters bring forth a pronounced intensification of production/ decomposition rates with an eventual increase in the sedimentation rate.

The above mentioned shallowing of the lake basin (which heralds the onset of the *second* stage of lake development or natural ageing) results in promotion of phosphorus release from bottom sediments back into bulk water due to enhancement of thermal convection and wind-induced turbulent mixing affecting progressively deeper and deeper layers. It further promotes production/decomposition rates and consequently pushes the trophic level of the lake upwards. The lake develops *mesotrophy*. Insofar as lake natural shallowing is virtually irreversible, lake ageing goes on unhindered with the result that increased decomposition rates entail biochemical oxygen demand (BOD) growth. This leads to a reduction in the dissolved oxygen content of bulk water. At a given moment, the level of primary production becomes so high that a considerable proportion of dissolved oxygen gets depleted, i.e. *anoxic* conditions set up in the hypolimnion whereas the lake upper layers easily replenish the loss of dissolved oxygen because of intense gas exchange continuously occurring at the air–water interface. The onset of anoxic conditions in lower hypolimnetic layers due to oxidation of copious dead matter by *aerobic* bacteria is conducive to development of conditions under which only *anaerobic* bacteria can exist. Although these bacteria are still capable of oxidizing the detritus, they release substantial amounts of such reduced gases as methane and hydrogen sulfide. The water quality becomes very low.

The depletion of hypolimnetic oxygen levels leads to a sharp intensification of phosphorus release from bottom sediments and, ultimately, to a further increase in the phytoplankton primary production, this time regardless of the actual phosphorus input provided by catchment. From this moment on, the lake is in transition to the *third*, final phase of its development. "Uncorking" a tremendous pool of phosphorus accumulated in bottom sediments results in the lake ecosystem ageing and deterioration gets an additional and this time irreversible acceleration. The lake develops *eutrophy*. Since changes in thermal and hydrodynamic regimes conducive to accelerating organic matter production only occur when a considerable proportion of the lake basin has become filled with sediments (and the sediment accumulation rate at the oligotrophic stage of the lake development is basically very low), the natural eutrophication of lakes is an extremely slow process going on for thousands of years. Accumulation of sediments and shallowing of the lake brings about a rapid growth of microphytes first in the littoral zone and then encroachment over the entire lake area. The lake gradually transforms into a swamp.

Anthropogenic eutrophication, being driven by multiply increased external nutrition loading, differs dramatically from the natural ageing of lakes, although the thermal regime in the lake remains an important player. It warrants a brief outline of water temperature patterns basically occurring in lakes throughout the year.

In a deep, cold temperate lake the vertical profile of temperature changes diurnally and seasonally. In winter time, the lake is thermally nearly homogeneous along the vertical. In spring, incoming solar radiation begins to increase. The excess energy is accumulated in the upper part of the lake and is mixed downwards due to wind-induced turbulent fluxes. However, the mixing only extends over a layer of finite thickness defined by the effectiveness of this mixing and, first of all, by the intensity of wind shear at the air–water interface. Being accompanied by a continuous rise in epilimnetic water temperature, a gradual deepening of this layer (the *epilimnion*) occurs during the summer period whilst the water at the bottom of the lake exhibits but marginal temperature increase. This is the hypolimnion where the vertical temperature gradient (dT/dz) is nearly zero. There is usually an intermediate layer (*metalimnion*) within the limits of which the vertical gradient dT/dz is very high. Although, strictly speaking, it is not correct, metalimnion is frequently associated with the *thermocline region*, which is in fact the horizon at which $\partial^2 T/\partial z^2 = 0$. In a stratified water body, the thermocline establishes almost a perfect barrier prohibiting any vertical fluxes of energy and matter from epilimnion to hypolimnion and *vice versa*.

Persisting during the summer heating period, this thermal stratification is then followed by a *convective overturn*, with the onset of autumn. The net energy balance at the air–water interface turns negative, the lake begins cooling, the upper water layers become negatively buoyant and sink. As a result, the lake becomes vertically isothermal, and a further lowering in bulk water temperature proceeds at an identical rate at all depths throughout the whole body of the lake until the temperature of maximum water density (*ca.* 277 K) is attained. A further cooling can make surface layers less dense, and a "reverse" stratification may be set up but the thickness of this region is considerably thinner than the epilimnion in summer time. Persistent sufficiently low temperatures may cause the lake to freeze over, whilst the temperature of bulk water remains usually in the range *ca.* 275–277 K.

Once the input of nutrients to the lake increases drastically, the concentration of phosphorus starts growing very rapidly with the result of forcefully stimulating algae multiplication—an excessive and rapid planktonic growth (Petrova *et al.*, 1987). Stemming from the anthropogenically driven acceleration of phytoplanktonic growth, a considerable increase of algal biomass in the water column results in deterioration of bulk water transparency. Such alteration to the water-column light regime restricts the vertical extent of comfortable habitat for phytoplankton to the uppermost layers. In pursuit of more favorable light conditions, the phytoplankton move upwards to get closer to the water–air interface. Having an avalanche-like nature, this process ultimately gives rise to development of phytoplankton-saturated "scum which is commonly associated with water "bloom" events. Water bloom is a relatively short-term event. Due to depletion of nutrients and dissolved oxygen, enhanced bacterial decomposition and zooplankton grazing, and associated release of toxic matter in the upper layers, the living conditions become unfavorable for algae. Phytoplankton death rates begin prevailing over phytoplankton growth rates, and the algal concentration drops. As a result, first the near-surface waters, and then the entire epilimnetic water column regains its initial transparency. Nothing on the water surface serves as a reminder about the past "drama". However, in concert with the high nutrition loading set up in the lake, the recovery of bulk water transparency creates the necessary preconditions for reiteration of the above dramatic events either during the same or forthcoming vegetation season(s).

The lake ecosystem becomes subject to changes proceeding at rates much higher than those characteristic of natural eutrophication processes. The time required for a lake to reach a higher concentration of phosphorus under conditions of powerful external nutrition proves to be determined by the lake-water mean residence time which is only 10–15 years for large lakes and several decades for great lakes, i.e. by time intervals incomparable with geological time scales distinguishing the analogous phase in natural ageing of lakes.

Excessive growth of phytoplankton and bacterial decomposition-driven depletion of dissolved oxygen in the lake bring about the onset of anoxia conditions in the near-bottom hypolimnetic layers. The replenishment of dissolved oxygen loss in the hypolimnion from the epilimnetic dissolved oxygen pool is barred by the thermocline. This results in an immense intensification of the nutrient (including phosphorus) release from the bottom sediments, and the autumn convective overturn brings about a nearly uniform dispersion of released nutrients throughout the vertical extent of the lake. However, unlike natural eutrophication, anthropogenic eutrophication is not irreversible at this phase. Moreover, anthropogenic eutrophication at this phase could be curbed and cut down provided the nutrient (phosphorus) input is radically reduced before the anoxic conditions have spread throughout the whole hypolimnetic zone. The reversibility of the anthropogenically driven eutrophication at this phase is totally due to the fact that, despite the increased trophic level, the anthropogenically afflicted large, deep and cold lake still retains its morphological features pertaining to an oligotrophic water body. Furthermore, its thermal regime, sun and sky radiation availability, and hydrodynamics also remain unchanged.

A regular taxonomic diversity in algal community indigenous to large oligotrophic lakes responds more or less uniformly to the phosphorus loading increase: no substantial enrichment of phytoplankton floristic composition is registered in anthropogenically "aged" lakes. However, dominant forms of algae expand considerably because dominant species characteristic of oligotrophic lakes are supplemented by some species which basically are dominant forms in the phytoplankton assemblage of naturally eutrophic water bodies (*Lake Ladoga ...*, 1992).

The competitiveness of a definite species under conditions of anthropogenic eutrophication of large, deep and cold lakes depends on its phosphorus uptake rate, production efficiency, water transparency/light availability, and vertical mixing and seasonal weather conditions. The same factors control the spatial distribution of phytoplankton in the lake. Almost invariably, dominant planktonic species characteristic of oligotrophic water bodies retain their leading role in pelagic, deep-water areas whereas in shallow coastal waters and bays they find themselves forced out or rather dominated by the algal species indigenous to naturally eutrophic waters. This, in its turn, largely determines a considerable spatial inhomogeneity in the distribution of organic matter production as well as in many other limnological parameters throughout the lake (Petrova *et al.*, 1987; Holopainen, Letanskaya, 1997).

Summing up the above, it could be stated that the development of anthropogenic eutrophication is conducive to a dramatic dilapidation of the lake ecosystem: it proves to be detrimental to useful lake productivity, and water quality (Viljanen *et al.*, 1996, Drabkova *et al.*, 1996). Sometimes it even causes a complete loss of natural resources of the lake. A massive growth of planktonic algae results in a distasteful odor of the water. It also brings about a pronounced increase of dissolved and suspended organic-matter content in the water. Excessive amounts of organic matter in the lake water stimulates a rapid and extensive growth of saprophytic bacteria, including pathogenic ones, as well as algal fungi. Metabolic processes of some algae, particularly blue–greens, result in release of toxic substances into the water. The toxicity of abundant planktonic species, such as blue–greens, is capable of causing a serious decrease or even death of animals (Gromov *et al.*, 1996). Sometimes people living on the shore contract toxicity- related diseases. Anoxia conditions occurring in the hypolimnetic layers due to an extensive demand for dissolved oxygen for mineralization of organic matter can provoke extinction of benthic animals, constituting the principal nutrition resources for fish. As a result, noble species of fish are supplanted by less valuable species (Koudersky *et al.*, 1996).

Nevertheless, toxic contamination of lakes does not arise exclusively from secondary effects of anthropogenic eutrophication. A direct admission of toxic matter (xenobiotics) into the lake from point and non-point (dispersed) anthropogenic sources residing on the shore is basically one of the most common pathways (Lozovik *et al.*, 1997; Volkov *et al.*, 1997).

Moreover, in many cases the atmosphere adds substantially (sometimes even predominantly) to the lake contamination: generally, a wide spectrum of xenobiotics are brought down to the lake by precipitation, solid and liquid aerosol fall-out, and snow deposition due to transboundary transport of pollutants from distant sources (Kenttamies, 1994; Lumme *et al.*, 1996). The list of pollutants recorded in different water bodies exceeds 300 000 chemical compounds, of which some 10 000 are considered xenobiotics (Froumin *et al.*, 1996; Slepoukhina *et al.*, 1997). For instance, about 600 different pollutants enter the largest fresh-water lake in Europe—Lake Ladoga. These include more than 300 toxic chemicals provided by anthropogenic sources (Roumyantsev *et al.*, 1997, Frumin *et al.*, 1997, *Ecological State ...*, 1996).

Sulphurous compounds, phenols, chlorophenols, organic acids, fatty and resinous substances, chlorine compounds are all generally added to the lake along with effluents from pulp and paper mills; pesticides enter the lakes through runoff from farms. The input of synthetic detergents to the lake is mainly due to domestic sewage and polycyclic aromatic hydrocarbons, crude and refined oil are admitted from heat and power plants by water transport, metal ions originate from metallurgic and related industries (Slepoukhina *et al.*, 1997, Kondratyev *et al.*, 1997). Many of the above pollutants are equally admixed to the lake water via downward transport of matter from the atmosphere (Doubrovina *et al.*, 1994).

The impact of xenobiotics on the lake biota and the entire lake ecosystem is multifaceted and invariably detrimental. One of the distinguishing features of inference of the anthropogenic chemical compounds with the lake ecosystem resides in arising multiple direct and feedback paths. As a result, the influence of chemicals on the lake biota is not confined exclusively to food chains, but also is exerted through indirect interactions (Rask, 1994; Sarvala, 1994; Koudersky *et al.*, 1996; Slepoukhina *et al.*, 1996; Verbitsky *et al.*, 1996; Kostamo *et al.*, 1997).

Some of the extraneous substances are capable of affecting the oxygen regime in the lake, as well as the active reaction of water (pH), while others initiate/cause a redistribution of xenobiotics in terms of their effective volume and surface phases rendering them more efficient in attacking the biota and the aquatic ambient environment.

Interestingly, chemical compounds containing heavy metals (e.g. copper, cadmium, cobalt) in small concentrations influence some algae (chrysophycean, cryptophycean) as growth stimulators, whereas higher concentrations of the same substances have inhibitory effect on the growth of algae. The inhibition and stimulation of algal growth are thought to be equally undesirable: any deviations in algal productivity or in other parameters appropriate to aquatic communities threatens the entire ecosystem. Furthermore, the swimming velocity in cell cultures becomes impaired by heavy metal ions at concentrations higher than a certain threshold. At metal ion concentrations exceeding the relevant threshold, some algae entirely lose their motility (Volosko, Titova, 1997). This entails in its turn a partial or complete loss of the cell's photo-orientation capability. Importantly, the suppression of phototaxis occurs even at low concentrations of heavy metals which may be explained by the existence of either a heavy-metal specific target in the cell's photoreceptor or an integral element in the sensory transduction chain.

Having been taken up by algae, some xenobiotics actively interfere with plant metabolic processes, altering the normal functioning of photosynthesis reaction centers in the cell.

Heavy metals are apt to accumulate in living organisms, particularly aquatic animals. Very distinctive signs of strong toxic pollution are deformities developing in some invertebrates, mainly oligochaetes. Toxicoses of various species of fish are also widespread in polluted lakes. Extractable organic halogens (EOH) and organic chlorine compounds are found to accumulate not only in fish but also in aquatic animals such as the ringed seal. Chemical analyses of Lake Ladoga seal blubber revealed high levels of EOH, DDT, and polychlorinated hydrocarbons (Kostamo *et al.*, 1997).

Acidification caused by air-borne pollutants, very much like toxic contamination originating from ground-based sources, may impact the lake in several ways (Mannio, 1994). It may produce direct toxic effects, e.g. through interfering with calcium uptake by living organisms, which are conducive to a decline in reproduction of normal species or may even cause death of individual organisms, particularly the youngest ones. Acidification is equally known to affect biota via indirect pathways by means of altering competition and predation that compose food-web interactions: some organisms are more sensitive than others, and the ensuing changes to the community structure can upset the balance between species. Admixture of airborne pollutants may impair the lake ecosystem through bringing about alterations to biogeochemical cycles: acidification changes water chemistry in lakes and streams both directly and indirectly consequent to changes in the drainage soil chemistry induced by atmospheric pollutants. The water chemism departures from the "normal" (or the established *status quo*) ultimately reflect damage on the lake biota. Extensive research in the north-western part of Russia and in Finland indicates that the abundance and biomass as well as species diversity of phytoplankon, bacterioplankton, macrophytes, periphytic diatoms, crustacean zooplankton and macrozoobenthos decrease dramatically mainly due to a selective impediment of metabolic processes. Notwithstanding the acidification impact, high humus levels in the lake seem to partly alleviate the above negative effects.

Acidification is reported to influence the lake's optical properties, and hence its light regime and productivity rates: with declining pH the algal size spectrum shifts to larger cell diameters causing an increased transparency of lake water (Havens, Heath, 1991).

Although many common fresh water species are moderately resistant to slightly enhanced water acidity, the most sensitive species begin disappearing at still relatively high pH levels. This may have very serious consequences on normal functioning of the lake food web. In this connection, the disappearance of some species of fish has an extremely dramatic influence on organisms pertaining to downward levels of the trophic pyramid (Kenttamies, 1994; Rask, 1994).

Probably the most dangerous aspect of acidification is its long-lasting impact on the lake ecosystem, since it produces irreversible or at least extremely slow reversible changes to the catchment biogeochemistry which are bound to continuously affect the lake biocoenoses via runoff and other pathways for many decades or even centuries (Sarvala, 1994).

1.2   INDIRECT ANTHROPOGENIC IMPACTS STEMMING FROM REGIONAL CLIMATIC CHANGE DRIVEN BY REGIONAL/GLOBAL CHANGES

The climate in eastern Fennoscandia, where the largest European lakes (LEL), Ladoga, Onega, Saimaa are located, is that of the Atlantic–Arctic temperate zone. It can be described as relatively mild, with not very severe although long winters, late springs with frequent cold sessions, cool and short summers, high relative humidity levels throughout the year and abundant precipitation. These features are due to the geographical vicinity of LEL to the Baltic, White and Barents Seas, as well as to the dominance of intensive cyclonic activity in all seasons.

Based on data from 1960–1990, the climatic map in Fig. 1.1 shows absolute maximum and minimum air temperature values from the beginning of observations up to 1990. For instance, the instrumental observations at the Petrozavodsk station on the Lake Onega shoreline started in 1815. Mean total precipitation for many years was calculated using the data on precipitation (in mm) from 32 hydrometeorological stations and 63 posts (Filatov, 1997).

Changes in diurnal sunshine duration (dsd) in the last 100 years are generally the same in various European regions. Time series of observations shows a maximum in the 1940s and 1950s with a notable decrease afterwards. This was caused by changes in specific features of the general atmospheric circulation, and the concentration of aerosols in the atmosphere. The data collected in Estonia and Finland since 1921 show the tendency for a reduction in dsd (Vehvilainen, Huttunen, 1997; *Estonia in* ..., 1996). The observations of dsd in the LEL area were initiated in the late 1920s–1930s. However, they were quite irregular and covered a short time period.

Temporal dynamics of dsd in the region demonstrates a tendency of dsd towards reduction, which has become particularly pronounced in the last 15–20 years.

A similar situation is characteristic of Europe as well (*Climate of Europe*, 1995; Kuusisto *et al.*, 1994). It can be explained by variations in atmospheric circulation, global climate changes, increasing cloudiness and the growing quantities of aerosols, primarily near large cities. The data on dsd obtained for the LEL area are valid not only for the relatively large city of Petrozavodsk located on the shore of Lake Onega, but also for the regions at considerable distances from cities, thus proving that the phenomenon is something larger than simply an impact (*Lake Onega* ..., 1990).

An analysis of mean annual data on air temperature reveals a positive trend with a 0.4°C growth in the last 115 years (Fig. 1.2). This value corresponds well to estimates of surface temperature changes in the northern hemisphere, which tends to grow by 0.5°C every 100 years. In eastern Fennoscandia (EF), in general, the mean

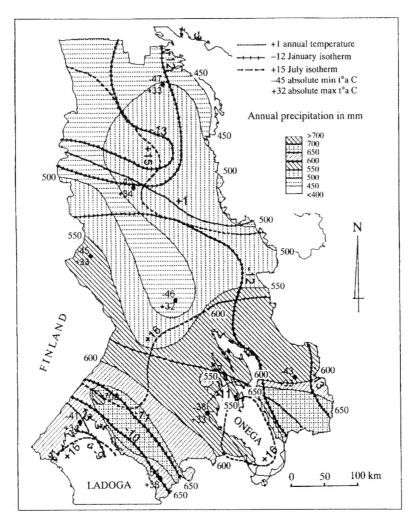

Fig. 1.1. Climatic map for the Karelia region (isolines show spatial distributions of mean annual values of air temperature and precipitation for a period 1960–1990; isotherms refer to July and January as well as to the air temperature absolute maxima and minima values). The points on the map denote the location of basic observational stations.

Table 1.1. Data on dsd (in hours) in the city of Petrozavodsk from 1951 to 1990.

Month	I	II	III	IV	V	VI	VII	VIII	IX	X	XI	XII	Year
Mean	22.3	62.7	132.7	185.3	261.0	289.8	289.0	216.0	126.6	60.0	23.2	8.3	1674
Max.	54.9	106.3	227.4	310.2	406.9	374.7	389.2	329.0	189.3	96.0	42.0	22.9	2008
Year	1963	1969	1963	1965	1978	1970	1973	1951	1975	1988	1953	1962	1963
Min.	2.9	15.2	36.4	108.6	134.0	131.3	191.2	128.7	57.4	13.0	3.3	0.7	1471
Year	1983	1974	1988	1970	1955	1976	1968	1978	1990	1952	1979	1984	1983

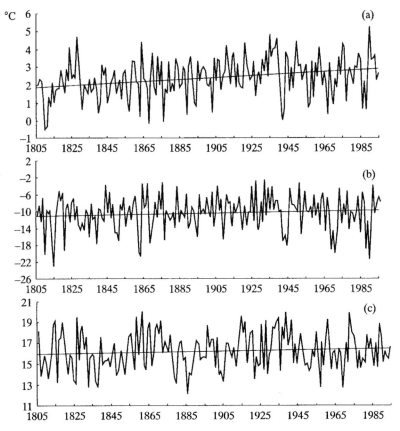

Fig. 1.2. Variability in the mean annual air temperature (a), as well as in winter (January) (b) and summer (July) (c) air temperatures.

annual air temperature has not increased continually throughout the century: the warm 1930s were followed by a cold period from 1965 to 1970 succeeded by an increase in the mean annual air temperatures in the late 1980s. Detailed analysis of seasonal temperatures recorded in the period of instrumental observations indicates that although the autumn and winter seasons in the second half of the 20th century grew colder, positive trends in the spring and summer temperatures resulted in a general increase in mean annual air temperatures (Fig. 1.2.).

The maximum mean annual temperature in the area was recorded in the 1990s. Caused by longer periods of westerly air transport over the LEL area, a notable positive trend (118 mm in 100 years) was also observed in the total annual precipitation series. However, it should be kept in mind that the absolute value of annual precipitation increase includes an error caused by the measurement method. It can thus only generally be stated that total annual precipitation in the study area has increased in the present century. A combined increase in precipitation and annual air temperature results as a rule in increased total evaporation. As a

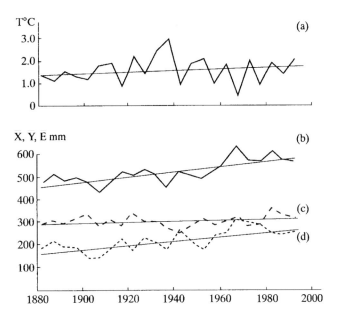

Fig. 1.3. Variability in air temperature (a) and in the elements of water balance: total runoff (b), precipitation (c), evaporation (d) in the LEL region. The period of smoothing is 9 years. Linear trends are designated by direct lines.

consequence, the total runoff in eastern Fennoscandia shows a slight upward tendency (13 mm in 100 years) (Fig. 1.3.).

Changes in general atmospheric circulation in the northern hemisphere in recent years has proved to be conducive to a considerable increase in the number of days with easterly air transport. This has become one of the reasons for an increase in mean annual temperature in the area mainly due to an increase in air temperature in the spring and autumn periods.

The duration and stability of easterly circulation in this area is usually lower than those of the westerly circulation. It manifests itself mostly in the winter period through low air temperatures, a smaller number of cloudy days and lower precipitation. This type of circulation loses much of its vigor by April when it is replaced by the meridional type. By the beginning of the summer period, westerly air transport takes over a dominant role. The dominant form of atmospheric circulation over the territory throughout the year is westerly transport (151 days), the easterly circulation lasts less (94 days), 120 days are governed by meridional circulation. An attempt has been made to establish the relationship between the annual values of meteorological elements (air temperature and precipitation) and the recurrence of the basic circulation types by analyzing the variations in these values in the period from 1945 to 1995, and by considering in detail the specific features of the formation of hydrometeorological conditions in the LEL area.

Instrumental observations extending over 100 years were used to calculate mean global air temperature changes and temperature changes in the LEL area. Data analysis of the growing deviations from the mean air temperature value, evidenced by a statistically reliable trend (*Climate Change* ..., 1996). The mean global air temperature is 15°C, the rate at which it changed was 0.50C in 100 years. An analysis of the data on the surface air temperature in the northern hemisphere in the last 100 years shows that intensive warming which started at the end of the last century actually ended by the 1940s. This warming was interrupted by a temperature reduction which lasted until the 1960s, and was followed in its turn by a new warming started in the mid-1960s. The mean annual air temperature T_a in the LEL area was as low as 1.6°C. The warmest years throughout the whole history of instrumental observations (i.e. 1881–1993) were the 1930s and the year of 1990 with the mean annual temperature at 4.5°C. Starting in the late 1970s, rather high amounts of precipitation and raised water levels in lakes were recorded in the LEL area and the northwest of Russia. Considerable climate changes in the region in those years are also manifested by a shorter period of snow cover of the catchment areas and a longer ice-free period on the lakes (*Climate Change* ..., 1995).

When comparing the values of mean annual deviations in the global and regional air temperatures for the LEL area, certain common features can be seen, and first of all—a statistically significant positive trend. The variance of fluctuation in global temperature mean levels for a large number of stations is much narrower than the variance of regional air temperature changes (ΔT) in the LEL area between 1881 and 1989 (Table 1.2).

Considerable differences are also observed in the spectral structure of the changes of global and regional air temperature. The spectrum of global variations in T_a resembles a "red noise" spectrum dominated by low-frequency secular variations. These fluctuations account for about 40% of air temperature variance. Minor maxima are also recorded in the range of frequences corresponding to temporal scales of about 30 and 5–7 years. The spectrum $S_T(\omega)$ (where ω is the frequency of air temperature oscillations, rad/year) for the area is different from the "red noise" spectrum. It has maxima with the same temporal scales in the considered frequency range of 0.1–0.5 cycle/year, but their contribution to total variance of air temperature changes is similar, so the spectrum cannot be qualified as a "red noise" spectrum, which is described by a low-order autoregression model (Privalsky, 1985; Rogkov, Trapeznicov, 1990). Regional changes in temperature anomalies seem, in the first approximation, to be directly proportional to the increase in

Table 1.2. Global and regional (*viz.* LEL) air temperature variations between 1881 and 1989.

Air temperature	Mean, °C	Variance, °C	Standard deviation, °C	Max., °C	Min., °C	Year
Global	−0.11	0.05	0.22	0.35	0.72	1885
Regional	1.57	1.19	1.09	4.10	1.1	1888

mean global air temperature. It is common knowledge that there is no direct proportional relationship between the changes in total precipitation and warming rate. Publications by some researchers (Kondratyev, 1992; Borisenkov, 1982) have shown that global warming could be caused by natural rather than man-made climate fluctuations. The most pressing are still the questions of whether global warming is a reality, and how strong it may become in the future? Global and regional climates should be taken into account when tailoring studies for the management of geosphere and biosphere processes, responsible for counteracted interactions between man and nature.

Eastern Fennoscandia is notable for a peculiar climate resulting both from specific features of atmospheric processes in the Atlantic Ocean, Arctic Ocean and Siberia, as well as from the effects of the LEL and the White Sea on the drainage basins. The climate formation here is influenced by the high percentage of coverage of the territory by surface waters (the lakes occupy 12% of the territory), forests and wetlands (Adamenko, 1987).

Because of low precision and low resolution of modern global climate models (GCM) applied to individual regions, hydrogeological consequences for regions and drainage areas can so far be estimated only roughly, and the models can hardly be used for forecasting. A common feature for all scenarios is a dramatic decrease in soil moisture in the summer, and a reduction in summer duration and increase in winter total runoff by approximately 30%. According to the forecasts given earlier by Shiclomanov (1988) a 12% decrease in annual total runoff in northwestern Russia is expected by the end of the century. Our data from the LEL area indicate that this forecast may not be quite accurate. Shiclomanov (1988) stated also that a dramatic increase in precipitation may occur in southern regions by 2020–2050 with the total runoff of northern rivers increasing all the time. This conclusion has already received some corroborating evidence from the changes in Caspian Sea water level. Several forecasts of climate change based on linear models, including that for the Caspian Sea (Privalsky, 1985, Ratkovich, 1993), proved erroneous. It is, therefore, very important to revise the theory and methods of forecasting the climate to enable proper decision-making in the control and management of water resources.

Hiltunen (1984) shows that the total runoff dropped throughout Finland in the first half of the century with its minimum in the 1940s and 1950s. Its spatial heterogeneity determined from 1939–1990 data turned out to be quite considerable. Total runoff for all seasons increased in the southern and central areas. The runoff in the north changed insignificantly with a slight decrease in the winter period. Variations averaged over the period of 11 years reached 0.5. The authors explain the total runoff fluctuations by natural climate variations of precipitation and atmospheric temperature. They also mention that the total runoff long-term time series tends to look quite stationary, while significant trends can only be observed in relatively short-time series. Heino (1994) also states that total runoff showed no notable changes which could be caused by the man-made component of climate fluctuations, and the recorded differences stay within the limits of natural fluctuations.

Investigations have shown that the 116-year series of the annual air temperature mean for the LEL area for the 1880–1995 period contains a positive linear trend equal to 0.4°C/100 years. Interesting in this context is the analysis of the spatial distribution of the trend value and the sign of annual and seasonal air temperature indices over the territory of the LEL area to estimate the potential effect of large water bodies (Lakes Onega and Ladoga) on the climatic characteristics of the territory.

In order to evaluate the possible mutual influence of LEL and the surrounding territory, spatial variations in air temperature and the tendencies in its changes were analyzed using data from 34 meteorological stations distributed fairly uniformly over the area.

Mean density of the observation posts was one post per 5070 km^2. The values of linear trends in the series of annual and seasonal air temperatures were calculated for the period between 1951 and 1992. The results of the calculations were used to analyze spatial variations in linear trends of annual and seasonal surface air temperature values for all 34 stations by determining the areas with identical signs of the trend.

The mean trend value for 34 stations was 0.62°C/42 years with a marked reduction in the south–north direction from the highest values of 0.80–0.97 at 61°N to 0.09–0.25°C/42 years at 64°N. Interestingly, the linear trend in annual air temperature series was positive at all 34 stations.

The time series of seasonal surface air temperatures shows considerable differences for stations located in different regions. Winter seasons (December–February) become warmer in the south of the area, in the northern part of the Ladoga Lake drainage area and near Lake Onega. The sign and values of the trends in the series of winter air temperatures change from −0.87°C to +1.01°C/42 years. The series of spring air temperatures (March–May) for all 34 stations contains positive values of the trends from 1.66° to 3.01–3.64°C/42 years in the south. The trend values were found to decrease from the south to north. The value of the spring air temperature trend mean for all stations is 2.58°C/42 years.

The trends of seasonal temperatures for the summer period (June–August) in the Ladoga and Onega Lake areas are positive with the highest value of 0.49°C/42 years. A characteristic feature of the spatial distribution of the autumn (September–November) temperature trends is the absence of a definite boundary between the regions of positive and negative values of the trends. Variability with regard to latitudes in the distribution of trend values was less than for other seasons. The distinctions in spatial distribution of seasonal trends of surface air temperature are significant. Nevertheless, the notable predominance of spring trends, which had positive sign and highest absolute value, resulted in generally positive annual trends for the whole LEL area. All seasonal and annual air temperature series for the Ladoga and Onega Lake areas demonstrated positive tendencies, which can be explained by the LEL effect on the climate in the adjacent drainage areas. Similar results concerning the effect of large lakes, this time the North American Great Lakes (NAGL), on the surrounding territory has been revealed for the NAGL area by Scott and Huff (1996).

1.3 DIRECT AND FEEDBACK MECHANISMS OF INTERACTIONS OF LAKES WITH AMBIENT ENVIRONMENTS (LAKES LADOGA AND ONEGA AS EXAMPLES)

The Lake Ladoga drainage basin belongs to the taiga subzone with its moderate climate, warm and moist summers, and cold and cloudy winters (Hydrological regime, 1966). The thermoregulating effect of the lake affects annual air temperature dynamics, especially in transitional/interseasonal periods. Thus, the mean air temperature in January on Valaam Island in the center of the lake is 2.7°C higher than that in the fjordic area in the north. The mean air temperature in July over the lake pelagic region is 0.9°C lower than that over the shore areas. The lowest annual air temperatures are recorded at the stations located on the eastern shoreline (+2.9°C), the highest ones on the southern shoreline (+3.8°C). The average duration of the ice-free period for the Lake Ladoga area varies from 103 to 181 days, these variations being mainly due to the effect of latitude and local conditions.

The wind regime formation within the Lake Ladoga area is affected by the atmospheric processes common for most of northwestern Russia. The distinguishing feature of Lake Ladoga's wind regime as compared with Lake Onega's is a relative homogeneity of the recurrence of wind direction. The dominant wind directions are southwestern, southern and southeastern. Western, northwestern and northern winds are also frequent. The most rare are eastern winds. The difference in recurrence between dominant and least frequent winds is only 2–3-fold.

According to the data available (Molchanov, 1946), cloudiness over Lake Ladoga is lower than that over neighboring areas. Mean annual values of total cloudiness over Lake Ladoga and the adjacent areas is about 4 degrees. The overcast skies probability in summer is 46–50%, in spring and autumn 65%, in winter up to 80%. Mean annual total cloudiness is rather evenly distributed over the lake area: the difference between cloudiness values from the offshore and shoreline stations is less than 0.3 degrees. The total cloudiness is an important factor affecting the amount of solar radiation reaching the underlying surface. With total cloudiness at 7 degrees, Lake Ladoga receives between 50 and 60% of the total incident radiation. The data from meteostations on the Lake Ladoga shoreline as compared with stations distant from the lake show that incoming solar radiation in the warm period increases due to lower cloudiness. Most of the total annual radiation (85–88%) is supplied in spring and summer.

The characteristics of the ice cover on lakes and climatic parameters are in a rather complicated correlation, e.g. ice thickness, and the time of ice break-up depend on a wealth of factors. In the case of Lakes Ladoga and Onega the ice-free period coincides with changes in air temperature; that is, there is a certain correlation between water temperature upward tendency and ice-covered area decrease.

When the southern part of Lake Ladoga begins to freeze up (December–January), the lake central part continues to cool down. In spring, ice melting proceeds in the south–north direction. The water in shallower areas warms up and cools down faster, hence freezing and ice break-up occur earlier in these areas.

Lake Onega is notable for high cloudiness throughout the year. The clouds are generally rather evenly spread over the water area. No big differences are observed in the distribution of total and low-altitude cloudiness in the center of the lake and in the shoreline areas. The mean annual total cloudiness over Lake Onega is 7–8 degrees. The highest values of total and low-altitude cloudiness throughout the year are recorded in autumn with the maximum in November. The recurrence of overcast days (8–10 degrees) in this period reaches 83–88%. Cloudiness decreases considerably between November and March, and never exceeds 4 degrees for low altitude and 6.5 degrees for total cloudiness after March. Spatially, the degree of cloudiness over Lake Onega is greatest over Petrozavodsk Bay and Povenets Bay, except for November–December when the highest cloudiness is observed over the lake rather than over the shoreline. Petrozavodsk is a rather large industrial town (population about 280 000) producing fairly high aerosol concentrations which are conducive to a considerable rise in cloudiness over the area, and as a consequence to a decrease in incident solar radiation.

Latitudinal differences in sunshine duration over the Lake Onega water area appear at the start of April, while in December–March the index is almost the same for the northern and southern parts of the lake. The amount of solar radiation reaching the underlying surface depends both on geographical latitude and on the specific features of atmospheric circulation processes. A large waterbody like Lake Onega transforms these processes thereby creating local circulation patterns which clearly apparent near the shoreline by causing frequent changes in wind direction. We have used the data from meteorological posts as well as satellite, ship and actinometric observations to characterize the specific features of this local atmospheric circulation. Very clearly seen is the local circulation demonstrated in the NOAA satellite images of Lake Onega.

The thermal status of the lake and its seasonal fluctuations affect a number of meteorological processes. The characteristic features of the temperature regime in the Lake Onega area have developed due to the effect of its huge water volume. Air circulation and ambient environments also affect the formation of atmospheric precipitation over the lake proper and the adjacent territories. Prevalent throughout the year is the marine air coming from the Atlantic, which is responsible for high precipitation. The mean annual precipitation over Lake Onega is 524 mm.

Regarding the ice cover on Lake Onega, in the period from 1880 to the present, the number of ice-free days reduced from 217 to 225 (Fig. 1.4a).

The longest series of observations (1693–1993) of ice cover on the Kemijoki River, Finland indicated that the date when the river becomes ice-free has moved from May 22 to May 10 (Kajander, 1993). Remote-sensing data (Kondratyev, Johannessen, 1995) also show a decrease in ice-covered area in the Barents Sea (Fig. 1.4b).

Lake Onega is currently at the stage of ecosystem destabilization and at an early stage of eutrophication. The lake bays and gulfs have already changed their initial oligotrophic status for a mesotrophic one, and they become an immediate source of eutrophication for the open lake areas. Lake eutrophication is a direct consequence of extensive pollution. As reported by *The Finnish Research ...* (1996), long-term observations (over 40 years) of Lake Onega indicate that climate warming did not produce significant effect on the ecosystem condition as had happened with small

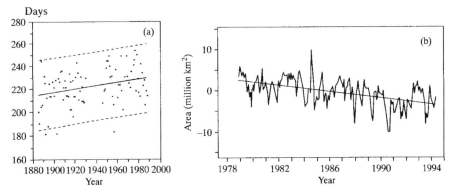

Fig. 1.4. (a) Ice-free period (in days) on Lake Onega (shown as dots) for the period 1880–1990. Linear trend is plotted as a direct line, confidential limits indicated as dotted lines. (b) The ice coverage area (km^2) in the Barents Sea for 1978–1994 as assessed from SAR remote sensing data.

water bodies in Finland and Karelia. The effect of the Great American Lakes on the regional climate has been assessed (Scott, Huff, 1996; Assel, Robertson, 1995). Climatic changes in the 20th Century may cause considerable changes in the LEL ecosystems and speed up eutrophication rates.

Comparison of the LEL ice-cover dynamics data and appropriate data on other lakes of the world shows that considerable changes are to be expected for boreal zone lakes due to a definite tendency towards warming. Observations in the North American lakes (*The Global Climate System Review*, 1995) demonstrated that ice cover is getting about 20 cm thinner. The ice on Lake Hoara in the Antarctic has been getting 20 cm thinner each year for the last two decades. At the same time, LEL are characterized by an increasing number of ice-free days. Ice cover on the lakes located at high latitudes is a good indicator of the climate changes.

Data presently available are not quite sufficient for identifying climate-induced changes in lake ecosystems. However, they are indicative of the fact (*Current Status*, 1987; Filatov, 1997) that the main cause of lacustrine ecosystem changes is man-made. Climatic effects, climate change and their impact on large lakes can be determined using the data on changes in water balance elements and lake level. It can be stated generally that the largest lakes produce a very significant effect on the features of spatial–temporal distribution of local climate characteristics. As illustrated, Lake Ladoga renders the climate more mild in the southwestern region, Lake Onega in the southeastern region, the White Sea in the northeastern part of the ambient area.

Hydrometeorological elements fluctuation spectra contain components with time scales ranging from a few seconds to several millennia. We consider here the seasonal fluctuations in water balance and water level. Multiyear fluctuations of these meteo-elements are still poorly studied, and hence poorly understood, whereas variability on this scale is of great interest for long-term forecasting over periods from several years to several dozen years.

Correlation and spectral analyses of stochastic processes have neen used to give structure to variances in hydrometeorological elements (such as river runoff, and lake water level). These can be represented as polycyclic parameters with a time scale of 2.5–7 and 25–30 years respectively. These results were obtained on the assumption that the processes are stationary in nature. Probability analysis methods for stationary series allow the structure of the process tobe studied, singling out the basic cyclic components. Still, they cannot be used to determine the variability of probability index estimates in time. The principal probability indices used for nonstationary processes are: the time-dependent mean value of the variable $m(t)$, auto- and cross-correlation functions $K(t, \tau)$, auto- and cross-time frequency $S(\omega, t)$ and double-frequency spectrum densities $S(\omega, \Omega)$ (where t, τ, ω, and Ω stand for time, lag of the correlation function, frequency of air temperature oscillations, frequency of energy temporal variations). Cross-correlation and spectral analyses have revealed that the multiyear variabilities in river runoff and lake water level are in agreement with fluctuations in air temperature, atmospheric precipitation and atmospheric circulation indices. Obtained for 128-year-long series segments, non-stationary spectra $S(\omega, t)$ demonstrate that the amount of variance occurring in the high-frequency range did not change throughout the observation period whereas the relatively short-term (10- and 30-year) fluctuations prove amenable for energy redistribution in time.

Let us now consider the characteristics of the time series of fluctuations of water level and water balance elements of large lakes. The nonstationary time-frequency spectrum corresponds to a polycyclic stochastic process, and is characterized by a multipeak structure. Fluctuations over periods of about 30 and 5–7 years are determined for all estimates of the spectral density of the water level in the lakes considered here. The amount of low-frequency fluctuations on a time scale of 20–30 years changes with time. This amount for the water level in Lake Ladoga varies from 28 to 54% with the mean value at 40%. In Lake Saimaa these figures are within the range 21–30% with mean values at 25 and 24%. The share of fluctuations in the frequency range 0.10–0.50 rad/year is lower than that of fluctuations considered above. It does not exceed 20% with the mean value for Lake Ladoga at 12%, and Saimaa 10%. The share of fluctuations on the time scale of about 5 years in total variability in Lake Ladoga is 23%, which proves that fluctuations play a rather significant role. The estimates of low-frequency fluctuations over 20–30 years in Ladoga, Onega and Saimaa lake water-level spectra from available data are not reliable enough. It appears expedient, therefore, to estimate the spectral density of changes in water level in Lakes Ladoga and Onega using a nonparametric method, *viz.* the maximum entropy method. One of the advantages of this method is a high-resolution capacity in frequency analysis and high reliability of the sought-after estimates. For instance, when the Fourier cosine transformation was applied to fluctuations in lake level, total runoff and atmospheric precipitation, the number of degrees of freedom v in the estimates of the spectral density was 10–15. When the spectral density is calculated using the maximum entropy method, the number of degrees of freedom is considerably higher (Fig. 1.5).

The spectrum calculated using this method has a number of peaks at frequencies 0.035, 0.16 and 0.21 l/year, which corresponds to fluctuations in time scale of 25–30, 10–11 and 5 years respectively. The total inflow in Lake Ladoga is characterized by

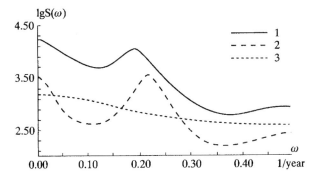

Fig. 1.5. The water-level spectra for Lakes Ladoga and Onega. Plots indicate as follows: 1—calculated with a model of autoregression of the fifth order, 2—calculated with a model of autoregression of the fifth order, 3—calculated with a model of autoregression of the first order.

fluctuations of different frequency ranges over a year: a sharp narrow-band maximum in spring is caused by the spring flood events; an autumn maximum is observed less regularly; some years also exhibit a spike connected with summer floods.

One of the characteristic features of the time series in question is the repeated recurrence of the observed process year after year when the modulation attenuation of the cycles over a year is very clearly seen in the lake data. It indicates that the process is nonstationary. Annual recurrence of seasonal fluctuations in its turn shows that this process may correspond to a periodically correlated stochastic process.

The spectral structure in the frequency range corresponding to multiyear variability is noticeably different. No definite narrow-band high-intensity maxima are observed here. A number of broad peaks with time periods of 5–7 and about 30 years are observed. The spectrum demonstrates that along with low-frequency modulation of fluctuations over the year, multiyear variability is formed by the overlapping of stochastic unordered fluctuations (such as the red noise or low-frequency trend) and cycles with time scales of 2 to 30 years. To be able to forecast the lake water level, it is important to relate it to the factors governing its variability. These factors for Lakes Ladoga and Onega, just as for other running-water bodies in the boreal zone, are surface inflow, outflow from the water body and, to a lesser degree, atmospheric precipitation and evaporation from the water surface. We will use the available time series of mean annual values of these processes to perform the cross-correlation and cross-spectral analysis between the fluctuations in lake water level and individual elements of the water balance.

Obtained on the basis of mean annual values, the results of analysis of the Lake Ladoga water level, total inflow from three main rivers, and the Neva River runoff reveal close cross-correlations ($r \geq |0.50|$). The maximal correlation coefficient is recorded at a zero shift which is an indication that the changes are synchronous with the factors that cause it (the total inflow and Neva River outflow). The correlation between changes in lake level on the one hand and atmospheric precipitation and total inflow on the other hand is low. Analysis of the estimates of cross spectral density of the changes in the discharge of rivers (Volkhov, Vuoksa, Svir, Neva) forming the basic surface inflow and outflow of Lake Ladoga and atmospheric precipitation showed that there were some significant maxima in the spectra. The variance within the main energy-bearing frequency zones (27–31, 6–7 years) is 50–60% for water level and inflow of three rivers, and it is about 80% for Neva River outflow.

High correlation coefficients for total inflow into the lake are observed for lake level and precipitation during the warm period of the preceding year. The ratio of water balance elements in Lakes Ladoga and Onega changes but little at different water content stages with only a slight decrease in the correlation coefficients for rise and recession phases. It is therefore quite sufficient to use the data from just the last cycle when building prognostic regression models for the lake level.

The presence of multiyear fluctuations with different sets of cycles proves that the process is nonstationary, and the annual recurrence is rhythmic. The results of analyzing spectral density estimates obtained for mean monthly data show that:

—the total atmospheric precipitation mean variance for the lake basin has a spectrum with a narrow distinct maximum on the 12-month time scale;
—the water level variance in the studied lakes has a spectrum with a maximum at low frequences and a second maximum with the 12-month time period. Characteristic of the total runoff is the spectrum with maxima at low frequences, on the time scale of 1, 3, 6 months.

The revealed peaks are much higher than is appropriate for red and white noises. Small broad peaks, their width pointing to their irregular nature, appear between narrow maxima. This irregularity with respect to both amplitude and cycle length can be detected in the time-frequency spectra. The following features are characteristic of Lakes Ladoga and Saimaa:

—identical progress of the cycles with 6–4-year time periods;
—well-expressed annual recurrence of maxima in the correlogram and their variability over a sequence of years which appear as greater or weaker spring–autumn maxima; it allows to qualify all three processes (i.e. river runoff, lake level, atmospheric precipitation) as periodically correlated stochastic processes. The multiyear variability in processes depends on the overlapping of multiyear modulation of the distribution over a single year and long-term fluctuations.

The most common model of nonstationary stochastic processes related to change in water balance elements (WBE) is annual periodicity against the background of nonstationary multiyear variability.

Let us estimate the dependence of the WBE and lake level change on the general atmospheric circulation. The development of the general atmospheric circulation includes prolonged periods when processes characteristic of one (or two) circulation patterns are well developed while the rest are slackened. Such periods dominated by certain circulation patterns have been named *epochs* (Girs, 1971). Recurrence of the processes associated with the C pattern (i.e. the pattern characterized by predominantly meridional transport of air masses) shows no trend, and can thus be used to delineate circulation epochs as well as to forecast the processes. Chronological disagreement between atmospheric circulation cycles and water content phases is due to some inertia in processes occurring in the drainage basin. Beginning with the 1940s, with increase in winter precipitation and augmentation of the temperature in the cold period as a background, a reduction is observed during the spring flood (the last phase). This situation can be explained by moisture losses from the snow surface at the end of winter under the dominance of cellular circulation.

When assessing climatic effect on water bodies, it is important to analyze fluctuations in water level as an integral index of system changes. Various models based on the analysis of the water balance linear equation were widely used to study water-level variations. These models were applied to closed water bodies (Caspian and Aral Seas, Lakes Balkhash and Issyk-Kul) by Privalsky (1985), Ratkovich (1993), to flowing waterbodies by Muzilev et al. (1982). Ratkovich (1993) convincingly demonstrated that while water body level reflects the integral effect of inflow, outflow, precipitation, evaporation and reaction of its own thermohydrodynamic system, its behavior is best described by probability characteristics of water balance. The water level has a very high degree of correlation, up to 0.9, which indicates the presence of long-range series of high and low levels. The conclusion is that there is no sense in trying to use, for forecasting purposes, long-range water-level observation series burdened with numerous errors. Data on water balance elements might be a much better candidate since they are more representative than the data on water level. Moreover, WBE are less autocorrelated. According to Ratkovich (1993), autocorrelation for evaporation is 0.4–0.5, for precipitation 0.1, for total runoff 0.3.

The stochastic nature of fluctuations of the elements only allows for a probabilistic description of the process. The values of inflow, precipitation and evaporation are set by stochastic models. This is the so-called class of dynamic-stochastic models. Models of this kind are categorized according to several characteristics: linear and nonlinear models, with one or several correlated or uncorrelated inputs.

Consider the results obtained using these models. When a time series of hydrological data is given a probabilistic description and different probability models are further used, it is difficult to see the physical meaning of many parameters of these models. Conversely, this meaning in the case of dynamic-stochastic models is clear. The linear models of this kind used to forecast the level of the Caspian Sea and the results obtained contradicted the sharp upward tendency in the water level. However, it can be admitted that the cycles and nature of the correlations used by some authors are not stable in time. At the same time, the statement that 30-year, secular and other cycles are not present in the series of hydrometeorological elements does not agree with numerous calculations based on analysis of long-range time series of observations of water level. Quasi-periodic fluctuations on the above time scales were found in changes in water level of not only the Caspian Sea but also in some large lakes of Europe (Malinina, 1966; Filatov, 1997) and North America (Kite, 1993; Bishop, 1990). The method of probabilistic forecasting based on linear water balance models developed by Privalsky (1985), and Ratkovich (1993) has been shown by Hublarjan *et al.* (1996) to be erroneous. The latter authors prove that application of the linear equation of water balance for forecasting fluctuations in Caspian Sea level resulted in an error as large as 1.67 m, though calculations by Ratkovich (1993) posited that the probability of such an error was very low—only 7×10^{-4}. It is evident therefore that this linear type of model cannot be applied to the processes. To estimate water-level variation, Hublarjan *et al.* (1996) studied nonlinear physical mechanisms in the system which considered the effect of random external forces on the behavior of nonlinear dynamic systems. They used the theory of so-called induced transitional noises. The water body was described as an unbalanced system with numerous nonlinear relations at which large fluctuations can emerge in the system. The authors mention that the water body in this case is an ideal object from the point of view of a synergetic approach. This gives an opportunity to envisage the probability of new stationary states forming in the system which is not possible within a linear approach. It also allows description of abrupt fluctuations in water level occurring against the background of slow climate change. The level of evaporation from the water surface expressed by water surface temperature depends on the water body depth. Bifurcations caused by thermally driven fluctuations in water level can occur in this case. As a result, a fundamentally important conclusion was made that because of the multiplicity of equilibrium levels of transition, water bodies may not be appropriate indicators of climate change at all, but reflect their own nonlinear water balance dynamics. The conclusions based on this model seem rather novel and can explain the long-lived "mystery of nature" and determine the mechanisms responsible for intensive fluctuations of water level occurring against the background of weak and slow climate change. According to this theory, the water level in water bodies differing in size and depth, but subjected to similar climatic conditions, e.g. Lakes Ladoga and Saimaa or Lakes Huron and Erie, should most probably demonstrate nonsynchronous fluctuations.

1.4　A COMPARATIVE ANALYSIS OF THE AMERICAN GREAT LAKES AND LAKES LADOGA AND ONEGA ECOLOGICAL STATUS

The cumulative erosive and in-filling impacts of the advance and retreat of the primordial Quaternary-era glacier has resulted in a wealth of large and small temperate lakes and interconnecting river systems throughout the northern hemisphere. The locations of most of the larger lakes were at the interface of neighboring crystalline plate and sedimentary rock provinces, predetermining pronounced inhomogeneities in their drainage areas. This resulted in lower mineral levels in northern tributary waters and greater mineral levels in southern inflow waters.

1.4.1　Lake Onega

Lake Onega is one of the great lakes in the world, and is the second largest freshwater body in Europe after Lake Ladoga (Molchanov, 1946). By its geographical position, Lake Onega is superior in the LEL system. It is located within the marginal zone of the Baltic crystalline shield where the shield borders the Russian plain (platform). The area of the lake is 9900 km^2, the water volume is 291 km^3. The lake's greatest length is 290 km, the lake width is 82 km. The mean depth is 30 m and the maximum depth is about 120 m. The drainage basin is 66 200 km2. The precipitation in the Onega Lake area is around 600 mm per year.

Besides its gigantic dimensions, Lake Onega is also remarkable in the sense that it is one of the least mineralized inland water bodies on the planet. The concentration of dissolved salts in the lake is under 39–46 mg/l. The mean concentration of nutrients mounts up to 10–14 µg/l for phosphorus (P), 0.52–0.65 for nitrogen (N), 0.3–0.5 for silicon (Si). The dissolved oxygen (O$_2$) concentration varies between 10.4–14.4 mg/l and is close to the relevant saturation level (*Lake Onega* ..., 1990).

Onega Lake phytoplankton are composed of more than 430 species: Bacillariophyta accounts for 35%, Cyanophyta for 19%, Chlorophyta and Chrysophyta for 7% each. Diurnal primary production in spring time may vary reaching 5–11 µg C/l.

The lake periphyton encompasses more than 506 types of algae, the major constituents of which are Bacillariophyta (60%) and Chlorophyta (28%). Annual periphyton production is 0.7–0.8 g C/m^2.

Lake Onega higher plants comprise 62 associations. *Phragmites australis* account for *ca.* 62% of the macrophyte stands area. *Nuphar lutea* is pretty abundant (6%) here for an oligotrophic lake. The total area of macrophyte stands is as high as 2.8 thousand hectares which is 0.24% of Lake Onega's water surface. Macrophyte annual production is 2.8 thousand tons.

The limnic heterogeneity of the lake affects the spatial distribution of bacterioplankton. The mean concentration of bacterioplankton is around $(0.1–0.3) \times 10^3$ cells/ml, and production levels are found at 0.05–0.2 mg C/l.

The protozoan plankton indigenous to Lake Onega comprise 130 types of organisms. Regardless of their small size, protozoan plankton play a very important role in recycling matter in Lake Onega. Protozoan plankton productivity is distributed nonuniformly throughout the lake: in the open parts the mean monthly biomass in June–October is as low as 0.01–0.04 g/m^3, whereas in July it is 0.05–0.092 g/m^3 which is still characteristic of an oligotrophic water body. In bays (which are almost invariably subjected to anthropogenic eutrophication) production levels are three to four times higher.

Lake Onega's zooplankton community encompasses 202 species including 90 crustaceous species and 112 rotifers. The production of prey and predatory zooplankton in the littoral zone over the summer period averages around 56.2 g/m^2, and 14.9 g/m^2, respectively. In the pelagic zone the prey and predatory zooplankton production is 19.3 and 3.4 g/m^2, respectively.

Bottom biotopes are composed of 530 types and forms of invertebrates, of which about 80% are mainly restricted to the lake littoral zone. The dominant group is constituted by oligochaetes. The pelagic-zone silt habitats accommodate assemblages of relict crustaceous species (*Gammaracanthus loricatus* Sars, *Pontoporeia affinis* Lindst., *Pallasea quadrispinosa* Sars, *Mysis oculata* var. *relicta* Loven).

With the exception of small shallow bays, Lake Onega was originally oligo-trophic. However, the industrial and agricultural impact has resulted over the last decades in a substantial violation of the initial trophic status of this water body. The first pronounced changes driven by eutrophication were reported back in the 1960s in the Kondopoga and Petrozavodsk bays. Since that time, the eutrophication process has been spreading further towards the pelagic zone. Presently, to some extent, the entire lake, except its central region, is consumed by anthropogenic eutrophication due to the impact of navigation, river discharge, runoff, industrial effluents, etc. Recent studies indicate that even in the pelagic part of the lake, there are changes indicative of anthropogenic eutrophication which, however, are still moderate, leaving the lake status within the margins of oligotrophy. These changes are primarily confined to a stable quantitative increase in benthic community parameters: during the last decade the benthos count increased by a factor of 3. Bacterioplankton concentration is presently assessed at $(0.4–0.7) \times 10^6$ cells/ml against 0.2×10^6 cells/ml reported in the 1960s. There are also alterations to the zooplankton community: the rotifer count rose significantly. In 1964–1967, rotifers accounted for 3% of the benthic biomass, in 1989–1993 their proportion increased up to 60%, and in 1995 up to 80%. The concentration of *Asplanchna priodonta*, which is a good indicator of trophy, increased from 0.13–0.9 thousands of organisms per m^3 to 2.8–6.4 thousands of organisms per m^3 over the last 35–40 years.

As already pointed out, the most significant alterations to the hydrobiological status took place in large bays of Lake Ladoga. The structural reconstruction of plankton communities is shown by reduction of biodiversity and changes in the ratios of major systematic groups. In the lake phytocoenosis, blue–greens, chloro-coccus and *Tribonema affine* become abundant. Eutrophication-driven alterations to the lake food-chain also brought about changes in the zoocoenosis status: the number of Calanoids diminished by comparison with Cyclops while the relative proportions of Cladocera and rotifers increased significantly. The role of oligo-chaetes in the benthic community becomes much more pronounced.

The bays become a secondary source of pollution of the lake. During periods of enhanced hydrodynamic activity in the lake, eutrophic waters from the bays are transported to the central regions of the lake. The most dangerous periods in this sense occur in spring and summer when the waters, accumulated in bays during the winter stagnation period, get involved in the all-lake water-exchange process.

The ichthyofauna of Lake Onega comprises about 50 species of fish. The most important marketable species are shallow-water cisco and smelt. The fishery productivity of the lake is rather low, about 1 kg/ha, and the total annual catch is under 3 thousand tons.

1.4.2   Lake Ladoga

The largest glacier-formed lake in Europe, and the sole source of fresh water for five million inhabitants of the city of St. Petersburg, Russia, Lake Ladoga (latitude 59–61°N), was created some 12 000 years ago. It measures approximately 220 km by 83 km with a water surface of *ca.* 17 800 km^2. With average and minimum depths of

51 m and 230 m, respectively, Lake Ladoga encapsulates almost 900 km^3 of fresh water. The Lake Ladoga watershed extends over 100 km from north to south and about 600 km from east to west, with an areal extent of about $260\,000 \text{ km}^2$ (Viljanen et al., 1996). The water renewal time of Lake Ladoga is 11 years, which indicates that the ecosystem is rather conservative (Petrova, 1990).

Morphologically and morphometrically, Lake Ladoga is inhomogeneous and can be subdivided into a northern and southern region, each characterized by distinctive coastline and distinctive littoral and pelagic bottom relief. Northern Ladoga is deep with a jagged coastline comprised of numerous long and narrow fjords. Bottom relief is markedly non-uniform with cleavages about 150–200 m deep neighbored by bottom elevations in the form of underwater hill chains that frequently surface to form more than 600 rocky islands in northern near-shore and pelagic lake areas. Southern Lake Ladoga is considerably less deep with depth progressively diminishing southward. The southern coastline is formed of a number of large open bays with depths <5–10 m.

The Lake Ladoga watershed also displays very pronounced north–south differences: forests with a predominance of coniferous species densely cover the northwestern, northeastern, and northern catchment areas; eastern and southern catchment areas are generally devoid of trees and are often swampy, but also arable in many areas due to artificial drainage.

Lake Ladoga receives water from over 30 large rivers as well as a host of smaller tributaries. At 74 km, the Neva River, however, is Lake Ladoga's singular outfall. The Neva River provides passage of Lake Ladoga water to the Gulf of Finland and, ultimately, to the Baltic Sea. The annual discharge of the Neva River into the Gulf of Finland is 80 km^3 with an average discharge rate of $2500 \text{ m}^3/\text{s}$.

The human population (exclusive of the five million inhabitants of St. Petersburg) residing within the catchment is slightly over one million with 75% of them living in towns. Local industry is largely oriented towards paper and cellulose production as well as towards the production of wood-based chemical compounds and merchandise. More than ten large pulp mills are located along Lake Ladoga's coastline, with many other industrial enterprises located on the banks of in-flowing rivers. Local industry is also focused on processing raw materials to produce aluminum, cement bricks, and chemical fertilizers. Several decades ago a number of metallurgic and oil refineries and industrial-machinery production plants were established, and industrial effluents are either being directly injected or carried as river discharge into the lake proper (*Anthropogenious* ..., 1982).

Local agriculture is mainly suburban in nature, oriented to supplying urban customers with milk and associated dairy products, poultry, beef and pork, processed meat, potatoes, and vegetables. Consequently, a large number of cattle and poultry collective farms have been established by the shore.

Being a large, deep, and cold body of water, Lake Ladoga's trophic status was predetermined to be strictly oligotrophic. Actually, it was from the time of lake formation until the latter part of the 1960s, when the trophic state of the lake began changing in a manner that could not be attributable to environmental influences of purely natural processes. The lake has ultimately developed a stable mesotrophy with extensive seasonal water-surface blooming events (Roumyantsev et al., 1997).

Until the early 1960s, the concentration of dissolved oxygen in the littoral area fluctuated from 9 to 15 mg/l (90–120% saturation). In 1968, water transparency, as measured by Secchi disk, was an average of 3.5 m. Concentrations of nutrients remained at low levels: total phosphorus was in the range 1–15 µg/l, nitrate loading varied between 0.1 to 1.2 mg/l, and ammonium (NH_4) content did not exceed a value of 0.24 mg/l (Viljanen *et al.*, 1996).

However, in 1987–1989, the mean phosphorus concentration in summer increased to 21 µg/l in the pelagic area and up to 32 µg/l in the littoral zone. In 1992–1993, total phosphorus concentration ranged from 15 to 29 µg/l: spatial and temporal variations in water quality distribution are very characteristic of the lake (Slepoukhina *et al.*, 1996).

Prior to the onset of anthropogenic eutrophication, the phytoplankton community from Lake Ladoga encompassed over 380 species, subspecies, forms and varieties: 45.5% Bacillariophyta, 33.2% Chlorophyta, 20% Cyanophyta, 3.7% Chrysophyta, 1.0% Pyrrophyta, 0.8% Xantophyta, and 0.8% Euglenophyta. (Petrova, 1990). This composition is appropriate for large temperate oligotrophic lakes.

However, by the late 1970s, the algal biomass rapidly began increasing attaining values considerably higher than those registered in the early 1960s. The algal biomass upsurge came about concomitantly with serious alterations to the composition of phytoplankton. Cyanophyta and Cryptophyta seized the leadership in summer phytoplankton. Concentrations of some algae (*Diatoma elongatum* var., *Oscillatoria*, *Microcystis*, and some others) associated with eutrophic waters have dramatically increased. Conversely, once abundant *Attheya zachariasii*, *Rhizosolenia eriensis* var. *morsa* and *Dinobryon* spp. have almost completely disappeared (Petrova, 1990). The annual primary productivity in Lake Ladoga exhibited a spectacular increase from about 15 g/m² in 1976 to ~140 g/m² in 1985 (Petrova *et al.*, 1987). The areal distribution of primary productivity in the lake acquires the features of pronounced inhomogeneity: coastal waters qualify as mesotrophic and even eutrophic types, whereas pelagic areas are still oligo-mesotrophic. Extensive blooms of blue–green algae in late summer have been observed throughout the whole lake area. However, in the late 1980s, the algal maximum biomass decreased again: it only reached half of that registered in the 1970s. This observation along with some others prompted a conclusion that lake trophy dynamics experienced variations which could be categorized as stages in the lake's recent history development. They are four in number (*Lake Ladoga ...*, 1992).

The *first* phase encapsulates a period of 5 years from 1976 to 1980. It is characterized by high phosphorus concentrations in Lake Ladoga (26 µg P/l, on average) as a consequence of strong phosphorus injection into the lake which was taking place in the late 1960s and early 1970s. That was a period of exponential rise in all quantitative indicators of the phytoplankton community, as well as a period when algae taxa characteristic of eutrophic water bodies became predominant in the phytoplankton assemblage. At the end of 1980, the floristic diversity (about 8–9 dominants) of phytoplankton in Lake Ladoga attained its peak while productivity rates largely became stable. Decomposition processes grew, bacterioplankton counts increased both in epilimnion and hypolimnion. For the first time, algal fungi in the littoral zone were reported to attain very high counts following the phytoplankton vernal peak in the area of warm stratified waters. This stage was marked by rapid accumulation of organic matter (primarily as detritus) in the lake. However, no alterations to the oxygen regime in the lake had yet been registered at that time.

The *second* phase (1981–1983) is characterized by a considerable decrease in the total phosphorus concentration (down to 23 µg P/l). Phytoplankton productivity became unprecedentedly (since the late 1960s) stable throughout the vegetation seasons. Decomposition processes gained acceleration: bacterial decomposition rates exceeded by 3–3.5 times the rates of primary productivity which was conducive to a considerable decrease in suspended and dissolved organic matter in the lake. However, at the end of 1983 the interannual fluctuations in dissolved organic matter content substantially diminished.

During the *third* phase (1984–1986), the total phosphorus concentration con-tinuously decreased (down to 22 µg P/l). The phytoplankton community displayed serious changes: oligotrophic forms began regaining their dominant role in the pelagic areas. Concentration of dissolved organic matter remained at 9.5–9.8 mg/l. Plots of kinetic biochemical oxygen demand (BOD) were transformed from exponential to linear, and BOD levels crawled down. The oxygen regime in the lake, for the first time, exhibited a lack of balance. Dissolved organic matter levels in near-bottom waters only reached 70–90%.

At the *fourth* stage (1986–1989), the total phosphorus concentration dropped to 21 µg P/l. Oligotrophic phytoplankton in pelagic areas became stably predominant, although productivity levels in the lake remained very high. Bacterial decomposition processes went down bringing about both a decrease in dissolved organic content in the water body and accumulation of organic matter in sediments. Depletion of oxygen levels was recorded in samples from the entire area of the lake. The lake ecosystem found itself profoundly destabilized.

In the years following the 1980s, Lake Ladoga's ecosystem remained essentially unbalanced. Provoked by eutrophication, the critical status of Lake Ladoga's ecology is further aggravated due to injection of chemical pollution. According to Iofina and Petrova (1997) Cd, Cr, Hg, V, Pb, Ce are lethal to many planktonic and algal fungi taxa. Moreover, the toxicity due to Hg and/or V suppresses blue–green algae dominance in summer plankton. In response to increasing toxic impact, primary productivity of algal diseases takes place.

The zooplankton community indigenous to Lake Ladoga is spatially hetero-geneous. Three different areas have been distinguished in the zooplankton spatial distribution: the northern archipelago, the deep pelagic zone, and the southern shallow area constituting the Volkhov River Bay. These areas are found to be specific not only in terms of zooplankton community features but also hydrophysical and water-quality parameters (Viljanen *et al.*, 1996). The pelagic zone and deep archipelago zooplankton exhibited remarkable stability throughout the entire monitoring period, i.e. 1948–1997. The biomass of zooplankton remains at low levels and no definite trends are yet noticeable in its time series variations. Copepoda have been prevalent in the zooplankton community over the observation period, which is pertinent to low productive (oligotrophic) water bodies (Slepoukhina *et al.*, 1997). However, the coastal region, with the highest water temperatures and highest nutritional levels, displays various degrees of eutrophication (from mesotrophic to eutrophic) evidenced by appropriate changes in zooplankton species composition and associations productivity (Andronikova, 1996a, b).

At least 385 species and forms of bottom-residing macroinvertebrates were identified in zoobenthos from Lake Ladoga in the 1950s and 1960s; about 85% of these species were collected within the littoral area. Since that time, no serious changes have been registered in zoobenthos community parameters. Species composition of macrobenthos and low levels of its biomass (the average values are less than $3 \, g/m^3$) still typify Lake Ladoga as an oligotrophic water body (*Ladoga Lake* ..., 1992).

The fish fauna of Lake Ladoga at the beginning of this century was reportedly rich. Lake Ladoga, at that time, provided perfect habitat conditions for altogether 48 species and forms of fish (Kudersky L.K., 1996). Of these, 25 species had commercial importance; salmonids, coregonids, pike, and perch being the most valuable. The construction of hydroelectric dams in rivers, timber floating, extensive fishing and other factors have drastically curbed the reproduction of some fish species. Atlantic sturgeon (*Acipenser sturio*) and Volkhov whitefish (*Coregonus lavaretus baeri*) became endangered species facing extinction. Due to irrational fishing, catches in Lake Ladoga experienced serious fluctuations. However, the measures undertaken in the early 1960s to control fishing resulted in a sensible increase in the total annual catch, although they proved to be insufficient to achieve conservation of such noble species as salmon, *Salmo salar* L. m. *sebago*, brown trout, *Salmo trutta* L., and migratory whitefish (Viljanen *et al.*, 1996).

Since 1990, it has become evident that anthropogenically driven deterioration of the aquatic environment has resulted in a considerable decrease in total fish catch. Worsening habitat conditions have led to a decline in whitefish stocks and, especially during the last few years, to a drastic decline in perch and pike. Anoxia in hypolimnetic waters affect fish as well as benthic populations. It means a dramatic reduction of food availability for major fish species and conducive to additional undermining of fish resources in Lake Ladoga.

Dedicated research indicates that extractable organic halogens and polychlorinated hydrocarbons as well as DDT accumulate in Lake Ladoga's fish-eating seals (Kostamo *et al.*, 1997). Heavy metal ions (Hg, Cd, Pb, Cu, Co, Cr, Zn, Mn, Al, V, Ce) are found in fish in quantities exceeding the relevant tolerable levels. As a consequence of this accumulation, many fish suffer from severe pathologies such as physical deformity, liver cirrhosis, and brain disease.

1.4.3   North American Great Lakes

Encapsulating over $22\,000 \, m^3$ of water, North American Great Lakes (NAGL)—Erie, Ontario, Huron, Superior, and Michigan—with total water area of *ca.* $300\,000 \, km^2$, are contained within a watershed which accommodates 10% of the

US population and 25% of the Canadian population. This amounts to 36 million people. For 75% of people residing within the NAGL watershed, these lakes are presently the sole source of drinking water. Numerous individual farms, farm settlements, towns and large cities as well as megalopolises (Buffalo, Cleveland, Chicago, Detroit, Toronto, Milwaukee and some others) are located on the shore area. This region accounts for more than 70% of US metallurgic production, and provides over 50% of Canada's overall produce. Excellent arable lands stretch out between the megalopolises, which account for more than 25 and 7%, respectively, of US and Canadian gross agricultural output. Swift development of this region was powerfully prompted by construction of a number of canals uniting the NAGL system into a single navigable water course which ends up on the Atlantic coastline. More than 70 thermal power stations withdraw water from NAGL. The NAGL are a large recreation center providing annually between 8 and 12 billion dollars to the US and Canadian state treasuries. On the whole, about 17 and 50% of the national annual revenue of the USA and Canada, respectively, comes from this region (Kondratyev, Pozdnyakov, 1993).

However, the NAGL ecological status has undergone a dramatic deterioration due to ploughing up virgin soils, developing agriculture and industry, increasing populace, changing the regions's hydrological conditions, and developing navigation and tourism. The anthropogenic impact has resulted, as in Lakes Ladoga and Onega, in two fundamental problems: lake water eutrophication and toxic contamination.

Geographical position and landscape characteristics largely favor the protection of NAGL against rapid natural ageing (eutrophication). The influence of erosion is low because of a relatively small water-catchment area, and due to the natural resistance of the water-catchment area bedrocks to erosion. Moreover, the NAGL watershed is densely covered with forests which are known to hinder soil erosion. In times before industrialization, only nitrogen was in sufficient amounts in the NAGL waters whereas phosphorus and silica were limiting nutritional elements. The NAGL water composition differed only very slightly from rainwater. The primordial oligotrophic status of NAGL is also favored by the lacustrine thermal regime: persisting for the major part of year, the vertical stratification prevents dispersion of nutrients along the vertical.

Anthropogenic increase in nutrient loading in the NAGL has resulted in the leading role in the summer phytoplankton assemblage moving from diatoms to blue–greens (Makarewicz, 1987). Residing at the water surface layer, the blue–greens reduce the intensity of solar radiation penetrating downwards through the water column. Light availability becomes a limiting factor for diatoms which are basically restricted to deep water . In its turn, the oppression of diatoms has brought about changes in the trophic chain (which, perhaps, has remained intact for the last 10 thousand years!). Zooplankton community became enriched with species residing in the upper layers. This led to relevant changes in ichthyofauna. While in the past century the NAGL were famous for rich resources in trout, cisco, umber, blue pike, and Atlantic salmon, nowadays dominance is seized by American herring, whereas sig, umber, and other species gradually find themselves forced out (Makarewicz, 1987; *Five Year* ..., 1988).

Anthropogenic eutrophication was particularly strong in Lakes Erie, Michigan, and Ontario. However, the measures undertaken in previous decades by the US and Canadian governments has efficiently improved the situation by drastic reductions in nutritional loading from anthropogenic sources (*A Citizen's* ..., 1988; *Environmental Quality*, 1990).

Unfortunately, apart from biogenic matter, a wide spectrum of xenobiotics also enter the NAGL. American and Canadian specialists have identified in NAGL waters over 300 highly toxic chemical compounds. The arising ecological problems also stem from phytoplankton: by the well-known mechanism of bio-amplification, the toxic substances taken up by phytoplankton increasingly accumulate in tissues of consumers at higher trophic levels. For instance, the concentration of toxic substances in tissue of birds settling in the NAGL region is 25 million times higher than in pelagic waters. Driven by deterioration of water quality, degradation of ichthyofauna has ked to almost complete extermination of mink and otter populations on shore areas of Lakes Michigan, Huron, and Superior. Birds feeding on fish also suffered a lot. Inadmissibly high concentrations of harmful substances have been found in embryos, the birth rates have gone down. Many waterbirds with unprecedented external and internal pathologies (e.g. absence of some jaw- and cranial bones, growing bald, liver decomposition, etc.) were encountered in the region (Gilbertson *et al.*, 1976; Hoffman, 1987). It is believed that polychlorinated biphenyls (the name encompassing more than 200 chemical compounds each displaying some individual pathogenic activity) as well as dioxins are particularly dangerous because of their capacity to accumulate in living tissue reaching high levels and interfere with gene mechanisms of living cells. Herbicide and pesticide runoff from arable lands has contributed profusely to lacustrine contamination. Heavy metals in industrial effluents and atmospheric pollutants mainly transported from western and southern regions of the USA have heavily aggravated the NAGL water contamination status bringing about alterations in aquatic environments similar to those summarized above for Lake Ladoga.

However, strict legislative measures undertaken by the US and Canadian authorities at municipal, provincial and federal levels during last decades, as well as implementation of new technologies have brought about a considerable reduction in admission of nutritional and toxic matters into NAGL. This resulted in amelioration of the general ecological situation in the basin of NAGL (*Environmental Quality*, 1990; *EPA* ..., 1989; *Five Year* ..., 1988; *US Progress* ..., 1989). Nevertheless, it is still too early to speak about a complete recovery of NAGL environments and restoration of oligotrophy in these lakes.

Interestingly, some common features are recorded in fluctuations of water level, elements of water balance and air temperature relating to LEL and NAGL. These fluctuations prove to be driven largely by contemporary climate change, and have a direct relevance to significant alterations in ecology of both systems of Great Lakes respectively in America and Europe.

The water level has noticeably increased since the beginning of the 1980s (Fig.

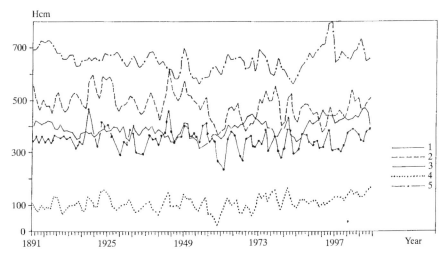

Fig. 1.6. Variability of mean annual water level in LEL and NAGL for 1880–1989: Huron (1), Erie (2), Saimaa (3), Onega (4), Ladoga (5).

1.6). Spectral analysis of observational data reveals both the existence of cycles on time scales of about 30 and 6 years, and a pronounced multiyear variability in seasonal fluctuations in the LEL and NAGL lake level fluctuations. The mean annual air temperature was also found to increase in the areas in question. The phases of variability of the water level and other hydrometeorological elements are not synchronous.

Specific features of climate changes in the NAGL area and fluctuations in water level of lakes are described by Bolsenga, Norton (1993), Kite (1992), Mortsch, Quinn (1996), Tushingham (1992). While these lakes and LEL have the same origin, the same time of formation, water-level variations may also be related to post-glacial uplifts (Saarnisto *et al.*, 1995). The data from a dense grid of water-level recording gauges set up in the NAGL indicates that isostatic uplifts were nonuniform. The rate of uplifting of the northern shoreline is about 400 mm/100 years, that of the southern shoreline is 100–200. The lake water level fluctuations due to climate change (quasi-periodic fluctuations within a century, and autoregressive components) are much more strong. Our analysis of the LEL and NAGL water-level mean annual fluctuations reveals a substantial degree of coherence in multiyear variability. It is easily discernible in mean annual water-level non-synchronous changes occurring in different limnosystems (Fig. 1.6).

Beginning from the 1940s, Lakes Ladoga, Onega and Saimaa have had a positive trend in water-level fluctuation. Changes caused by regulation of water-level regime in lakes Saimaa and Onego account for only a few centimeters over the period from the early 1950s onwards, whereas the amplitude of natural fluctuations recorded in the three lakes is about 100 cm. Cross-correlation and spectral analysis of pairs of time series for water levels in Lakes Ladoga, Onega, and Saimaa indicate the presence of identical cycles on the time scales of about 30 and 6–7 years.

Cross-correlation functions $K(\tau)$ reveal a marked synchronism of water-level fluctuations expressed via the maximum shift (lag) $K(\tau)$ found at a zero shift τ. As this shift increases, the cross-correlation values decrease tending towards zero. Cross-correlograms contain quasi-periodic components. For example, when the Lake Ladoga's water-level interannual fluctuations increase, analogous fluctuations in water level of Lakes Onega and Saimaa increase as well.

In 1980–1996, superimposed on cyclic fluctuations, a positive trend was recorded in water-level variations in both LEL and NAGL (Filatov, 1997; Kite, 1993). Water-level mean annual values of all the studied lakes in the late-1990s were very high. Mean monthly data from a number of the lakes in this period contain maximum values of water levels over the whole period of observations. Thus, the value $H = 192\,cm$ was recorded in Lake Onega in July 1995. A close value of $H = 189\,cm$ was recorded in July 1924. By comparing the absolute minimum for the same month observed in July 1940, $H = 33\,cm$, we see that it is 1.6 m lower. In 1995, the Lake Onega water level was the highest it has been over the last 45 years of observations since the water-level regime of the lake was regulated. In 1995, the mean annual water level in Lake Saimaa was 397 cm, which is 50 cm higher than the relevant mean value for many years. It should be mentioned that the water level in this lake is thoroughly monitored and controlled by a special automated system. Data on average, maximum and minimum mean annual water levels of LEL and two American lakes (one shallow lake—Erie, and one deep lake—Huron) are considered. These two lakes can to a certain degree be accepted as analogs of their European counterparts—Lakes Saimaa and Ladoga or Saimaa and Onega in terms of their morphometric parameters. Data on the NAGL were taken from the literature (Bishop, 1990; Kite, 1992; *Living with the Lakes*, 1993).

Cross-analysis of time series of water level in two North American Great Lakes, Huron and Erie, indicated a high degree of correlation. Cross-correlation functions show that at the multiyear variability scale the water levels in these lakes have high correlation $r \geq |0.72|$ at low τ values. Dominant cycles on time scales of about 30 years appear in correlograms. When Lake Huron's water-level grows, the water level of Lake Erie is observed to grow synchronously. Both the results of estimates by Kite (1992, 1993) and our own estimates show that the fluctuation spectra of levels in these lakes contain some quasi-periodic components on time scales of about 30 and 6–7 years. Quite significant as well is a low-frequency component described by the red noise spectrum or a first-order autoregression model (Kite, 1993). The Report *Living with the Lakes* (1989) describes the time series of mean annual fluctuations in NAGL water levels throughout the period of instrumental observations, showing that just like the LEL, the NAGL fluctuations are highly synchronous, and that each system of lakes has dominant quasi-periodic fluctuations characteristic of this concrete system.

Cross-analysis of the fluctuations in NAGL and LEL water levels proves that the changes in these systems are not synchronous.

The above results indicate that the nature of variability is different. The correlation function shift for the maximum $K(\tau)$ value is 6 years, with the relatively high value of the cross-correlation function equal to nearly |0.50| at the time shift of about 6–8 years.

Cross-correlation functions for other lake pairs are similar: Saimaa–Erie, Ladoga–Huron. Cross-spectral analysis (cross-spectrum $\lambda(\omega)$, coherence $F(\omega)$, and the phase lag $\theta(\omega)$) has shown that most significant values are recorded at frequencies corresponding to long period fluctuations (30 and 6 years).

Cross-spectral analysis of lake water level indicates that fluctuations in time scales of 30 and 6 years characteristic of the two (LEL and NAGL) systems are relatively highly synchronized ($F(\omega) = 0.50$), and the phase lag is about 5 years. Relatively short 6-year fluctuations of water level have a phase lag of about a quarter of the period, i.e. 1–2 years, and rather low spectral correlation. Water-level rise in lakes due to growing atmospheric precipitation in the region by 129 and 146% due to changes in the general atmospheric circulation will result in a number of problems. Some traits can already be seen, e.g. the NAGL water level, particularly Lake Superior in July 1995 and 1996, was close to its maximum values throughout the history of measurements (July 1985). However, this increase does not coincide with the response of the NAGL system to climate warming. Comparison of the data on water levels in the NAGL and LEL in the last high-water periods shows that the maximum value for Lake Onega was recorded in July 1995, for Lake Superior in July 1996. This lack of synchrony between water levels in the NAGL and LEL is caused by changes in atmospheric circulation in the American and Eurasian sectors of the northern hemisphere as well as in the air temperature in the NAGL and LEL areas.

Evaluation of changes in ecosystems, water balance components and water level using empirical data seems quite problematic. Therefore, these estimates in recent studies of lake-water level in North America, Finland and other countries have been made using the global climate model (GCM). The most widely known models used in the study of changes in lake ecosystems are those by Goddard of Space Flight Center (GSFC), Geophysical Hydrodynamics Laboratory (GFDL), Oregon State University (OSU), Canadian Climate Center (CCC) and some others (see, e.g., Kite, 1993; Mortsch, Quinn, 1996). Given stationary conditions and relatively insignificant isostatic fluctuations of the shores, modelling data (Mortsch, Quinn, 1996) show the following effects of climate change on precipitation, evaporation, river runoff and lake water level.

Air temperature increase due to growing concentration of atmospheric CO_2, i.e. the greenhouse effect, will result in converse changes in the regions in question. According to calculations using global climate models with various scenarios (doubling of CO_2) will cause an upward tendency of air temperature in the NAGL area in all seasons, while atmospheric precipitation will go down, and evaporation from water bodies and drainage basins will grow, and hence lake water levels will drop. These scenarios suggest that the changes in Lake Superior, which is located at higher latitudes, will be less pronounced than they are in Lakes Erie and Ontario. An increase in lake water level causes increased erosion of the shoreline and deterioration of water quality followed by increasing costs of water supply to the population while the water needs to be additionally treated. The first floors of houses located near the shore may be flooded.

Calculations for Finland and Karelia using the models show that atmospheric precipitation in the LEL area will grow. Doubling the CO_2 concentration in the LEL area may cause a 5–15% increase in precipitation resulting in elevation of the level of the lakes (Carter et al., 1995).

Temporal fluctuations in amounts of precipitation lead to variations in river runoff, in time of occurrence and intensity of floods, lake water temperature, evaporation and eventually affect the ecosystem at large. It is difficult to reflect these changes in the GC models correctly both because their rather low resolution for drainage basins, water bodies and individual areas, and because of the response of water ecosystems which is rather difficult to explain. Let us give several examples as an illustration. On-site observations and known scenarios of global and regional climate changes carried out jointly for drainage basins and water bodies using GCM suggest that the greatest variations in lake level are to be expected in northern regions. The amount of atmospheric precipitation will grow in Canada, Finland, Northwest Russia (Climate Change, 1996; Hiltunen, 1984; Kuusisto, 1994; Kuusisto et al., 1996; Scott, Huff, 1996; The State of Canada's Climate, 1995; Vehvilainen, Huttunen, 1997) causing an increase in the water level of the largest European lakes and the lakes of Canadian provinces Ontario and Manitoba. Long-term observations of the lake water level in semiarid and some tropical regions reveal both intra- and multiyear variability. This is especially pronounced in lakes with a relatively short conditional water exchange. These lakes are very sensitive to variations in atmospheric precipitation. Reconstruction of lake water level variations in these regions with respect to climate changes was made using paleolimnological methods involving the analysis of diatoms in bottom sediments (Davidova et al., 1993; Prediction, Detection and ..., 1996; Scott, Huff, 1996; Tushingham, 1992). Different lakes respond to climate variations individually. Water level variations in the Great African lakes and in Lake Titicaca in South America have their own characteristic features. The mean annual values of water levels of lakes located in different parts of the world, e.g. Michigan (North America), Titicaca (South America) and Kinneret (Middle East), demonstrate considerable variability within this century and across the years (Climate Change, 1996).

Water level in Lake Titicaca has risen by 6.3 m since the 1940s. Within the same period, Lake Kinneret experienced, simultaneously with a significant increase in water level, considerable (over 2 m) variations in its amplitude. Even Lake Michigan, located in the moderate zone of North America, demonstrated very high (about 1.7 m) fluctuations in water level for the period from 1960 to 1993 (Privalsky, 1992). We further mention the considerable changes of lake water level in high latitudes which occur during warming. The water level of closed lakes fed by glaciers in the large Antarctic desert MacMerdo increased by 10 m in 20 years, which means it grows almost 0.5 m every year (Climate Change, 1996; Gornitz et al., 1997). Naturally, the use of GCM must not limit application of other possible techniques for forecasting changes in WBE and lake level. Quite acceptable are palaeo- and dendrochronological analyses, the method of historical analogs and probability analysis. GCM are now widely used to estimate changes in water balance elements and lake level. Below are some known estimates of changes in NAGL obtained GCM.

If the atmospheric CO_2 concentration doubles, the NAGL area at 42–47°N will experience a decrease in precipitation and total runoff; if the temperature in the lower atmosphere grows then evaporation will increase in the area. As a consequence, water level will drop and lake surface temperature will grow. Lake Superior will probably experience the least changes. As the water temperature in the lakes increases, considerable changes will be seen in their ice cover. According to the results of modeling, the greatest increase in lake water level will be observed in Lakes Huron, Michigan and Erie (1–2.5 m). The NAGL area welfare is largely based on water resources, utilization of lakes for energy production, water supply, transport, recreation, mining of mineral resources and as collectors of waste water including heated water from nuclear power plants. Thus the economy closely depends on fluctuations in lake water level. If the lake water level drops by more than 0.5–1 m, which may occur if CO_2 concentrations double, the region's economy will suffer a heavy blow.

Some scenarios may result in a 1 to 2.5 m lowering of the water level in Lake Michigan. Though climate change accompanying greenhouse effect is relatively insignificant in tropical lakes, the latter may also experience strong fluctuations in water level, e.g. the level of Lake Victoria in Africa grew unexpectedly in the early 1960s though precipitation stayed at its average level (Johnsson, 1995). Considerable variations in water level are often recorded against the background of relatively insignificant climate change, reporting badly on the economy of the regions, especially in the basins of the Rivers Neva and St. Lawrence.

As shown earlier, changes in precipitation, total runoff and water level north of 60°N will be opposite to the changes in the NAGL area. This is the fundamental difference between changes in water level and WBE in the LEL and NAGL. Temperature change in both water systems will have the same sign, but surface temperature amplitude will be half as small in the LEL system when compared with NAGL. Considerable changes will also be going on in drainage basins in terms of soil geochemistry and hydrology, the quality of surface and ground water. The NAGL area will face serious problems with reorientation of agriculture, location of water intake facilities, utilization of lake and ground water for drinking water supply. If CO_2 concentration in the atmosphere over the LEL area doubles, surface air temperature will grow by approximately 2–3°C. Precipitation will grow by about 4–5% with the greatest increase in winter 2%/10 years, and 1%/10 years in summer. Computations using global climate models yield less stable estimates for summer temperatures than for winter.

This chapter primarily addresses the specific features of the NAGL and LEL systems response to global climate changes. Changes in water level in lakes located in other climate zones have a number of similar features. Natural climate changes in each system are quasi-synchronous, the amplitude of these fluctuations is much greater than the amplitude of controlled variations in water level. Consider the regularities in the LEL and NAGL water-level variations with respect to natural climatic and man-made factors. Neither our data nor data reported by other authors hardly leads to the conclusion that changes in water level and WBE depend on climate change caused by the "greenhouse" effect. Shnitnikov (1966), Bishop (1990), Tushingham (1992) have shown that present-day water-level values were also observed in the past, and so, cannot be a result of anthropogenic forcing. Preliminary results have demonstrated that the characteristics of fluctuations recorded in water levels of large lakes were caused by climate fluctuations on the time scale of many centuries.

Similar regularities were obtained for lakes in the boreal zone of North America (*The State of Canada's Climate*, 1995). Surface water regime will vary almost simultaneously with climate dynamics. The effect of climate change on the boreal zone will in general be favorable for the biosphere, energy resources created by autotrophic plants will grow.

In some scenarios (*Climate of Europe*, 1995; *Climate Variations in Europe*, 1994), warming at the LEL latitudes may reach 1°C by 2010 and 2°C by 2030 with air temperature increase higher in winter than in summer, with the amplitude of fluctuations of surface air temperature δT (mean square deviation) retaining its present level. Climate in the LEL area as well as in Finland will become more maritime, and in northern areas more continental. As air temperature increases, water temperature in lakes will grow, the ice-free period will last from 7 to 9 months, and atmospheric precipitation will increase. This will result in rising water level in lakes. As these processes develop, erosion of the shores will intensify, more nutrients and organic matter will be washed out of the catchment area, and eutrophication of the basin will grow.

The above-described tendencies in climate change and their possible consequences should be considered when developing strategies for reasonable economic use and protection of natural resources. Climate fluctuations in Fennoscandia will need modification of agricultural practices in the region. Some additional measures on reducing the transport of biogenic matter from non-point sources are needed to attenuate the effect of eutrophication of water.

1.5 FEASIBILITY, MERITS AND LIMITATIONS OF USING A LARGE, DEEP LAKE AS A PHYSICAL MODEL OF THE OCEAN

Investigations of lakes are important not only in terms of limnology *per se* and/or local ecology. Large and deep lakes are considered as physical models of oceans, and as such could be used as test-sites for studying many attributes of marine and oceanic environments through *in situ* and remote observations as well as via mathematical modelling (Kondratyev *et al.*, 1990).

Presently, special emphasis is being placed on studying global oceans because of their exceptional role in forming climatological conditions on our planet. El Niño events, increasing levels of carbon dioxide in the atmosphere, and many other effects are thought to be directly related to the global ocean system activity and its dynamic interactions with the Earth's atmosphere. Understanding of this fact has led to perception and eventual realization of a series of programs, such as the World Ocean

Circulation Experiment (WOCE), the Tropical Ocean Global Atmosphere (TOGA) 10-year experiment, the North Atlantic Community Project, the Fine Resolution Antarctic Model (FRAM), as well as conjugated ocean–atmosphere numerical experiments including the Russian "Sections" Program based on the concept of so-called energy-active zones (EAZO), Joint Global Ocean Flux Study (JGOFS) (1988), the International Geosphere-Biosphere Program (IGBP) (1988), (see e.g., Kondratyev, 1992).

At the same time, recent developments of the World Climate Research Programme (WCRP) and the IGBP exhibit tendencies towards regional aspects of global change (*Global Change*, 1993). Some national and International regional programs have consequently been launched: in USA BOREAS (as a follow up to FIFE), in Asia GAME, in Europe ECHIVAL, HAPEX-MOBILITY, LOTREX-HIBE, EFEDA, NOPEX, etc. Within these regional research programs specific emphasis is now being placed on inland and coastal waters, especially because of their influence upon the global climate and *vice versa*, as well as in the context of their ecological importance. The United Nations Environmental Programme (UNEP launched a world-wide Environmentally Sound Management of Inland Water Programme (EMINWA). GEWEX plans a series of dedicated projects (International Continental Scale Project), covering the Mississippi drainage area, the Mackenzie River Study, and the GEWEX Asian Monsoon Experiment (GAME), and finally, the Land–Ocean Interactions in the Coastal Zone (LOICZ) (Kondratyev, Pozdnyakov, 1996). In Europe, in addition to GEWEX and IGBP activities, the Baltic Sea Experiment (BALTEX) is established. The Helsinki Commission (HELCOM) has also put forward the Baltic Joint Comprehensive Environmental Action Program, which, apart from addressing purely geophysical problems, is strongly oriented towards regional needs.

Thus, even a mere enumeration of scientific goals and relevant national/international programs clearly indicates the multifaceted importance of studies of inland waters both in their own right and in relation to global oceans.

Oceans are a "flywheel" of the planetary climatic system. Due to their gigantic cumulative surface, they accumulate enormous amounts of heat (acquired mainly through absorption of solar radiation) and release it continually throughout the year. They are also source and sink pools for many gases controlling the Earth's climatic status.

In view of their large latitudinal extent, oceans do not display pronouncedly hydrological seasons. However, in mid-latitudes such seasons are discernible, although there is a phase lag of about 6 weeks in arrival of the seasons (with respect to their austral counterparts). This phase lag is driven by the ocean's thermal inertia which also brings about a substantial decrease in the amplitude of annual water-temperature variations (Kondratyev *et al.*, 1990).

At least at latitude $40°N$, oceanic hydrological seasons could be specified from water surface-temperature (T_s) variations: mid-March to late June correspond to the hydrological spring, July–August to the summer, September–November to the autumn, and December–mid-March to the winter.

The circulation of oceanic waters is intimately related to the atmospheric general circulation. Under the influence of trade winds, water currents are formed in the equatorial zone which upon reaching the eastern coasts of America and Asia, are divided into two branches, i.e. northern and southern. The northern branches of the currents driven by the trade winds work their way along the eastern coasts far into the high latitudes. With the prevalence of west–east winds in high latitudes, the waters are further driven eastward and find themselves entrained in large-scale anticyclonic movements (Robinson, 1995).

Global ocean thermodynamic features are best exemplified in the Atlantic Ocean. It is most interesting to compare the thermohydrodynamics in the northern Atlantic with the relevant features of large and deep temperate lakes, using Lake Ladoga as a highly representative example. On the left side of the Gulf Stream core, cyclonic gyres are formed propagating down to depths exceeding 50 m; on the right side anticyclonic movements are generated with sizes twice as large as the cyclonic ones above. Within the frontal zones, some meso-circulation cyclonic and anticyclonic patterns arise measuring from 50 to 2000 km. These originate in meanders detaching from the Gulf Stream (Parker, 1971).

Unlike the oceans, in large temperate lakes the hydrological seasons are distinctly shown, although the aforementioned phase lag is basically also present (Ryanzhin, 1995). The timing of season transition is significantly influenced by both the atmospheric general circulation and local weather conditions (Adamenko *et al.*, 1991).

In large temperate lakes like Lake Ladoga, cyclonic circulation of water is most characteristic. Not related to persistent winds, it arises in spring when the water temperature in the coastal zone increases more rapidly than in hypolimnetic waters. As the spring heating period progresses, the coastal waters stratify rapidly, although the deep waters remain isothermal until the autumn when a lake-wide stratification is established. The frontier (so-called "*thermal bar*") separating these two thermally different regions in the lake provides horizontal temperature (and hence density) gradients from which pressure-induced currents originate: the water in the peripheral part of the lake starts moving off the shoreline, and being affected by the Coriolis force, deviates to the right, thus forming a cyclonic circulation. With progressive heating of the lake, the thermal bar displaces towards the deepest (and hence the coolest) part of the water body. The horizontal velocity of the thermal bar displacement is directly proportional to the intensity of heat fluxes coming into the lake and inversely proportional to the slope of the bottom. The strength of the cyclonic circulation gradually increases with enhancement of the radial thermal gradient in the peripheral part of the lake. Meanwhile, in the central cold-water region the temperature rises but very slowly approaching 4°C which corresponds to the highest density of water. Thus a density current also forms in the cold-water region of the lake driving waters from the center to the thermal bar. This current, deviating to the right due to the Coriolis force, forms an anticyclonic inner circulation. However, with spring warming and thermal bar evolution, the anticyclonic water movement domain progressively shrinks and finally fades away totally surrendering the lake to cyclonic circulation. It is noteworthy that cyclonic and anticyclonic circulations bordering the thermal bar on both sides cause formation of meanders and ring-like water motions of different spatial scales which accompany the evolution of the lake thermal bar (Tikhomirov, 1982; Filatov, 1983).

At the beginning of the hydrological summer the thermocline resides in near-surface waters in the pelagic region and at a depth of 8–10 m in the coastal zone. The thermocline acquires a dome-like form. In the second part of the hydrological summer, the dome of cool water begins sinking, whereas t_s becomes rather uniform throughout the whole lake.

By the fall, the winds become stronger causing wind-driven local circulation patterns superimposed on previously established but essentially subdued general cyclonic circulation. Wind-driven circulation patterns acquire the form of rings of different sizes, sometimes reaching 20 to 30 km in diameter. A further cooling of the lake results in gradual decrease in t_s accompanied by transfer of heat down to deeper layers.

At the beginning of winter cooling (in late December for Lake Ladoga), t_s in the pelagic region becomes less than 4°C. This coincides in time with the formation of a still ice edge along the lake coastline. Soon, winter homothermal conditions throughout the lake become established.

Regardless of the aforementioned differences in "genetics" of water motion patterns in oceans and large deep lakes, they bear a lot of similarities apart from tides and planetary Rossby waves which are only characteristic of oceans (Table I.3).

Consequently, in many respects a large deep lake can be taken as a scaled model of an ocean. Indeed, lake-wide cyclonic circulation discussed above might be considered as a model of oceanic synoptic gyres. Thermal fronts frequently arising within oceanic jet currents and synoptic gyres are very similar to the thermal bar in lakes which usually manifests itself very distinctively on the water surface because of much stronger (than in oceans) thermal gradients in the horizontal plane. Displacement of the thermocline surface, due to surface wind stress, may result in the thermocline intersecting the water surface. At such times of *upwelling*, deep water may be brought to the surface. It occurs both in oceans and large and deep lakes. The vertical velocities in central areas of lake circulation patterns as well as within the lake coastal upwelling zones are an order of magnitude higher than in the oceans, which makes it possible to assess more reliably the role of regular vertical motions and advection in forming the upper layer of the thermocline.

Numerous results of numerical experiments and field observations conducted in lakes and oceans suggest that the decisive role in setting up one or another type of general circulation in deep natural water bodies belongs to baroclinicity and bottom relief, although the near-surface wind also plays a substantial role in quasi-periodic fluctuations in the vigor of the general circulation. The near-surface wind is also conducive to the emergence of currents in the upper layers. Since circulation patterns in deep lakes are relatively simple, while lake dynamics is closely analogous with that of oceans, it is possible to test the reliability of oceanic thermohydrodynamic models using lakes as test-sites. For example, autonomous buoy stations (ABS) could be deployed over the entire area of a large deep lake such as Ladoga, Onega or one of the American Great Lakes in order to adequately assess the interactive effect of

Table 1.3. Major dynamic processes and phenomena basically occurring in oceans and lakes (Kondratyev *et al.*, 1990).

Process/phenomenon	Characteristic dimensions of water basin (L = horizontal; H = vertical)	
	Ocean (L = 10^3 km, H = 4×10^3 m)	Deep lakes (L = 10^2 km, H = 10^2 m)
Rossby planetary waves	$\lambda = 10^5$ m, $V_{orb} = 50$ cm/s	—
Inertial gravity waves	+	+
Solitons	($V_{ph} = 1$ cm/s)	($V_{ph} = 1$ cm/s)
Kelvin waves	($V_{ph} = 10$ cm/s)	($V_{ph} = 10$ cm/s)
Topographic waves	+	+
Standing waves (seiches)	$\lambda = 10^3$ km at the Equator	$\lambda = 10^2$ km
Acoustic waves	+	+
Capillary and gravity waves	+	+
Tidal motions	$V = 15$ cm/s	—
Synoptic gyres	$V_{orb} = 150$–200 cm/s	$V_{orb} = 20$ cm/s
Coastal jet streams	$V = 50$ cm/s	$V = 10$ cm/s
Small-scale turbulence	+	+
Microinhomogeneities	+	+, but devoid of mechanism of dual diffusion
Stratification	+	+, but devoid of constant thermocline
Fronts	+	+
Coastal upwelling	+	+

Notes: λ = the wavelength, V_{orb} = the orbital velocity, V_{ph} = the phase velocity, V = the linear velocity.

water motion on various spatial/temporal scales. Understandably, a strictly analogous experiment could not be conducted in oceans because of presently insurmountable logistic problems.

In view of considerable water surface-temperature gradients appropriate to lakes, a combined use of ABS, shipborne and remote sensing data, *internal waves* (i.e. waves arising at the interface separating water layers of different density which move with respect to each other) can be studied in more detail. Such studies would also give better insight into the nonlinear interaction of internal waves with large-scale water motion including general circulation.

If the lake water is fresh, i.e. practically devoid of dissolved salts, a whole variety of inhomogeneities in the spatial distribution of physico-biochemical properties could be studied under conditions when the interference of one of the controlling factors in oceanic water (water salinity) could be completely eliminated. Since, in the absence of water salinity, the mathematical formulation of many thermodynamical processes is facilitated, it would on one hand considerably simplify the analysis, and on the other, highlight the actual role of salinity in forming oceanic physico-biochemical fields.

A large deep lake, considered as a model of the ocean, is a suitable water basin in which to conduct comprehensive field studies relating to the thermodynamic interaction between the atmosphere and the water body. One of the major advantages of this approach in further developments of studies under such programs as "Sections is the use of a scaled ocean model with precisely controlled internal parameters of both atmospheric and aquatic boundary layers, as well as external factors determining system boundary conditions. Consequently, a large deep lake can be an "open-air" laboratory where sophisticated field experiments with well-controlled boundary conditions can be conducted.

Investigations of basic regularities in spatial distribution and temporal variability of hydrometeorological elements and energy exchanges between the atmosphere and water bodies is a common problem in oceanography, limnology and climatology. Development of the theory and adequate mathematical models of the atmosphere–water body interactions is only possible when based on coupling reliable submodels simulating thermal processes and circulation mechanisms in these two media. The energy exchange between the atmosphere and the ocean takes place across the respective boundary layers, which, in their turn, arise from this exchange.

Revealed through theoretical and field investigations on lakes, specific features of momentum, heat and water vapor redistribution at the air–water interface as well as related processes with pronounced feedback paths provide the grounds for considering the lake and atmosphere over it as a single thermodynamic system functioning due to one and the same energy source, which is incoming solar radiation. Obviously, there is a strict analogy between the oceans and lakes in this sense: oceans, like lakes, impact atmospheric processes via heat and water-vapor exchange whereas the atmosphere influences in return the dynamics of underlying water bodies. However, such a statement does not imply necessarily the validity of direct application of relevant results obtained in lakes to oceans. It should be borne in mind that, on one hand, the ocean is invariably stratified and reacts to atmospheric impacts as a two-layer system: changes induced by atmosphere in the upper layer are of seasonal nature, whereas in the lower layer they have a climatic character. Conversely, lakes, particularly shallow ones, may remain unstratified for long periods, sometimes even the whole year round, and as such respond to atmospheric forcing as a unity. Such differences should be taken into account when considering the possibility of using a lake as a scaled physical model of the ocean. Moreover, respective dimensionless criteria determining the extent of similarity between oceanic and lake processes under study should be closely examined (Kondratyev *et al.*, 1990).

Scale analyses and field observations indicate that vertical transport of heat and momentum arising from atmosphere–water body interactions is driven by turbulent mixing (Monin, Ozmidov, 1981). In both lakes and oceans, a thermally quasi-homogeneous layer forms near the air–water surface interface. The investigation of processes taking place within the upper wind mixed layer and in the zone of thermocline is important in terms of assessing the potential sources of energy accumulated in the water basin. From this point of view, a study of lake mixed-layer development can provide a more adequate insight into relevant oceanic processes than laboratory experiments because of an unconditionally closer physical similarity between oceans and lakes than between oceans and laboratory tanks.

Apart from the turbulent entrainment issue, a number of other important problems of the atmosphere–ocean interaction can be solved more accurately and in a simpler way by using a large deep lake as a natural model for the ocean. For instance, the behavior of coefficients of dynamic interaction at high wind speeds is poorly understood largely because of considerable technical difficulties associated with conducting large-scale experiments in stormy oceans. Obviously, lakes are better suited as test sites with a system of anchored stations for continuous observations during such extreme events.

Two other problems associated with the ocean surface become tractable when looking at opportunities provided by lakes. These are the quantitative assessment of the capillary waves role in the formation of microscale inhomogeneities (structures), and visualization, perhaps with the help of bubble clouds, of internal waves emerging onto the water surface, a frequently occurring but poorly studied phenomenon in oceans.

In studying energy transformation at the air–water interface, additional problems come into being with the formation of ice cover: the up and down transfer of heat, moisture and momentum becomes dependent on the status and spatial distribution of ice. Understandably, it is much easier to conduct such studies on lakes due to their limited areas. To answer the question to what extent one may use large fresh-water lakes like Ladoga, Onega or NAGL as models to study ice growth/decay in the ocean, it is necessary to consider together ice-cover formation processes in large lakes and oceans (Chizhov, Borodoulin, 1984).

The onset of a positive net heat flux coming up to the atmosphere in the fall–winter time leads to cooling and subsequent freezing of the water surface. The formation of ice begins with the appearance of ice needles, i.e. small, needle-shaped crystals suspended in water. The ice needles rapidly grow both in number and size to form nonfreezing clusters called ice greese. If accompanied by snowfall, a snow mass piles up on the water surface gradually transforming into the sludge. Under windy conditions, the greese, snow sludge and in-water ice together produce the shuga—an amassment of loose, porous, whitish lumps of ice. In the fall, within the coastal zone, under calm and cold weather conditions, a fragile crust of ice (ice rind) grows but gets easily broken by winds and waves. These initial phases of ice growth in oceans and lakes are very similar. However, the enhanced salinity of oceanic water compared with fresh water slows down ice-growth rates and delays freezing: although fresh water freezes at $0°C$, in typical sea-water conditions of 36 ppt salinity the freezing point is $-1.9°C$ (Shouleykin, 1968). Besides, since the temperature of the highest sea-water density is greater than the temperature of freezing, a strong convective mixing in the upper-water layers arises which also delays the onset of freezing. Moreover, sea-water freezing is further delayed due to salinity enhancement in near-surface waters: upon sea-water freezing, only a fraction of dissolved salts find their way into the ice, while the remaining part increases the local water salinity, and hence, the density of the surface-water layer. This mechanism leads not only to lowering the freezing temperature, but also promotes and intensifies convective mixing which slows down the ice-growth rates.

The subsequent development phases of ice cover in oceans and fresh-water lakes differ considerably due to salinity. Sea ice, starting from dark nilas, has limited thickness and is characterized by a specific coloration arising from brine migration through the bulk ice.

In fresh-water bodies, apart from ice in initial development phases, which is similar to the sea ice, dark and light nilas can also be identified as resembling sea nilas very closely. Here the similarity between formation of fresh and sea-water ice-cover ends. Fresh-water ice further develops according to its own laws, which are different from those appropriate to sea-water ice. As a result, it does not change colour either with age (in the absence of melting) or in the course of ice thickness building up.

However, despite the aforementioned specific features distinguishing the intrinsic mechanisms of growth of ice in lakes and oceans, the common property of ice of both origins resides in the analogy of processes of ice/snow cover interaction with the atmosphere. Therefore, many problems relating to heat exchange between ice-covered ocean areas and the atmosphere should be closely studied on lakes in winter time, e.g. in regions of expanding ice edges, or in constant unfrozen patches of water in the midst of the icebound lake.

There are many similarities in the characteristics of heat balance in oceans and lakes except those relating to fetch, wave effects and large-scale horizontal advection of heat.

The net energy N supplied to the ocean is given by (Henderson-Sellers, 1984)

$$N = R - (P + LE) \tag{1.1}$$

where R is the net radiation (shortwave plus net longwave), P is the sensible heat loss, and LE is the latent energy loss.

To estimate the temporal variability of the sum $P + LE$ and of the resulting heat exchange N in the ocean, it is necessary either to define the air–water surface temperature increment (Δt) or separately give the air and water temperature for all the sites under study.

Similar relationships have been suggested for inland water bodies. The major difference between oceans and lakes is that in oceans the total flux $P + LE$ and the resulting heat exchange N are always inversely proportional to the temperature and directly proportional to the temperature gradient across the air–water interface whereas in lakes there is a period (December–March) when the total flux $P + LE$, unlike the rest of year, grows linearly with rising air temperature.

In oceans, the surface temperature gradients of latitudinal water are rather substantial. For instance, in the North Atlantic the water surface temperature varies by 7, 10 and 5°C upon sequential transition through latitudinal zones 0–35, 35–45, 45–65°N, respectively. The mean meridional water temperature gradients in the near-equatorial, subtropical and temperate zones are 0.18, 1.0, and 0.25°C/100 km, respectively. Interestingly, the temperature gradients of latitudinal water decrease with depth whereas the temperature gradient of meridional water, peaking in the subtropical zone retains this peculiarity throughout the entire quasi-homogeneous layer down to a depth of about 0.5 km (Filippov, 1984).

In large lakes, the temperature gradients of latitudinal water are two to three orders of magnitude higher than in the ocean (Ryanzhin, 1994). This gives us an opportunity of choosing in a large lake, such as Ladoga, areas analogous (in terms of component values of heat balance) to almost every region in the North Atlantic or Pacific Oceans.

The temperature gradients of vertical water in Lake Ladoga can reach values of about 10°C per 100 m in spring time. It indicates that, at least in this season, lake-water stability either equals or even exceeds that observed at the Equator. Moreover, within the zone extending from the coastline to the thermal bar, lake-water stability is pretty close to that typical of the Atlantic temperate zone. Apparently, it is possible to identify in a large lake some areas which, in terms of vertical stability of water, are analogous to oceanic EAZO.

It is very important to study the impact produced by monomolecular films of natural and anthropogenic origin as well as by wind-induced foam and white-caps on mass and energy exchange in oceanic upper layers. These processes are difficult to study in oceans. Conversely, in lakes such observations and dedicated experiments would not pose serious logistic and technical problems. *In situ* measurements from captive balloons of the vertical profiles of air temperature and humidity, and near water-surface wind would provide valuable data on turbulent fluxes of momentum, heat and water vapor in the boundary layer. Moreover, such data would furnish answers to many methodological questions related to measurements of relevant coefficients of exchange and to the impact made by large research vessels on the results of *in situ* measurements in the atmospheric boundary layer.

Finally, the identification of benefits arising from using a large deep lake as a model of the ocean would not be complete without mentioning the hydro-optical aspect of this issue.

The composition of ocean water is similar to that of lakes: both water types basically contain dissolved and suspended matter. The dissolved component encompasses mineral salts, gases, organic substances. The suspended fraction (so-called hydrosol) is composed of mineral, mainly terrigenous, matter, and living and dead organic matter (plankton and detritus). Although they have similar qualitative chemical composition, oceanic and lake waters seriously differ in terms of salinity.

The salinity of ocean water varies along the meridian: it is minimum near the equator and maximum at latitudes 20°, averaging 35 ppt. Lake-water salinity varies from 10 to 250 g/m^3. Lake Ladoga's water salinity is *ca.* 50 g/m^3, which is two orders of magnitude less than the average salinity of the ocean water (Shouleykin, 1968).

In the open ocean, concentrations of dissolved organic and suspended minerals are fairly low, whereas phytoplankton abundance can be quite considerable, particularly during blooming events. Although dissolved organic levels in the open ocean are low, they basically correlate with phytoplankton content.

In lake water, dissolved and suspended fractions may be found in large and not mutually correlating amounts (e.g. as observed in Lake Ladoga). In this sense, the hydrooptical situation in lakes strongly resembles the hydro-optical situation appropriate to oceanic coastal zones which are basically affected by river discharge, runoff, erosion of the coastline, eolian fallout and many other factors apart from the anthropogenic influence produced by riparian industrial and urban sewage effluent, agriculture and transboundary transport.

In general, phytoplankton communities in oceans and lakes in each vegetation season are very much alike in terms of belonging to the same major taxonomic groups, but they differ at the level of concrete algae species indigenous to ocean and fresh water (Kissilev, 1969).

Substantial spatial inhomogeneity in horizontal and vertical planes is characteristic of both ocean and lake waters. This is a result of various processes taking place on the water surface, and in deep water. Generally speaking, lake water is optically more complex than ocean water because of the variety and amounts of co-existing optically active components (e.g. phytoplankton, suspended and dissolved mineral and organic matter, etc.). Inhomogeneities in their spatial/temporal distributions are more pronounced and, hence, more readily traceable down to the relevant driving mechanisms and phenomena. That is why their investigation in lakes is less complicated and more promising in terms of attaining research goals.

There is also an important methodological aspect to this issue: since in a big and deep lake, such as Ladoga, almost the entire variety of hydro-optical conditions of the ocean coastal zone could be encountered, dedicated field and numerical experiments can be carried out to develop techniques of retrieving water quality parameters from remote (space-, air-, and shipborne) data as well as for conducting appropriate training and validation experiments. In many instances, such research is logistically more readily accomplished in lakes than in marine/oceanic coastal waters, and the results thus obtained are perhaps more adequate since input data for development of the retrieval algorithm and numerical modelling could be obtained for well-known boundary and in-water conditions (Kondratyev, Pozdnyakov, 1988).

Closing this discussion, it should be pointed out that the outlined merits of using a large deep lake as an analog for oceanological studies in reality are much more numerous and versatile extending far beyond the scope of our book. Our vision of the book format and composition precluded a more detailed and exhaustive examination of this issue.

REFERENCES

A Citizen's Agenda for Restoring Lake Ontario. 1987. New York: Great Lakes United, 87 pp.

Adamenko V.N. 1987. *Lakes and Climate*. Leningrad: Nauka Press, 263 pp. (in Russian).

Adamenko V.N., Kondratyev K.Ya., Pozdnyakov D.V., Chekhin L.P. 1991. *Radiation Regime and Optical Properties of Lakes*. Leningrad: Gidrometeoizdat., 300 pp. (in Russian).

Andronikova I. 1996a. Zooplankton characteristics utilizable in Lake Ladoga monitoring. *Hydrobiologia*, **322**, 173–179.

Andronikova I.N. 1996b. *Structural and Functional Organization of the Lake Ecosystem Zoopankton*. St. Petersburg: Nauka Press, 190 pp. (in Russian)

Anthropogenious Eutrophication of Lake Ladoga. 1982. Leningrad: Nauka, pp. 51–39 (in Russian).

Assel R., Robertson D. 1995. Changes in winter air temperatures near Lake Michigan, 1851–1993, as determined from regional lake-ice records. *Limnology, Oceanology*, **40**, No 1, 165–176.

Bishop C. 1990. Historical variation of water levels in Lake Michigan–Huron. *J. Great Lakes Res.*, **16**, No. 3, 406–423.

Bolsenga S., Norton D. 1993. Great Lakes air temperature trends for land stations, 1901–1987. *J. Great Lakes Res.*, **19**, No. 2, 379–388.

Borisenkov E.P. 1982. *Climate and Human Activities*. Moscow: Nauka Press, 131 pp. (in Russian).

Carter T., Posch M., Tumenvirta H. 1995. *SILMUCEN and CLIGEN User's Guide*. Helsinki: SILMU, 62 p.

Chizhov A.N., Borodoulin V.V. 1984. On the nomenclature of ice types. In: *Proceedings Hydrological Institute*. Leningrad: Gidrometeoizdat., pp. 47–51 (in Russian).

Climate Change. 1996. Impacts, adaptation and mitigation of climate change. In: Watson R., Zinyowera M., Moss R., Dokken D. (eds.), *Scientific–Technical Analyses*. Cambridge: IPCC, 412 pp.

Climate Change of Europe in History. 1995. Krenke A.N. *et al.* (eds.) Moscow: Nauka Press, 224 pp. (in Russian).

Climate of Europe. 1995. First European Climate Assessment. The Netherlands: ECSN, p. 73.

Climate Variations in Europe. 1994. Helsinki: Academy of Finland, Vol. 3, 385 pp.

Current Status of Lake Ladoga. 1987. Petrova N., Terjevik A. (eds.). Leningrad: Nauka Press, 203 pp. (in Russian).

Davydova N., Kalmykov M., Sandman O. *et al.* 1993. Recent palaeolimnology of Kondopoga Bay, Lake Onega, reflecting pollution by a large pulp mill. *Verh. Int. Verein. Limnol.*, **25**, 1086–1088.

Doubrovina L.V., Fedorova N.V., Pomazovskaya I.V. 1994. Experimental research on heavy metal toxicity in an acidic environment. In: Mononen P., Lozovik P. (eds.), *Acidification of Inland Waters. Proceedings of the 3rd Soviet–Karelian–Finnish Symposium on Water Problems, Joensuu, Finland, 3–7 June 1991*. Helsinki: Finnish National Board of Waters and the Environment, pp. 117–129.

Drabkova V., Roumyantsev V., Sergeeva L.V., and Slepoukhina T.D. 1996. Ecological problems of Lake Ladoga: causes and solutions. In: Simola H., Viljanen M., Slepoukhina T., Murthy R. (eds.), *Proceedings of the 1st International Lake Ladoga Symposium, St. Petersburg, Russia, 22–26 November 1993*. Dordrecht: Kluwer Academic, pp. 1–7.

Ecological Problems of Northwestern Russia and Ways of their Solution. 1997. Inge-Vechtomov S.G., Kondratyev K.Ya., Frolov A.K. (eds.). St. Petersburg: Nauka Press, 528 pp. (in Russian).

Ecological State of Water Bodies and Water Courses Pertaining to the Neva River Basin. 1996. Alimov A.F., Frolov A.K. (eds.). St. Petersburg: Nauka Press, 224 pp. (in Russian).

Environmental Quality. 1990. Twentieth Annual Report. Washington, D.C.: The White House, 495 pp.

Estonia in the System of Global Climate Change. 1996. Punning J-M. (ed.). Tallinn, Vol. 4, 205 pp.

Filatov N.N. 1983. *The Lake Dynamics.* Leningrad: Gidrometeoizdat., 165 pp. (in Russian).

Filatov N.N. 1991. *Hydrodynamic of Lakes.* St. Petersburg: Nauka Press, 196 pp. (in Russian).

Filatov N.N. 1997. *Climate Change and Variability of Eastern Fennoscandia and Water Level Fluctuations of Largest Lakes of Europe.* Petrozavodsk: Karelian Research Centre, 148 pp.

Filippov D.M., 1984. *A Climatic Analysis of the Physical Fields in the Atlantic and Pacific Oceans.* Leningrad: Gidrometeoizdat., 213 pp. (in Russian).

Five Year Strategy for Great Lakes National Program Office. 1988. Chicago: EPA, 96 pp.

Froumin G.T., Chernykh O.A., Krylenkova N.L., Scherbak. 1996. Lake Ladoga: chemical pollution and biochemical self-purification. In: Simola H., Viljanen M., Slepoukhina T., Murthy R. (eds.), *Ecological Problems of Lake Ladoga: Causes and Solutions. Proceedings of the First International Lake Ladoga Symposium, St. Petersburg, Russia, 22–26 November, 1993.* Dordrecht, *et al.*: Kluwer Academic, pp. 143–147.

Frumin G., Soussareva O. Barkan L. 1997. Water quality in Lake Ladoga. In: Simola H., Viljanen M., Slepoukhina T. (eds.), *Proceedings of the 2nd International Lake Ladoga Symposium, Joensuu, Finland, 26–30 August 1996.* Joensuu: Joensuum juliopisto, pp. 47.

Gilbertson M., Morris R.D., Hunter R.A. 1976. Abnormal chicks and PCB residue levels in eggs of colonial birds on the Lower Great Lakes, 1971–1975. *Auk*, **93**, No. 5, 434–442.

Girs A.A. 1971. *Many Years Variability of Atmospheric Circulatons and Long Term Hydrometeorological Forecasting.* Leningrad: Hydrometeorological Press, 280 pp. (in Russian).

Gornitz V., Rosenzweig C., Hillel D. 1997. Effects of antropogenic intervention in land hydrological cycle on global sea level rise. *Global Planetary Change*, **14**, 147–161.

Gromov B.V., Vepritsky A.A., Mamkaeva K.A., Voloshko L.N. 1996. A survey of toxicity of cyanobacterial blooms in Lake Ladoga and adjacent water bodies. In: *Ecological Problems of Lake Ladoga: Causes and Solutions. Proceedings of the 1st International Lake Ladoga Symposium, St. Petersburg, Russia, 22–26 November 1993.* Dordrecht: Kluwer Academic, pp. 149–151.

Guidelines of Lake Management. Vol. 1. *Principles of Lake Management.* 1989. Jorgensen S.E., Vollenweider R.A. Shiga, Japan: International Lake Env. Committee Foundation, 200 pp.

Havens K.E., Heath R.T. 1991. Increased transparency due to changes in the algal size spectrum during experimental acidification in mesocosms. *J. Plankton Res.,* **13**, No. 3, 673–679.

Heino R. 1994. Climate in Finland during the period of meteorological observations. Finnish Meteorological Institute Contr., No.12, 120 pp.

Henderson-Sellers B. 1984. *Engineering Limnology.* Boston: Pitman Advanced Publishing Program, 356 pp.

Henderson-Sellers B., Markland H.R. 1987. *Decaying Lakes.* Chichester: John Willey & Sons, 279 pp.

Hiltunen T. 1984. *What Do Hydrological Time Series Tell About Climate Change.* Helsinki: Water and Environmental Institute, 41 pp.

Historical Runoff Variations in the Nordic Countries. 1995. Hisdal H., Erup J. (eds.).

Hoffman D.J. 1987. Teratogenicity and aryl hydrocarbon hydroxylase activity in Forster's terms on Green Bay, Lake Michigan. *Environ. Res.,* **42**, No. 2, 176–184.

Holopainen A.-L., Letanskaya G.I. 1997. Effects of nutrient load on phytoplankton species composition and productivity in Lake Ladoga. In: Simola H., Viljanen M., Slepoukhina T. (eds.), *Proceedings of the 2nd International Lake Ladoga Symposium, Joensuu, Finland, 26–30 August 1996.* Joensuu: Joensuum juliopisto, pp. 87.

Hublarjan M.G. Naidenov V.IO., Krutova N.M. 1996. Analysis of water level by nonlinear models. *Rep Acad. Sci.,* **35**, No. 4, 539–543 (in Russian).

Hydrological Regime and Water Balance of Lake Ladoga. 1966. Leningrad: Nauka Press, 323 pp. (in Russian).

Iofina I., Petrova N. 1997. The impact of toxic metals on the structure of planktonic communities from Lake Ladoga. In: Simola H., Viljanen M., Slepoukhina T. (eds.), *Proceedings of the 2nd International Lake Ladoga Symposium, Joensuu, Finland, 26–30 August 1996.* Joensuu: Joensuum juliopisto, pp. 114–119.

Johnsson T. 1995. Paleoclimatic studies suggest that lake Victoria was dry. *Bull. IDEAL,* 1.

Kajander J. 1993. Methodological Aspects on river cryophenology exemplified by a tricentennial break-up time series from Tornio. *Geophisica,* **29.** No. 12, 73–96.

Kapoustina L. 1996. Bacterioplankton Response to Eutrophication in Lake Ladoga. 1996. In: Simola H., Viljanen M., Slepoukhina T., Murthy R. (eds.), *Ecological Problems of Lake Ladoga: Causes and Solutions. Proceedings of the 1st International Lake Ladoga Symposium, St. Petersburg, Russia, 22–26 November 1993.* . Dordrecht: Kluwer Academic, pp. 17–22.

Kenttamies K. 1994. Effect of acid deposition on waters. Acidification of inland waters. In: Mononen P., Lozovik P. (eds.), *Proceedings of the 3rd Soviet–Karelian–Finnish Symposium on Water Problems, Joensuu, Finland, 3–7 June 1991.* Helsinki: Finnish National Board of Waters and the Environment, pp. 9–41.

Kissilev M.A.,1969. *Marine and Inland Fresh Water Plankton*. Leningrad: Nauka Press, 657 pp. (in Russian).

Kite G. 1992. Spectral analysis of selected Lake Erie levels. *J. Great Lakes Res.*, **18**, No. 1, 207–217.

Kite G. 1993. Analysing hydrometeorological time series for evidence of climatic change. *J. Nordic Hydrol.*, **24**, 135–150.

Kondratyev K.Ya. 1992. *Global Climate*. Leningrad: Nauka Press, 357 pp. (in Russian).

Kondratjev K.Ya., O. Johannessen.1995. *Arctic and Climate*. St. Petersburg: Nauka Press, 139 pp. (in Russian).

Kondratyev K.Ya., Pozdnyakov D.V., Filatov N.N. *et al.* 1990. Using large lakes as analogues for oceanographic studies. *Modelling Marine Systems*. Boca Raton: CRC Press, Vol. 11, pp. 341–384.

Kondratyev K. Ya., Kantor R.R., Kargin B.A., Pozdnyakov D.V. 1987. *Numerical Modeling in Developing Optical Remote Sensing of Inland Water Bodies*. Leningrad: Nauka Press, 62 pp. (in Russian).

Kondratyev K.Ya., Adamenko V.N., Henderson-Sellers B., Pozdnyakov D.V. 1990. Using large lakes as analogues for oceanographic studies. In: *Modelling Marine Systems*. Boca Raton: CRC Press, Vol. 2, pp. 299–344.

Kondratyev K.Ya., Pozdnyakov D.V. 1988. *Optical Properties of Natural Waters and Remote Sensing of Phytoplankton*. Leningrad: Nauka Press, 185 pp. (in Russian).

Kondratyev K.Ya., Pozdnyakov D.V. 1993. Ecology of North American Great Lakes: problems, solutions and perspectives. *Water Resources*, **20**, No. 1, 113–122 (in Russian).

Kostamo A., Pellinen J., Kukkonen J. Viljanen M. 1997. EOX and organic chlorine compounds in fish and seal samples from Lake Ladoga, Russia. In: Simola H., Viljanen M., Slepoukhina T. (eds.), *Proceedings of the Second International Lake Ladoga Symposium, Joensuu, Finland, 26-30 August 1996*. Joensuu: Joensuum juliopisto, pp. 130.

Kudersky L.K., Jurvelius J., Kaukoranta M. 1996. Fishery of Lake Ladoga—Past, Present and Future. In: Simola H., Viljanen M., Slepoukhina T., Murthy R. (eds.), *Ecological Problems of Lake Ladoga: Causes and Solutions. Proceedings of the 1st International Lake Ladoga Symposium, St. Petersburg, Russia, 22–26 November 1993*. Dordrecht: Kluwer Academic, pp. 57–64.

Kuusisto E. *et al.* 1994. *Climate and Water in Europe: Some Recent Issues*. Geneva: WMO, 8 pp.

Kuusisto E., Kauppi L., Heikinheimo P. (eds). 1996. *Ilmastonmuutos ja Suomi*. Helsinki: Helsinki University Press, 265 pp.

Lake Ladoga: Ecosystem Status Criteria. 1992. Petrova N.A., Torzhevik A.Yu. (eds.). St. Petersburg: Nauka Press, 325 pp. (in Russian).

Lake Onega Water Ecosystem and Tendency of Their Variability. 1990. Leningrad: Nauka Press, 264 pp. (in Russian).

Living with the Lakes: Challenges and Opportunities. 1989. A progress report to the IJC. The water level, pp. 16–18.

Lozovik P., Niinioja R., Raspletina G. 1997. Anthropogenic and natural loading in lake Ladoga. In: Simola H., Viljanen M., Slepoukhina T. (eds.), *Proceedings of the 2nd International Lake Ladoga Symposium, Joensuu, Finland, 26–30 August 1996.* Joensuu: Joensuum juliopisto, pp. 26–32.

Lumme I., Arkhipov V. 1996. Atmospheric Deposition of Sulphur, Nitrogen and Base Cations in Scots Pine Stands of Southeastern Finland and the Karekian Isthmus, NW Russia in 1992. In: Simola H., Viljanen M., Slepoukhina T., and Murthy R. (eds.), *Ecological Problems of Lake Ladoga: Causes and Solutions. Proceedings of the 1st International Lake Ladoga Symposium, St. Petersburg, Russia, 22–26 November, 1993.* Dordrecht: Kluwer Academic, pp. 223–226.

Makarewicz J.C. 1987. *Phytoplankton, and Zooplankton in Lake Erie, Huron and Michigan*, Rep. No. 06 EPA—905/2-87-002. Chicago: EPA, 137 pp.

Malinina T.I. 1966. Lake Ladoga water balance. In: *Hydrological Regime and Water Balance of Lake Ladoga.* Leningrad: Nauka Press, pp. 182–203 (in Russian).

Mannio J. 1994. Acidification and trace metals in lakes. In: Mononen P., Lozovik P. (eds.), *Acidification of Inland Waters. Proceedings of the 3rd Soviet–Karelian– Finnish Symposium on Water Problems, Joensuu, Finland, 3–7 June, 1991.* Helsinki: Finnish National Board of Waters and the Environment, pp. 109–117.

Molchanov I.V. 1946. *Lake Ladoga.* Leningrad: Hydrometeorological Press, 559 pp. (in Russian).

Monin A.S., Ozmidov R.V. 1978. Turbulence in the oceans. In: *Physics of the Ocean.* Moscow: Nauka Press, Vol. 1, 148 pp. (in Russian).

Mortsch L., Quinn F. 1996. Climate change scenarios for Great Lakes Basin ecosystem studies. *Limnology, Oceanography*, **41**, No. 5, 903–911.

Muzilev S.V., Privalsky V.E., Ratkovich D.Ja. 1982. *Stochastic Models in Engeneering Hydrology.* Leningrad: Nauka Press, 184 pp. (in Russian).

Parker C.E. 1971. Gulf Stream rings in the Sargasso sea. *Deep-Sea Res.*, **18**, 981–993.

Petrova N.A. 1990. *Phytoplankton Succession Driven by Anthropogenic Eutrophication of Large Lakes.* Leningrad: Nauka Press, 200 pp. (in Russian).

Petrova N.A., Goussakov B.L., Stravinskaya E.A. 1987. Specific features of anthropogenic eutrophication of large lakes. In: *Present Status of the Lake Ladoga Ecosystem.* Leningrad: Nauka Press, 213 pp. (in Russian).

Privalsky V. 1992. Statistical analysis and predictability of Lake Eriewater level variations. *J. Great Lakes Res.*, **18**, No. 1, 236–243.

Privalsky V.E. 1985. *Climate Variability.* Leningrad: Nauka Press, 183 pp. (in Russian).

Rask M. 1994. Effects of acid deposition on fish populations of small lakes in Finland. In: Mononen P., Lozovik P. (eds.), *Acidification of Inland Waters. Proceedings of the 3rd Soviet–Karelian–Finnish Symposium on Water Problems, Joensuu, Finland, 3–7 June 1991.* Helsinki: Finnish National Board of Waters and the Environment, pp. 81–97.

Ratkovich D.Ja. 1993. *Hydrological Basis of Water Supply.* Moscow: Nauka Press, 429 pp. (in Russian).

Robinson I.S. 1995. *Satellite Oceanography.* Chichester: Wiley-Praxis, 455 pp.

Rogkov V.A., Trapeznikov J.A. 1990. *Stochastic Models of Oceanographic Processes*. Leningrad: Hydrometeoizdat., 271 pp. (in Russian).

Roumyantsev V., Viljanen M., Slepoukhina T. 1997. The present state of Lake Ladoga, Russia—a review. In: Simola H., Viljanen M., Slepoukhina T. (eds.), *Proceedings of the 2nd International Lake Ladoga Symposium, Joensuu, Finland, 26–30 August, 1996*. Joensuu: Joensuum juliopisto, pp. 11.

Ryanzhin S.V. 1994. Latitudinal-altitudinal interrelationships for the surface temperatures of the Northern Hemisphere freshwater lakes. *Ecol. Modelling*, **74**, No. 3–4, 231–253.

Ryanzhin S.V. 1995. Interrelationships for spatial and temporal variability of thermohydrodynamic processes in freshwater lakes. DSc. Thesis, St. Petersburg State University, St. Petersburg, Russia, 232 pp.

Saarnisto M., Gronland T., Ekman I. 1995. Late glacial of Lake Onega—contribution to the history of the Eastern Baltic Basin. *Quaternary International*, **27**, 111.

Sarvala J. 1994. Effects of acidification on the biota of freshwater ecosystems. In: Mononen P., Lozovik, P. (eds.), *Acidification of Inland Waters. Proceedings of the 3rd Soviet–Karelian–Finnish Symposium on Water Problems, Joensuu, Finland, 3–7 June 1991*. Helsinki: Finnish National Board of Waters and the Environment, pp. 63–71.

Scott R., Huff F. 1996. Impacts of the Great Lakes on regional climate conditions. *J. Great Lakes Research*, **22**, No. 4, 845–864.

Shiklomanov I.A. 1988. *Investigations of Water Resources*. Leningrad: Hydrometeorological Press, 203 pp. (in Russian).

Shnitnikov A.V. 1966. Inter-century variability of water resources of Lake Ladoga basin. In: *Hydrological Regime and Water Balance of Lake Ladoga*. Leningrad: Nauka Press, pp. 557 (in Russian).

Shouleykin V.V. 1968. *Marine Physics*. Moscow: Nauka Press, 1084 pp. (in Russian).

Slepoukhina T. D., Belyakova I.V., Chichikalyuk Y.A. 1996. Bottom sediments and biocoenoses of northern Ladoga and their changes under human impact. In: Simola H., Viljanen M., Slepoukhina T., Murthy R. (eds.), *Ecological Problems of Lake Ladoga: Causes and Solutions. Proceedings of the 1st International Lake Ladoga Symposium, St. Petersburg, Russia, 22–26 November 1993*. Dordrecht: Kluwer Academic, pp. 23–28.

Slepoukhina T.D., Froumin G.T., Barbashova M. 1997. Concepts for ecological monitoring of Lake Ladoga. In: Simola H., Viljanen M., Slepoukhina T. (eds.), *Proceedings of the 2nd International Lake Ladoga Symposium, Joensuu, Finland, 26–30 August 1996*. Joensuu: Joensuum juliopisto, pp. 16–25.

The Finnish Research Programme on Climate Change. 1996. Roos, J. (ed.), Final Report of SILMU. Helsinki: SILMU, 507 pp.

The State of Canada's Climate. 1995. Monitoring variability and change. SOE Report No. 951, 51 pp.

Tikhomirov A.I. 1982. *Thermal Regime of Large Lakes*. Leningrad: Nauka Press, 232 pp. (in Russian).

Tushingham A. 1992. Postglacial uplift predictions and historical water levels of the Great Lakes. *J. Great Lakes Res.*, **18**, No. 3, 440–455.

Vehvilainen B., Huttunen M. 1997. Climate change and water resources in Finland. *Boreal Env. Res., Helsinki*, No. 2, 318.

Verbitsky V.B., Tereshchenko V.G. 1996. Structural phase diagrams of animal communities in assessment of freshwater ecosystem conditions. In: *Ecological Problems of Lake Ladoga: Causes and Solutions. Proceedings of the 1st International Lake Ladoga Symposium, St. Petersburg, Russia, 22–26 November 1993.* Simola H., Viljanen M., Slepoukhina T., Murthy R. (eds.). Dordrecht: Kluwer Academic, pp. 277–282.

Viljanen M., Roumyantsev V., Slepukhina T., Simola H. 1996. Ecological state of Lake Ladoga. In: Varis E., Porter S. (eds.),   *Karelia and St. Petersbourg: from Lakeland Interior to European Metropolis.* Joensuu: Joensuum juliopisto, pp. 107–125.

Volkov I. V., Zaicheva I.N., Moisseeva V.P. 1997. Aquatic ecotoxicology problems on a regional scale. *Water Resources*, **24**, No. 5, 5 (in Russian).

Volosko L.N., Titova N.N. 1997. Effects of heavy metal ions on the motility and phototactic orientation of chrysophycean and cryptophycean algae from Lake Ladoga. In: Simola H., Viljanen M., Slepoukhina T. (eds.), *Proceedings of the 2nd International Lake Ladoga Symposium, Joensuu, Finland, 26–30 August 1996.* Joensuu: Joensuum juliopisto, pp. 220.

2

Fundamental limnological processes and relevant indicators used in contemporary lake studies

2.1 HEAT FLUX, STORAGE AND BUDGET OF A LAKE WATER– SEDIMENTS SYSTEM

The main principals of freshwater lake thermics were first studied and established by Forel in his pioneer work (1892). Then, the basics of lake heat storage and budget were formulated by Birge (1915), advanced and summarized by Hutchinson (1957). Later the relevant problems have been reviewed, studied and considered in monographs (Frey, 1966; Kirillova, 1970; Pivovarov, 1972; Wetzel, 1975; Lerman, 1978; Straškraba, 1980; Tikhomirov, 1982; Serruya and Pollingher, 1983; Henderson-Sellers, 1984; Straškraba and Gnauck, 1985; Adamenko, 1985; Adamenko *et al.*, 1991; Ryanzhin, 1989, and others).

Here we consider and evaluate heat flux components and heat budget of a lake. Contribution of heat flux components into a lake heat budget, simplifications and simplified equations, as well as some thermal phenomena revealed by remote sensing are also discussed. Besides, GIS WORLDLAKE constructed to study the general zonal and azonal relationships for lake thermal and water regimes is also presented.

2.1.1 Thermal characteristics of lakes. Global inventories, database and GIS

2.1.1.1 *Lake surface temperature and remote sensing*

Lake surface temperature reflects a total heat flux and heat budget of a lake surface. In turn, thermal processes developed in a water body, such as thermal bar, organized gravity convection, internal waves, Langmuir circulation, etc., reveal themselves at the lake surface through either surface temperature or visually. The most successful remote sensing temperature data concerning lake thermal phenomena have been obtained for lake thermal phenomena characterized by relatively strong either horizontal or vertical gradients of surface temperature (thermal bar, internal waves, etc.; see Figures in § 2.7). For example, the vast relevant data obtained for lakes from satellite IR images have beenstudied and generalized (Kondratyev *et al.*, 1986; Filatov *et al.*, 1990; Malm and Jonsson 1993; Malm, 1994, and others). Thus, the surface temperature is one of the important and informative parameters to be studied. On the other hand, lake surface temperature is the most easily, reliably and often-measured parameter in remote sensing.

2.1.1.2   *Global lake inventories and database*

Natural lakes are one of the most easily available water resources, and contain considerable amounts of the Earth's water. However, although there are approximately 10 000 000 lakes with area >0.01 km^2 (Meybeck, 1995), it seems that more or less detailed limnologic studies including at least a survey of lake depths, have been carried out in approximately 40 000 of them. Moreover, the closed (with no outflow) and many open lakes are excellent climatically sensitive objects. Remote sensing and long-term monitoring of these lakes can provide useful information for study of climatic change. Particularly, satellite radar altimetry offers the advantages of day/night and all-weather capability in the production of relative lake level changes on a global scale.

However, inventories of lakes have been mainly developed for morphometric characteristics of regional (Murray and Pullar 1910; Rihter, 1959; Yakushko, 1983, 1985, etc.) and national (Domanitskii *et al.*, 1971; Nikitin, 1977; *Lakes in China*, 1995, etc.) scales. Inventories on a global scale presented in (Halbfass, 1922; Herdendorf, 1982, 1990; Ryanzhin, 1989) are also far from completed. Herdendorf (1982, 1990), e.g. only collected data on "large lakes" with surface area >500 km^2.

The Mullard Global Lakes Database (MGLD) has been constructed in Mullard Space Science Laboratory (MSSL) and University College London, UK (Birkett and Mason, 1995; Birkett, 1995, 1998). MGLD mainly assesses the number and global distribution of closed lakes and accepts altimeter data from the TOPEX/POSEIDON satellite. MGLD contains information on lake name, type (closed, open, reservoir), location, approximate area, satellite TOPEX/POSEIDON altimeter coverage, data on lake level and anthropogenic activity. MGLD holds records on 1415 lakes, with areas ≥100 km^2, 329 of which have been preliminarily designated as closed. The original database of MGLD was compiled from lake inventories (Halbfass, 1922; Herdendorf, 1982, 1990, and others). Unfortunately, the orbit of the TOPEX/POSEIDON satellite is such that it only passes over 53 identified closed lakes compared with, e.g. 92 by Geosat and 152 by ERS-1 during its 35-day repeat phase. This is due to the relatively shorter frequency of repeat measurements (10 day) compared with Geosat (17 day) and ERS-1 (35 day). Preliminary analysis of altimeter data on selected lakes, rivers and wetlands from MGLD have been recently given (Birkett, 1995, 1998). However, MGLD does not contain data on thermal characteristics of lakes.

2.1.1.3 GIS WORLDLAKE

To derive general zonal and azonal relationships for thermal and water regimes of the world lakes, geo-information system GIS WORLDLAKE in MapInfo has been developed in the Limnology Institute, St. Petersburg, of the Russian Academy of Sciences. The original database (in Windows Excel) of WORLDLAKE contains the following lake data:

Geographic: (1) lake name (including local and/or alternative) in English, Russian and Latin form; (2) continent, country, administrative region and district, geographical region; (3) co-ordinates, above sea level (asl) altitude, snow line elevation for a given latitude (for high elevated and mountain lakes);

Morphometric: (1) origin of lake basin (glacial, tectonic, etc.); (2) lake type by outflow (closed/open) and by level variation (regulated/not); (3) average and maximum depth and their ratio; area and volume; length of the shoreline and shoreline development; (4) lake catchment and specific catchment areas;

Hydrological: (1) range of seasonal and long-term (year-to-year) water-level variation; (2) annual average water balance components (precipitation, evaporation, river and groundwater inflow/outflow); (3) retention (residence) time; (4) specific discharge;

Meteorological and climatological: (1) annual average (avg), extreme and range (rng) values of (a) air temperature, (b) global radiation, (c) cloudiness, (d) integral albedo, (e) absorbed radiation, (f) effective back radiation, radiation balance, turbulent flux, latent heat due to evapo/transpiration and ice formation/melting, total heat income due to rivers inflow/outflow, (g) Bowen ratio, (h) total surface heat flux; (2) aridity index (precipitation to evaporation annual mean ratio); (3) ice thickness and duration of ice cover;

Hydrochemical: (1) lake type by salinity (fresh, salt, brackish, etc.); (2) average salinity and its characteristic vertical gradient; (3) chemical composition of water (for saline lakes); (4) pH;

Hydrothermal: (1) annual average, extreme and range values of (a) surface temperature, (b) heat storage in water and in sediments, (c) heat flux at water–sediment boundary; (2) mixing lake type (moderate dimictic, cold monomictic, polymictic, etc.) real and predicted by the model (Ryanzhin, 1994); (3) seasonal (midsummer) thermocline depth and corresponding values of wind speed, fetch, Secchi depth and extinction coefficient; (4) time interval between spring and autumn thermal bar; (5) water and sediments heat budget in a year cycle;

Others: (1) comments; (2) references.

Original data for WORLDLAKE were collected from the above publications as well as from (1) UNEP/ILEC and IAHS publications (*Data*, 1988–1993, etc.); (2) world, national and regional lake catalogs and inventories; (3) general limnologic monographs listed above (e.g. there are data on ~820 lakes in Hutchinson, 1957), (5) proceedings of national and international conferences, as well as from (6) relevant publications in international and national journals for the last 40–50 years.

Currently WORLDLAKE covers data on more than 11 000 freshwater and saline lakes. The data are distributed as follows: Americas 1800 lakes (USA, Canada 1300; Central and South America 150, 350 respectively), Africa 300 lakes, Australia, New Zealand and Oceania 400 lakes, Antarctica 70 lakes, Asia 1700 lakes, Europe more than 8000 lakes (Russia *ca.* 5000). The database of WORLDLAKE is open for further compilation and is supposed to be set at Internet. Additional data, particularly, for such exotic regions as Antarctica, Amazon, Tibet, Himalayas, etc. are going to be received from satellite data quite soon. Current and previous information, as well as suggestions can be received/sent from/to by e-mail ryanzhin@lake.spb.su.

2.1.2   Equations for heat storage and budget

Although, the main amount of energy comes to a lake from the sun (from the atmosphere–lake boundary), generally a lake should be considered as a part of the *lake + atmosphere + sediments + watershed* system. The thermal and hydrodynamic state of a lake as well as its temporal-spatial variations are governed by matter and energy fluxes at lake boundaries. Variation in fluxes affects the thermal regime and energetic budget of a lake. Watershed effects are usually parameterized as lateral boundary conditions imposed through river inflows/outflows. Then the system is reduced to *lake + atmosphere + sediments*, where the effects of sediments become considerable in relatively shallow lakes and lake sites (Ryanzhin, 1997). Seasonal (annual) variations energetically predominate in moderate lakes. So, we here consider temporal–spatial variability of heat fluxes and factors affecting heat storage and budget of a lake mainly in a year cycle.

Heat storage of lake water body

The thermal status of a lake can be described through lake-water storage (enthalpy) E_w or heat budget of a lake water body $\partial E_w/\partial t$. In general, a temporal variation of E_w for a 2-D lake is expressed through the heat flux at lake boundaries and internal heat source (Fig. 2.1), that can be expressed after integration of the heat conductivity equation over the water body as:

$$\frac{\partial E_w}{\partial t} = \frac{\partial}{\partial t} \int_0^H \int_0^l c_p \rho_w \theta_w(t, z, x)\, dz\, dx = H l c_p \rho_w \frac{\partial \bar{\theta}_w}{\partial t} = F_0 + F_H + F_1 + P_w \quad (2.1)$$

where t is time; (x, z) is Cartesian right-hand side co-ordinates with z-axis directed downward with $z = 0$ at air–lake boundary; t is time; H, l are lake depth and length; $c_w \cong 1.00\,\text{cal}\,\text{g}^{-1}\,°\text{C}^{-1} \cong 4.19\,\text{kJ}\,\text{kg}^{-1}\,\text{K}^{-1}$, $\rho_w \cong 1.0\,\text{g}\,\text{cm}^{-3} \cong 1 \times 10^3\,\text{kg}\,\text{m}^{-3}$ are specific heat capacity and density of water; $\theta_w(t, x, z)$, $\bar{\theta}_w(t)$ are instant and averaged over the lake volume $(V = Hl)$ water temperature; F_0, F_H, F_l are vertical and horizontal (lateral) components of a total heat flux at lake boundaries; P_w is integral internal (spatially distributed) heat source (consumption/production per unit time) in lake water.

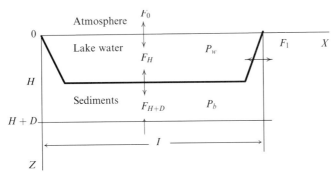

Fig. 2.1. Heat flux components at lake boundaries. 1 is water body and lake sediments length; H is lake depth; $D, H + D$ are sediments thickness and lower boundary; F_0, F_H, F_{H+D} are vertical heat flux at air–water, water–sediments, and the lower sediments boundary; F_{H+D} equals negative geothermal flux F_G; lateral heat flux F_l is inflow–outflow of heat advected by rivers, ground waters and horizontal diffusive heat exchange between lake water and sediments; P_w, P_b are internal (spatially distributed) heat sources (production/consumption) in water and sediments.

Water density and heat capacity spatially–temporally vary, being dependent on temperature θ_w, salinity s and pressure p. Variation of ρ_w are described by the equation of state $\rho_w = \rho_w(\theta_w, p, s)$. In turn, c_w decreases slightly with increase of θ_w from $0°$ to $50°C$. Variations of s are considerable in saline and, particularly, in hypersaline lakes. However, under natural lake conditions the above variations are negligibly small compared with variations of θ_w. Therefore, they are neglected when (2.1) is derived.

Each of the flux components F_0, F_H, F_l in (2.1) combines heat flux induced by different mechanisms: radiate, advective, diffusive and phase transformation. However, the former and the latter are usually meaningful only at air–water boundary. Hence, they are included only into F_0. In turn, lateral flux F_l mainly characterizing inflow–outflow of heat by rivers and ground waters, also describes horizontal diffusive heat exchange between a lake and surrounding sediments, e.g. at relatively high slopes of the lake bottom. Similarly, although F_H mainly describes diffusive heat exchange between water and sediments, it may also define a vertical advection of heat due, e.g. to ground waters.

Thus, (2.1) simply states that variation in heat storage of a lake is governed by sum of heat flux at the boundaries and internal heat source of water. Equation (2.1) can be detailed:

$$\frac{\partial E_w}{\partial t} = Hlc_p\rho_w\frac{\partial \bar{\theta}_w}{\partial t} = \underbrace{I_0(1-a) + B_e}_{R=Radiation\ balance} + \underbrace{LE + Q_t + Q_{pr} + Q_{ice} + Q_{moon}}_{air-water\ boundary}$$

$$+ \underbrace{Q_{io}}_{\substack{inflow/outflow \\ by\ rivers}} + \underbrace{Q_{gr} + F_b - F_G}_{\substack{water-sediments \\ boundary}} + \underbrace{P_w}_{\substack{production-consumption \\ in\ water}} \qquad (2.2)$$

here I_0 is global irradiance of lake surface, i.e. direct + scattered short-wave (0.35– 0.75 μm) insolation with absorbed part $I_0(1 - a)$; a is integral albedo; B_e is effective back long-wave radiation of lake surface; $R = [I_0(1 - a) + B_e]$ is radiation balance; LE, Q_T are vertical heat flux caused by evaporation/condensation and sensible heat flux at air–water boundary; $L \approx 595\,\mathrm{cal\,g^{-1}} = 2.50\,\mathrm{MJ\,kg^{-1}}$ is latent heat of freshwater vaporization that slightly decreases with an increase of θ_w as $L \approx 2.50\text{–}2.51 \times 10^{-3}(T - 273°)$, where $[L] = \mathrm{MJ\,kg^{-1}}$, T is absolute temperature of evaporated water; E is water vapor vertical flux; Q_{io}, Q_{pr}, Q_{gr} are advection of heat (inflow/outflow) due to rivers, precipitation and ground waters; Q_{ice} is latent heat of ice formation/melting; Q_{moon} is solar radiation reflected from moon and starlight; F_b is vertical fluctuating diffusive heat flux at water–sediment boundary; F_G is steady-state geothermal flux.

Heat storage and budget of vertical water column

Balance equation (2.2) is reduced for vertical water column. Then, advection of heat by rivers Q_{io} and other lateral heat flux components should be replaced with an integral horizontal advection of heat caused by lake currents Q_{adv}. In turn, the vertical component of Q_{gr} should be introduced to integral vertical heat flux at water–sediment boundary. Then, (2.2) is rewritten as:

$$\frac{\partial E_w}{\partial t} = Hc_p\rho_w\frac{\partial \bar{\theta}_w}{\partial t} = \underbrace{I_0(1 - a) + B_e}_{R=\text{Radiation balance}} + LE + Q_t + Q_{pr} + Q_{ice} + Q_{moon}$$

$$\underbrace{}_{\text{air−water boundary}}$$

$$+ \underbrace{Q_{adv}}_{\substack{\text{horizontal advection} \\ \text{by currents}}} + \underbrace{Q_{gr} + F_b - F_G}_{\substack{\text{water−sediments} \\ \text{boundary}}} + \underbrace{P_w}_{\substack{\text{production−consumption} \\ \text{in water}}} \qquad (2.3)$$

2.1.3 Heat flux components. Evaluation and simplification

Flux components in (2.2) and (2.3) reveal temporal–spatial variability. Thus, their contribution to lake heat storage can be considered in terms of either annual range and extremes, or annual average magnitudes. Although some annual mean flux components in (2.2) and (2.3) equal zero, mean values of $I_0(1 - a)$ and LE, Q_T are the main positive and negative parts of heat budget, respectively. However, the role and relative contribution of all flux components varies in a year cycle.

Radiation reflected from moon and heat of starlight

Since the moon reflects sunlight, the effect and corresponding term Q_{moon} in (2.2) and (2.3) are regulated by the timetable of the moon. However, even radiation from the full moon is a very small fraction of that from the sun, i.e. $Q_{moon}/I_0 \sim 2\text{–}3 \times 10^{-4}$ (Hutchinson, 1957). In turn, the intensity of starlight is very much less than Q_{moon}. Thus, although these sources of heat may effect an aquatic ecosystem to a certain degree, they are generally neglected when calculating heat balance of a lake.

Internal heat source in lake water

The term P_w in (2.1)–(2.3) includes production (consumption) of heat per unit time in lake water (Fig. 2.1). It parameterizes production of heat from (1) dissipation of mechanic energy, (2) destruction of organic matter, and consumption of heat due to photosynthesis. Measurements of mechanic energy dissipation carried out at the Caspian Sea, Lawrentian Great Lakes, Lake Tahoe, Lake Ladoga, Lake Onega and others (Palmer, 1973; Filatov *et al.*, 1981; Ryanzhin *et al.*, 1981; Filatov, 1983, 1991; Imberger, 1985) give small values of $\sim 10^0\text{–}10^{-5}\,cm^2 \times s^{-3}$. Destruction of organic matter mainly occurs in sediments and not in water itself. Thus, this source of heat is indirectly involved in flux F_b rather than in P_w. Next, although consumption of heat by photosynthesis is poorly studied, this part of heat loss seems to be very small compared with the main terms in (2.2) and (2.3). Besides, the energy consumed by photosynthesis is transformed into organic matter. Therefore, a release of heat by destruction of organic matter should be of the same order as that consumed by photosynthesis. Hence, these very small terms are seemingly annually balanced. Thus, P_w is very small compared with other main terms in (2.2) and (2.3) and is generally neglected. It should be underlined, however, that in transparent lakes P_w includes distributed absorbed radiation. Then P_w is considerable and is not omitted in (2.2) and (2.3), as discussed below.

Heat caused by precipitation

Variation of lake enthalpy E_w due to heat income with precipitation Q_{pr} depends on lake water to precipitation enthalpy ratio, i.e. on specific enthalpy (enthalpy in a unit volume): lake to precipitation volume and temperature ratio. Generally, contribution of Q_{pr} in the thermal budget of a lake tends to increase with increase of precipitation contribution into the water balance of a lake. Therefore, Q_{pr} is meaningful for the thermal regime of shallow lakes with high variation in lake level during the rainy season. Note that Q_{pr} also includes latent heat flux if, e.g., snow falls on ice-free, or rain falls on ice-covered lake surface. However, in moderate lakes of considerable volume the flux Q_{pr} is usually small compared with major flux terms, and is commonly neglected in a heat budget.

Heat due to ground waters

Value of Q_{gr} is significant for heat regime of lakes with considerable ground water feeding, e.g. for karst lakes. Contribution of Q_{gr} to the heat budget of a lake can be estimated through specific enthalpy in a similar manner as it is for Q_{pr}. It should be pointed out, however, that since ground water feeding remains the worst-studied component of water balance of a lake of any origin, Q_{gr} is one of the most uncertain flux components in lake heat budget.

Advection of heat by rivers

Similarly, as for Q_{pr} and Q_{gr}, variation of lake enthalpy E_w caused by rivers $Q_{io} = (Q_i - Q_o)$ is governed by the lake-to-river inflow/outflow enthalpy ratio, i.e. on lake to inflow/outflow volume and temperature ratio. River inflow tends to increase E_w if specific enthalpy of river water is higher than that of a lake. For closed lakes (with no river outflow) $Q_{io} \neq Q_i$. However, it may be $Q_i \neq Q_o$ and $Q_{io} \neq 0$ even if a volume of river inflow equals outflow. The reason is that the temperature difference between lake and inflow/outflow water can vary in a year cycle. Hence, Q_i, Q_o and Q_{io} are temporally variable and their annual mean values may decline from zero. In Fig. 2.2 the long-term (year-to-year) average annual course of Q_i, Q_o and Q_{io} is given for the largest European lake, Lake Ladoga, 60.9°N, 31.4°E, 4.8 m asl, NW Russia, characterized by many inflows and only drained by the River Neva. It is seen that the values vary in a year cycle ranging within $\sim \pm 10\,\mathrm{W\,m^{-2}}$ and the net flux Q_{io} is positive in spring and negative from summer to the fall. It is shown in §§ 2.6 and 2.7, when discussing hydrodynamic effects of river inflow on lake hydrodynamics, that inflow–lake temperature difference often seen from IR satellite images (see,

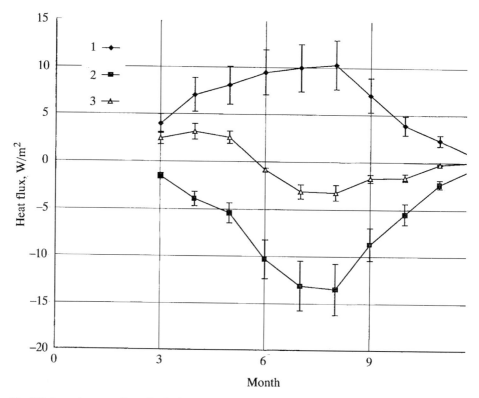

Fig. 2.2. Annual course of heat flux input Q_i, output Q_o, and net heat flux $Q_{io} = (Q_i + Q_o)$, due to rivers at Lake Ladoga, 60.9°N, 31.4°E, NW Russia. Values recalculated from original data (Veselova, 1968); points averaged for 1957–1962 are given with std values. 1—heat income by rivers, 2—heat outcome by River Neva, 3—net total heat flux.

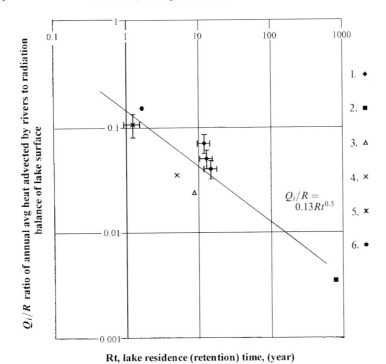

Fig. 2.3. Ratio of annual average heat advected to the lake by rivers to radiation balance of lake surface Q_i/R *versus* lake retention time $R_T = V/V_{inf}$ for selected lakes. Data included in GIS WORLDLAKE were recalculated from original values given in cited sources. Data for Lake Ladoga show year-to-year variability. Approximated curve $Q_i/R \sim 0.13 R_T^{-1/2}$ is shown. Lakes: 1—Ladoga, Russia, 2—Issik-Kul, Kirgizstan, 3—Mergozzo, Italy, 4—Valencia, Venezuela, 5—Punnus-Jarvi, Russia, 6—Mead, USA.

Figures in §§ 2.6 and 2.7) in some cases leads to front formation. In contrast to lowland moderate lakes, in high elevated lakes the temperature of inflow rivers with glacial feeding can be much lower than a lake's all-year-round temperature.

It is reasonable to compare annual average values of Q_i and that of the radiation balance of a lake surface. A ratio Q_i/R likely depends inversely on lake retention (residence) time $R_T = V/V_{inf}$, where V_{inf} is river inflow for a taken time interval, i.e. $Q_i/R \propto R_T^{-1}$ with the following asymptotes:

$$Q_i/R \to 0 \qquad \text{at } R_T \to +\infty \text{ (lake with no inflow)},$$
$$Q_i/R \to +\infty \quad \text{at } R_T \to 0 \text{ (river)}. \tag{2.4}$$

It is seen from Fig. 2.3, where data on Q_i/R *versus* R_T^- taken from GIS WORLDLAKE are represented for selected lakes, that despite data scatter, the above tendency is adequate and the relationship can be approximated as:

$$Q_i/R = 0.13 R_T^{-1/2} \tag{2.5}$$

R_T characterizes the transitory degree of lake water. Then, under an increase of R_T any lake could be treated first as a lake itself, then as a reservoir, and finally as a river, i.e. as *lake → reservoir → river*. Thus, assuming, e.g., threshold value of $Q_i/R \sim 1$, a boundary between lake and reservoir can be estimated from (2.5) as $R_T \approx 1.7 \times 10^{-2}$ year (~6–7 day).

Advection of heat to vertical water column

For a separate vertical water column the advection of heat by lake currents Q_{adv} should be considered instead of Q_{io}, Q_i, Q_o and other lateral heat flux components. Then Q_{adv} includes both lateral advective and poorly studied diffusive heat flux. Q_{adv} varies in a year cycle, being important at a point of time of both the spring and fall thermal bar in moderate dimictic lakes. In many other cases, in contrast to oceanic conditions, the effects of Q_{adv} on a heat budget of lake water column can be neglected. Moreover, on average Q_{adv} is annually balanced to zero in a lake. It should also be noted that $Q_{adv} \sim 0$ if horizontally (spatially) averaged vertical water column is considered.

Heat flux at water–sediment boundary

Diffusive heat flux at the water–sediment boundary F_H consists of time-independent (steady-state) geothermal flux F_G that is negative, i.e. directed from sediments to a lake water body to increase water temperature and time-fluctuating flux $F_b(t)$. Thus:

$$F_H = F_G + F_b(t) = -\chi_b G - \chi_b \frac{\partial \theta_b}{\partial z} \tag{2.6}$$

here $G = \partial\bar{\theta}_b/\partial z > 0$ is always a positive steady-state geothermal gradient; θ_b is sediment temperature; $\chi_b = c_b\rho_b\nu_b$ is molecular bulk thermal conductivity of sediments; c_b, ρ_b, ν_b are specific bulk heat capacity, density and molecular thermal diffusivity of sediments, respectively.

202 and 446 probing of G and F_G obtained from 78 European, American, Asian and African lakes have been compiled and processed in Ryanzhin (1992) to evaluate geothermal effects on lake heat budget and to calculate global average estimates. It was shown that G varies from 0.009 K m^{-1} in Africa's Lake Nyasa (Malawi), 12°S, 34.9°E, Malawi/Tanzania/Mozambique, to 0.28 K m^{-1} in Lake Windermere, 54.3°N, 3°W, England giving the global average $0.9(\pm0.05)$ K m^{-1} where $\pm$ std value is shown in brackets. In turn, F_G ranges over two orders from 6.28×10^{-3} W m^{-2} in Lake Malawi to 3.12×10^{-1} W m^{-2} in Lake Baikal, 54.6°N, 106.8°E, Siberia, Russia, yielding the global mean $7.45(\pm4.31) \times 10^{-2}$ W m^{-2} that is close to the global average terrestrial value. A more recent review of relevant measurements of F_G in lake sediments is presented in Ryanzhin (1997). It should be stressed that in balance equations (2.2) and (2.3) only two terms, absorbed radiation and geothermal flux, always tend to lake heating in a year cycle, i.e.

$$I_0(1 - a) \geq 0 \quad \text{at } z = 0 \quad \text{and} \quad F_b < 0 \quad \text{at } z = H \tag{2.7}$$

Since there is no zonal trend of F_G, its contribution to lake heat budget increases on average with a decrease in either absorbed radiation, or radiation balance of lake surface, i.e. with an increase in both latitude or lake altitude. Therefore, in spite of its relatively small value, F_G should contribute considerably to heat budget of some lakes, e.g. cold amictic (perennially ice-covered) lakes (Ryanzhin, 1992, 1997).

The role of seasonal heat exchange between lake water and sediments in lake heat budget were first stressed by Birge *et al.* (1927) during excellent field studies at Lake Mendota, Wisconsin, USA. A review of relevant studies and suitable parameterizations as well arc presented in Ryanzhin (1997). The fluctuating diffusive heat flux at water–sediment boundary F_b is strongly governed by either an amplitude of seasonal temperature variation at lake surface or an intensity of vertical mixing in a water column. Particularly, it was shown that a seasonal amplitude of F_b increases rapidly with decrease in lake depth H (Ryanzhin, 1995, 1997).

Heat of ice formation/melting

Heat flux due to ice formation/melting can be estimated from the simplest 1-D equation of ice thickness h_i as:

$$\rho_i L_i \frac{\partial h_i}{\partial t} = Q_0 - Q_h \qquad (2.8)$$

where $\rho_i \approx 0.917\,\mathrm{g\,cm^{-3}} = 9.17 \times 10^2\,\mathrm{kg\,m^{-3}}$, $L_i \approx 80\mathrm{cal\,g^{-1}} = 0.334\,\mathrm{MJ\,kg^{-1}}$ are freshwater ice density and latent heat of ice formation/melting; the former slightly increases with decrease in ice temperature; Q_0, Q_h are heat fluxes at ice–water and ice–air boundaries.

According to (2.8), the ice thickness h_i increases and decreases when $Q_0 > Q_h$, and $Q_0 < Q_h$, respectively. Instead of (2.8), so-called *freezing degree day* semiempiric models are often used to calculate growth of ice thickness (e.g. Pivovarov, 1972). However, they are essentially bulk thermodynamic with little or almost no connection to lake processes, such as vertical mixing occurring beneath the ice. On the other hand, (2.8) is enlarged to develop more detailed and complicated ice + snow cover models (Patterson and Hamblin, 1988; Carmack and Farmer, 1982, and others). However, it should be noted that the thickness of snow overlying the ice cover is restricted by the ability of floating ice to support the snow weight. Corresponding balance equation for maximum snow thickness gives (Pivovarov, 1972):

$$h_{s(max)} = h_i \frac{\rho_w - \rho_i}{\rho_s} \qquad (2.9)$$

where $\rho_s \cong 0.33\,\mathrm{g\,cm^{-3}} = 3.3 \times 10^2\,\mathrm{kg\,m^{-3}}$ is average snow density.

Substitution of characteristic values into (2.9) yields the critical ratio of $h_{s(max)}/h_i \sim 2.20\text{–}0.25$. The excess snowfall may break the ice cover and substantially change and complicate heat flux at the air–water boundary. Q_{ice} is positive and negative during ice formation and melting respectively; that is especially the case in the heat budget of shallow lakes and lake sites. For example, formation of 20-cm-thick ice during one month releases heat at an average rate $\sim 25\,\mathrm{W\,m^{-2}}$ (annual mean $\sim 2\,\mathrm{W\,m^{-2}}$). However, freezing and melting are usually balanced in a year cycle and

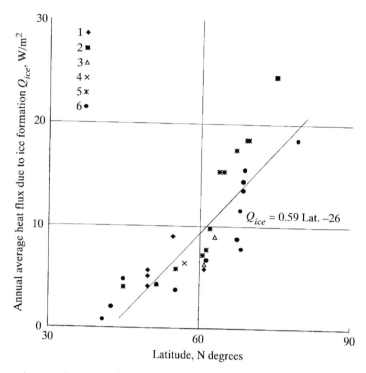

Fig. 2.4. Annual average latent heat flux due to ice formation/melting Q_{ice} in lakes located close to sea level ($Z \leq 300$ m asl) depending on lake latitude φ. Flux values attributed for annual time interval. Data included in GIS WORLDLAKE were recalculated from maximum ice thickness data given in cited works. Note the almost linear relationship Q_{ice} over φ with latitudinal gradient $\partial Q_{ice}/\partial \varphi \cong 0.59$ W/m^2 dgr (Lat.). 1—ELA, Canada; 2—Meretta, Canada; 3—Baikal, Russia; 4—Small Karelian lakes; 5—lakes of NW Canada; 6—North American lakes.

annual mean Q_{ice} is zero in temperate lakes. The exceptions occur in both high latitude and altitude lakes where ice cover does not disappear completely during spring–summer heating. High-latitude Arctic Lake Char, 74.7°N, 94.8°W, 34 m asl, Canada, and Lake Taymir, 73.6°N, 102.5°E, 6 m asl, Siberia, Russia, are good examples. Numerous similar examples for small Canadian Northwestern Territories (NWT) lakes have recently been given by Pienitz *et al.* (1997a, b). Thus long-term variability in Q_{ice} can exist and for such a situation the year mean value is $Q_{ice} > 0$.

Ice-cover duration and ice thickness vary zonally, depending particularly on latitude φ. In Fig. 2.4 only the positive part of Q_{ice} recalculated from ice thickness and yearly averaged (data taken from GIS WORLDLAKE) is given *versus* φ. As is seen, the flux Q_{ice} almost linearly increases with increase of latitude with approximated regression:

$$Q_{ice} = \frac{\partial Q_{ice}}{\partial \varphi} \varphi - 26 \, \text{W m}^{-2} \qquad (2.10)$$

where $\partial Q_{ice}/\partial \varphi = 0.59$ W m^{-2} dgr (Lat.)$^{-1}$ is average latitudinal gradient of heat flux Q_{ice}.

Thus, the contribution of Q_{ice} in lake heat budget increases on average with increase of latitude.

Temporally averaged lake heat budget equation

Consider the heat budget of a vertical water column. Then assuming Q_{gr}, Q_{pr} and P_w to be small compared with other heat flux components, eq. (2.3) is reduced to:

$$\frac{\partial E_w}{\partial t} Hc_p\rho_w \frac{\partial \bar{\theta}_w}{\partial t} = \underbrace{I_0(1-a) + B_e}_{R=Radiation\ balance} + LE + Q_t + Q_{ice} + \underbrace{Q_{adv}}_{horizontal\ advection} + \underbrace{F_b - F_G}_{water-sediments}$$

$$\underbrace{\qquad\qquad\qquad\qquad\qquad\qquad\qquad\qquad}_{air-water}$$

$$(2.11)$$

If heat storage of a lake has no long-term (year-to-year) variation then, as shown above, the flux components Q_{ice}, Q_{adv} and F_b are annually balanced and after integrating (2.11) with respect to t, for annually averaged heat flux components one has

$$\underbrace{I_0(1-a) + B_e}_{R} + LE + Q_T - F_G = R + LE + Q_T - F_G = 0 \qquad (2.12)$$

where R and F_G are the main positive and LE and QT are main negative parts of annual heat budget of a lake. As shown above, F_G excepting some specific cases, is negligibly small compared with other heat flux terms in (2.12). Then only annual average fluxes R, LE and Q_T have to be considered to evaluate spatial variability of lake heat status.

2.1.4   Heat flux and budget of a lake surface

Heat budget of air–water surface

The equation for heat budget or heat balance of the atmosphere–lake boundary is derived from (2.2) or (2.3) at $H \to 0$. Since $\partial\bar{\theta}_w/\partial t$ is a limited value, instead of (2.2) and (2.3) at $H \to 0$ one has:

$$F_w = -c_p\rho(K_Z + \nu_w)\frac{\partial\theta_w}{\partial z} = \underbrace{I_0(1-a) + B_e}_{R=Radiation\ balance} + LE + Q_T \quad \text{at } z = 0 \quad (2.13)$$

where $F_w = -c_p\rho(K_Z + \nu_w)\,\partial\theta_w/\partial z$ is the near-surface vertical heat flux in water; $\nu_w = \chi_w/c_w\rho_w \cong 1.2 \times 10^{-3}\,\text{cm}^2\,\text{s}^{-1} = 1.2 \times 10\,\text{m}^2\,\text{s}^{-1}$, $\chi_w \cong 1.2 \times 10^{-3}\,\text{cal s}^{-1}\,{}^\circ\text{C}^{-1}$ $\text{cm}^{-1} = 5.0 \times 10^{-1}\,\text{W m}^{-1}\,\text{K}^{-1}$ are molecular thermal diffusivity and conductivity of freshwater; K_z is effective (turbulent) vertical thermal diffusivity in a water column that parameterizes vertical mixing of different origins, such as turbulent stirring at stable and/or neutral buoyancy flux, gravity convection, Langmuir circulation, etc. Note that the flux Q_{ice} should be added to the right side of (2.13) during the ice formation/melting and ice-cover period.

Sensible and latent heat flux

Sensible Q_T and latent heat flux caused by evaporation (condensation) LE are the main negative parts of the heat budget of a lake surface. Calculation of Q_T and LE are based on either semiempiric bulk formula or the Monin–Obukhov similarity theory. A review of bulk-formula approaches for lakes is given in Henderson-Sellers (1984). The Monin–Obukhov theory states that the turbulent regime in the horizontally homogeneous near-surface layer of the atmosphere is fully described by friction velocity u^*, buoyancy parameter $b = g/T$ (g is gravity acceleration; T is absolute air temperature) and normalized vertical fluxes of water vapor E/ρ_a, and of sensible heat $Q_T/c_a\rho_a$. Then the vertical gradients of wind speed u, air temperature T and specific humidity $q = 0.622e/p$ (e, p is vapour and atmospheric pressure) are given:

$$\frac{\partial u}{\partial z} = \frac{u_*}{kz}\psi_u(z/L) \tag{2.14}$$

$$\frac{\partial T}{\partial z} = \frac{-Q_T}{kzu_*c_a\rho_a}\psi_T(z/L) \tag{2.15}$$

$$\frac{\partial q}{\partial z} = \frac{-E}{kz\rho_a}\psi_q(z/L) \tag{2.16}$$

$$L = \frac{-u_*^3}{kg(Q_T/Tc_a\rho_a + 0.608E - /\rho_a)} \tag{2.17}$$

where k is the von Karman constant; L is the Monin–Obukhov scale; ψ_u, ψ_T and ψ_q are dimensionless universal functions of dimensionless height z/L.

The form of ψ_u, ψ_T and ψ_q for stable and unstable ($z/L > 0, < 0$) stratification over a lake have been derived from field lake data by Panin (1985) as

$$\psi_u = \begin{cases} (1 - 13z/L)^{-1/4}, & \text{at } z/L \leq 0 \\ 1 + 6z/L, & \text{at } z/L \geq 0 \end{cases}$$

$$\psi_T = \psi_q = \begin{cases} (1 - 6z/L)^{-1/2}, & \text{at } z/L \leq 0 \\ 1 + 9z/L, & \text{at } z/L \leq 0 \end{cases}$$

Mironov (1991) later carefully reviewed available field studies of the form ψ_u, ψ_T and ψ_q over the natural water bodies and additionally examined some special cases of free convection in near-surface air layer at light winds.

In moderate lakes the maximum latent and sensible losses of heat occur in summer or late summer and minimum in winter and fall. Ice cover considerably prevents losses. Annual average values of LE reveal latitudinal trends at sea level. For example, according to Adamenko (1985, see his Fig. 1.14), $-LE$ increases from $\sim 60\,\text{W m}^{-2}$ at $40°$–$45°$N to maximum $\sim 150\,\text{W m}^{-2}$ at the tropical cycle and then slightly drops to $\sim 125\,\text{W m}^{-2}$ for equatorial lakes. However, latitudinal trends of Q_T seem not to have been carefully studied so far. It also seems no relationships for altitudinal variations of Q_T and LE have been established so far.

Effective back radiation

Effective back long-wave radiation B_e with maximum emission at 8–15 μm is a sum of both net lake surface and atmosphere long-wave radiation. Although the atmosphere does not radiate as a black body, B_e is traditionally approximated as:

$$B_e = -\sigma\delta(T_w^4 - T^4)f(n,q) \tag{2.18}$$

where $\sigma = 5.67 \times 10^{-8}\,\mathrm{W\,m^{-2}\,K^{-4}}$ is the Stefan–Boltzmann constant; $\delta \cong 0.90\text{–}0.98$, T_w are darkness coefficient and absolute temperature of lake surface; $f(n,q)$ is empirical function of cloudiness n and specific humidity q of the atmosphere. Forms of $f(n,q)$ used in limnology are discussed in Kirillova (1970); Pivovarov (1972); Henderson-Sellers (1984) and Panin (1985).

Since $(\theta_w - \theta_a)/T_0$ is a small value ($T_0 = 273\,\mathrm{K}$), the linearized equation is used instead of (2.18) as:

$$B_e = -\sigma\delta T_0^3(\theta_w - \theta_a)f(n,q) \tag{2.19}$$

Temporal–spatial variability of B_e, particularly its zonal trends, are more worthy of study when compared with those for Q_T and LE. However, in moderate lakes the annual mean values and a range of B_e values are considerably less varied compared with I_0, Q_T and LE.

Global solar radiation

Global short-wave solar radiation reaching the top of the atmosphere I_∞ at a given geographic latitude φ and instant of time is accurately specified by List (1951):

$$I_\infty = I_s(r/r_0)^2 \cos\left(\frac{\pi}{2} - h_\odot\right) \tag{2.20}$$

where $I_s = 8.3\,\mathrm{J\,cm^{-2}\,min^{-1}} = 1.4 \times 10^3\,\mathrm{W\,m^{-2}}$ is solar constant; $h_\odot$ is sun elevation (sun height); $r/r_0 = [1 + 0.0335\cos(2\pi d/365)]$ is Earth-to-Sun separation; d is day of a year; $\cos(90 - h_\odot) = \cos\varphi\cos\delta + \sin\varphi\sin\delta\sin h_s$; h_s and $\delta = \arcsin(0.3978\sin\varepsilon)$ are solar hour angle and declination; $\varepsilon = 2\pi(d - 80)/354 + 0.0335 \times \sin(2\pi d/365) - \sin(2\pi80/365)$ is ecliptic longitude of the earth in its orbit.

Figure 2.5 shows that at the top of the atmosphere the annual average I_∞ is distributed around the maximum at the geographical equator and minimum towards the poles. Calculated regression is also shown in Fig. 2.5. However, the annual range of global insolation is minimal at $\varphi = 3.4°\mathrm{N}$ called the *meteorological equator*, rather than at $0°$ (Prescott and Collins, 1951). Therefore, in studying seasonal processes of lake thermics in relation to latitude, particularly within the tropics, it may be useful to adjust φ to the *corrected latitude* $\varphi' = (\varphi - 3.4°)$ as done in Straskraba (1980, 1993), Straskraba and Gnauck (1985), Lewis (1987), and others. It should be stressed that the above meaning of meteorological equator differs from that used to position the equatorial trough, a zone of low pressure caused by convergence of tradewinds leading to higher formation of clouds.

The daily and monthly sum of global irradiance I_∞, being independent of the atmosphere quality, have also been well calculated and tabulated for selected dates

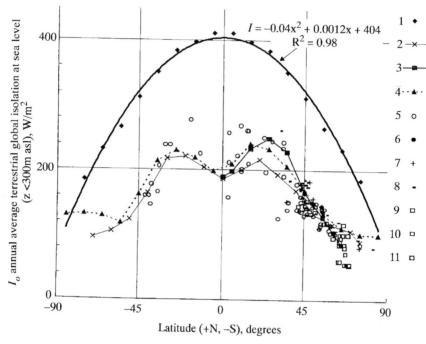

Fig. 2.5. Annual average global insolation measured at 151 lakes located at sea level ($z < 300\,$m asl) depending on latitude φ. Values at the top of the atmosphere with corresponding regression and global terrestrial insolation observed at sea level are also shown. Lake data included in GIS WORLDLAKE were originally taken or recalculated from cited studies. 1—top atmosphere; 2–4—terrestrial observed; 5—world lakes (wl) see *Data* (1988); 6—Volgograd res.; 7—world lakes (Straškraba); 8—wl (Adamenko, 1985); 9—European lakes; 10—Kola Peninsula lakes; 11—High Arctic lakes.

and latitudes by List (1951), Sellers (1965), Gates (1980) and other workers. Note that at latitudes higher than 10°N and 5°S, daily insolation reveals a single annual maximum near the time of solstice. Whereas, within 10°N–5°S there are two peaks, at the end of both March and September, and correspondingly two minima. The lowest one coincides with the hemispheric winter.

In the Earth's atmosphere, insolation is considerably attenuated being absorbed and scattered. Under cloudless conditions the integral insolation I_0 being a sum of direct and scattered radiation, at sea level is described:

$$I_0 = I_\infty \exp(-\eta M) \tag{2.21}$$

here η is the extinction coefficient referred to one standard atmosphere; M is air mass number = length of insolation path in the atmosphere.

Global trends of insolation

Kondratyev (1965) derived a theoretical relationship for short-wave radiation reaching sea level under cloudless conditions based on the rigorous equation for transformation of radiation in the atmosphere. Observed global latitudinal

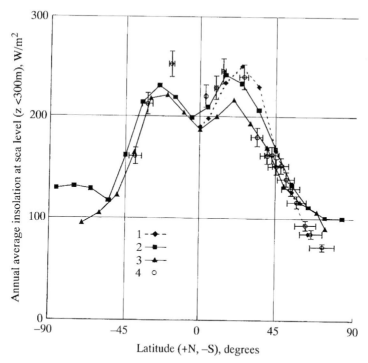

Fig. 2.6. Similar to Fig. 2.5, but averaged over the closest located lakes. Note the visible decrease of insolation due to higher cloudiness within the Tropics close to the Equator. 1–3—terrestrial observations (Landsberg, 1961; Berlyand, 1961; Budiko, 1956); 4—world lakes.

distribution of annual mean global terrestrial values for sea level was presented by Budyko (1956), Landsberg (1961), and Berlyand (1961). In Figs. 2.5 and 2.6, these trends are plotted using field data taken from GIS WORLDLAKE (see § 2.1.1) for lakes located close to sea level ($z < 300$ m asl). It is seen that in the northern hemisphere the annual mean lake-surface irradiance increases from $\sim$80–90 W m^{-2} at the polar circle to $\sim$220–250 W m^{-2} at the tropical circle, and then reveals lower values of $\sim$190–200 W m^{-2} for equatorial lakes (Fig. 2.6). The latter is obviously explained by higher average cloud cover at the equator. It should be noted that mean data for lacustrine and terrestrial trends are close to each other on average (Fig. 2.6). However, average irradiance for high-latitude lakes (at $\varphi > 60°$N) is visibly lower than terrestrial values given by Budyko (1956), Landsberg (1961), and Berlyand (1961), which may be caused by higher annual mean cloudiness over the lakes as compared with surrounding territory.

Latitudinal global trends for annual range (max–min) of insolation for lakes located at sea level is presented in Fig. 2.7. It is seen that extreme ranges of $\sim$50 W m^{-2} and $\sim$220–240 W m^{-2} are attributed to equatorial and middle latitude ($\sim$45°N) lakes, respectively. It should be emphasized that in moderate latitudes the range values for lakes are lower than observed terrestrial flux. Similarly, as pointed

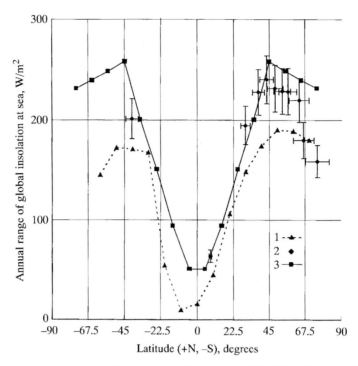

Fig. 2.7. Annual range (max–min) of insolation for lakes at sea level (<300 m asl) *versus* latitude φ. Lake data averaged over 122 lakes of the world. Original data are included in GIS WORLDLAKE. Global trends for observed terrestrial values are also shown. 1—terrestrial observed; 2—for 122 lakes located at sea level; 3— theoretical terrestrial at atmospheric transmissivity 0.7.

out for annual mean values, the differences may result from the lower summer and higher winter cloudiness over temperate lakes compared with terrestrial values, that was derived from the field studies on Lake Sevan (Ryanzhin and Rubtzov, 1986), Siberian lakes (Adamenko, 1985; Adamenko *et al.*, 1991) and Great Lawrentian Lakes (Scott and Huff, 1996). Obviously, additional field lake data are needed to examine spatial trends both for the southern hemisphere and high elevated lakes.

Effects of cloudiness

Equations (2.20) and (2.21) give a potential (maximum possible) irradiance of a lake surface at sea level under cloudless conditions. However, as shown in Fig. 2.8, cloudiness considerably modifies insolation at lake surface. Moreover, cloudiness highly complicates remote sensing in IR and visible spectral bands. To parameterize effects of cloudiness on I_0 the empirical relationships are usually derived in the form:

$$f(n) = 1 - cn \qquad (2.22)$$

here n is cloudiness (in points); c is empirical coefficients. For example, data given in Fig. 2.8 are approximated by (2.22) at $c = 0.08$.

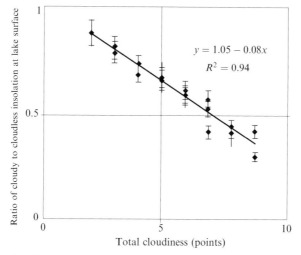

Fig. 2.8. Cloudy to clear-sky radiation ratio at lake surface depending on total cloudiness n (points) at lowland Lake Punnus-Jarvi, Karelia, Russia. Mokiyevskii's field data have been modified from Andronikova (1980). Appropriate regression $f(n) = 1 - cn$, at $c = 0.66$ is shown.

Non-linear empirical relationships for $f(n)$ with coefficients tabulated for selected latitudes have also been approached instead of (2.22) (e.g. Berlyand, 1961; Kirillova, 1970; Jirka *et al.*, 1975). In some more careful studies $f(n)$ is detailed in a manner similar to (2.22) to assume clouds of lower, middle and higher level (see review in Pivovarov, 1972). It should be noted that $f(n)$, being empirical, is not universal and reveals considerable temporal–spatial variability (Berlyand, 1961; Andronikova and Mokiyevskii, 1984). Hence, the corresponding empirical approximations are only locally adequate.

Integral albedo

According to Fresnel's law, albedo for direct radiation depends on angles of incidence and refraction. However, since I_0 consists of direct and scattered radiation characterized by different spectra, the integral albedo a depends on wave-length, sun elevation $h_\odot$, cloudiness, sea state and physical properties of lake water.

Direct radiation having diurnal and seasonal course highly predominates in I_0 at high sun angles and low values of cloudiness. In Fig. 2.9 average field data obtained at low cloudiness and moderate state of a lake surface are presented for separate temperate lakes (data were taken from **GIS WORLDLAKE**). As is seen from Fig. 2.9, a strongly governed by $h_\odot$ only slightly decreases at high sun elevation and is almost invariably small ($a \sim 0.04$–0.08) at $h_\odot \geq 40°$. Then it increases rapidly to 0.20–0.30 with a decrease in $h_\odot$ to $\sim 10°$–15°. However, an opposite tendency of a decrease in a and an increase in data scatter are clearly seen at lower sun height $h_\odot < 10°$–15°. One of the reasons is that at low sun angles a is mainly dependent on sea state rather than on $h_\odot$, whereas lake state could significantly differ in separate lakes. Another reason for data scatter could be low accuracy of albedo measurements at low values of $h_\odot$. Thus,

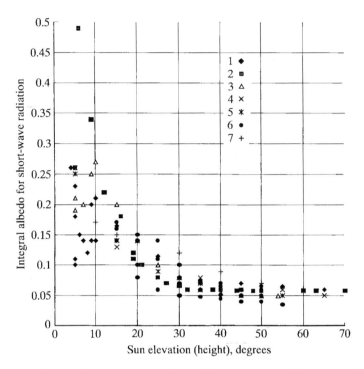

Fig. 2.9. Integral albedo *a* for short-wave insolation of ice-free surface of temperate lakes *versus* sun elevation (height) $h_\odot$. Clear sky and moderate sea state of lake conditions. Field data averaged over the sets of measurements. Note that albedo is almost invariable at $h_\odot \geq 30°$. Tendency of decrease in albedo and increase in data scatter are clearly seen at $h_\odot < 10–15°$.

an empirical relationship in the form submitted by Pivovarov (1972), can be derived from the data of Fig. 2.9 for clear sky and moderate sea state conditions as:

$$a = b/(\sin h_\odot + b) \quad \text{at } h_\odot > 10° \tag{2.23}$$

here $b \approx 0.038$ is the empirical coefficient calculated from the data of Fig. 2.9 that is close to $b \approx 0.040$ found by Pivovarov (1972) for description of Black Sea data at moderate sea state (3–4 balls).

Since sun elevation $h_\odot$ has an expressed seasonal and diurnal variability, similar variations are revealed in albedo (Figs. 2.10 and 2.11). $h_\odot$ and consequently *a* are strongly controlled by latitude φ (zonality). Minimal (summer) value of *a* in a year cycle increases with increase in latitude (Table 2.1).

The models have been developed to describe these variations for lakes and reservoirs (Kirillova, 1970; Pivovarov, 1972, and others). However, it was shown (Schindler *et al.*, 1974; Henderson-Sellers, 1984; Andronikova and Mokiyevskii, 1984; Adamenko, 1985; Kondratyev *et al.* 1986; Chekhin, 1987; Adamenko *et al.*, 1991, and others) that variations in cloudiness, lake water colour and transparency considerably complicate spatial–temporal variability of albedo predicted by the models.

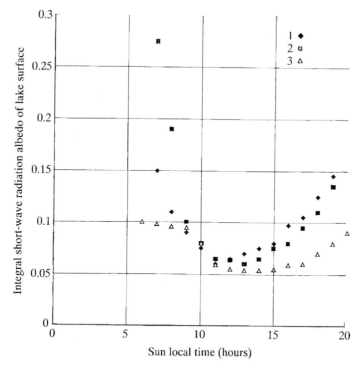

Fig. 2.10. Diurnal course of integral albedo *a* for a short-wave insolation of the temperate lake surface in mid-July. Note that daily variation in *a* is considerably smoothed for cloudy days compared with clear-sky days. 1—clear sky, Sevan; 2—clear sky, Issik-Kul; 3—cloudy sky.

Effects of ice cover

Another aspect of the albedo problem is ice cover. Field data on *a* for ice and snow cover are shown in Table 2.2. It should be pointed out that ice and snow structure and quality are not often specified and indicated by the authors. Therefore, some uncertainty and significant scatter of the field data in Table 2.2 are understandable.

Table 2.1. Range of maximum mid-day sun elevation $h_\odot$ and minimal year-cycle values of integral short-wave radiation albedo *a* at clear sky and ice-free period of lakes at selected latitude φ.

Latitude	0°–20°	30°–40°	50°–60°	70°–80°
Sun elevation $h_\odot$	65°–88°	60°–80°	40°–60°	20°–45°
Albedo *a* according to:				
Pivovarov (1972), Eq. (2.23) at $b = 0.040$	0.04	0.04	0.04–0.06	0.05–0.10
Adamenko (1985)	0.04	0.04	0.04–0.07	0.06–0.14
This study Eq. (2.23) at $b = 0.038$	0.04	0.04–0.05	0.04–0.06	0.06–0.11

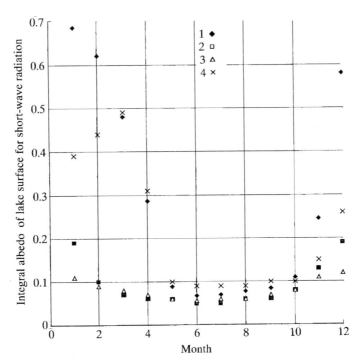

Fig. 2.11. Annual course of integral albedo a for short-wave radiation of temperate lakes. Lakes Punnus-Jarvi (1) and Ladoga (4) are ice-covered from December to May, and Lakes Sevan-3 and Issyk-Kul (2) are not.

It seems the values in Table 2.2 should be considered as minimal ones. a values for ice-covered lake considerably differ from that given by (2.23) being on average higher than that for ice-free lake surfaces. On the other hand, albedo for ice-cover, as opposed to ice-free lake surfaces, is more closely correlated with ice and snow physical properties. Thus, ice-cover significantly biases an annual variation of a as seen from Fig. 2.11 where field data on annual course of albedo are given for freezing in a year cycle for Lakes Punnus-Jarvi and Ladoga and for not-freezing Lakes Sevan and Issyk-Kul. In a year cycle the values of a for different lakes are low and close to each other for the ice-free period (May–October). However, they rapidly increase during the ice-cover period (November–May) in freezing Lakes Punnus-Jarvi and Ladoga.

Penetration of solar radiation

Vertical penetration of solar radiation into a lake water body is parameterized as:

$$I(t, z) = I_0(1 - a) \exp(-\beta z) \tag{2.24}$$

where β is the integral extinction coefficient (m^{-1}). In some detailed studies eq. (2.24) is considered in its differential form with a spectral expansion of a, I_0 and β (e.g. Pivovarov, 1972, 1979; Panin, 1985).

Table 2.2. Field data on integral albedo of ice and snow.

Ice/snow quality	Destruction of ice/snow (%)	Albedo	Reference
ICE			
Intensive melting ice partly covered with snow	50	0.41	Bogorodskii and Gavrilo, 1980[1]
Clear new formatted ice		0.35–0.40	Piotrovich, 1958
Melting ice	70	0.33	Bogorodskii and Gavrilo, 1980[1]
—		0.25	Grenfell and Maykut, 1977
Dirty melting ice		0.10–0.15	Piotrovich, 1958
SNOW			
Clear new fallen snow		0.95-0.98	Piotrovich, 1958
—	0	0.88	Bogorodskii and Gavrilo, 1980[1]
—		0.85–0.90	Smirnova, 1968
New, cold, freezing		0.85	Grenfell and Maykut, 1972
Old melting		0.60	Grenfell and Maykut, 1972
Dense snow on ice	0	0.77	Bogorodskii and Gavrilo, 1980[1]
Dirty wet snow		0.20	Piotrovich, 1958

[1] Data compiled by Bogorodskii and Gavrilo (1980) from different sources as well as sun elevation at the moment of measurement are not often indicated by authors.

As seen from (2.24), the vertical penetration of solar radiation is strongly governed by integral extinction β. The depth of photic layer, an important parameter of lake ecological system, is also closely defined through β. It was first shown in oceanic studies that β widely varies, being $\beta = 2.6 \times 10^{-2}$, $1.0 \times 10^{-1}\,m^{-1}$ for pure freshwater and oceanic water, respectively (Jerlov 1976; Pivovarov 1979). Detailed studies of β have been carried out in lakes of the former USSR (Kirillova, 1970; Andronikova, 1980; Andronikova and Mokiyevskii, 1984; Panin, 1985; Chekhin, 1987) and reviews are also presented for American and English (Henderson-Sellers 1984) as well as world (Panin, 1985; Adamenko et al., 1991) lakes. It was shown that β reveals a great spatial–temporal variability reaching values of $5-10\,m^{-1}$ or even more for very turbid lakes. β is governed mainly by dissolved matter, water colour, lake trophic status, etc., being on average much higher than those given above for pure freshwater and ocean water. In moderate lakes β reveals seasonal and diurnal variability due to chemical and biological cycling, being higher in midsummer and lower in winter (see the annual courses of β presented in Andronikova and

Mokiyevskii, 1984; Kondratyev *et al.*, 1986). According to Kirillova (1970), V. Shuleikin first noticed that β inverse correlates with Secchi depth S (see, e.g., the plots in Pivovarov, 1972) as:

$$\beta \sim cS^{-1} \tag{2.25}$$

where $c \sim 2$ is the empirical coefficient.

Relationships similar to (2.25) are used for numerical modelling when direct data on β are absent. Thus, the incident radiation term $I(z)$ in (2.24) has the asymptotes:

$$I(t, z) \rightarrow (1 - a)I_0(t, 0) \quad \text{at } \beta \rightarrow 0, \qquad \text{and} \tag{2.26}$$

$$I(t, z) \rightarrow 0 \qquad\qquad \text{at } \beta \rightarrow +\infty \tag{2.27}$$

Equation (2.26) characterizes a *thoroughly transparent lake* with no vertical attenuation of solar radiation. In turn, (2.27) presents a *fully turbid lake* where radiation is completely absorbed at air–water surface. It is pointed out above that for a transparent lake the solar radiation term is considered as an internal (spatially distributed) heat source (P_w in (2.1)–(2.3)) and is not included in total heat flux at the air–water boundary. However, in order to ascertain the influence of solar radiation on a heat regime of a relatively shallow lake it is helpful to introduce dimensionless optical depth of a lake defined as $H\beta$ (Ryanzhin, 1995, 1997) instead of (2.26) and (2.27). Then, at $H\beta > 1$ the water–sediment boundary should not be affected by incident radiation, whereas at $H\beta < 1$ it is subjected to direct solar heating. The latter was recently precisely measured at the small Karelian Lake Uros, Russia (Malm *et al.*, 1996). In that case the role of penetrating solar radiation should be included in the total heat flux F_H at the water–sediment boundary. For pure freshwater a threshold value of $H\beta = 1$ corresponds to lake depth $H \approx 39\,\text{m}$. Hence, the condition:

$$I(t, z) \rightarrow 0 \quad \text{at } \beta \rightarrow +\infty, \text{ or } H\beta > 1 \tag{2.28}$$

describes the effect of penetration of solar radiation on a heat budget of lake water more accurately compared with (2.26) and (2.27). Thus, (2.13) for the total heat flux at the air–water boundary is modified for very transparent lakes as:

$$F_w = -c_p\rho(K_z + \nu_w)\frac{\partial\theta_w}{\partial z} = B_e + LE + Q_T \quad \text{at } z = 0 \tag{2.29}$$

Equation (2.29) states that for completely transparent lakes the main heat flux components B_e, LE and Q_T balance vertical heat flux close to the lake surface.

2.2   TEMPORAL–SPATIAL VARIABILITY OF LAKE SURFACE TEMPERATURE. GLOBAL APPROACH AND CLASSIFICATION

Global zonal trends of the atmospheric and hydrospheric temperature are mainly caused by trends of insolation, the dominant incoming part of the heat budget of the Earth's surface. Detailed zonal climatologic temperature trends based on the vast field data have been calculated for the world ocean– and air–surface temperatures (e.g. Pivovarov, 1979; North *et al.*, 1981). However, similar trends for lake surface temperatures have been calculated only recently (Straškraba, 1980; Ryanzhin, 1989). These trends are evidently basic for thermal classification of lakes.

2.2.1 Thermal classification of lakes

2.2.1.1 Forel's classification

Forel (1892) subdivided lakes according to their surface temperature annual course and temperature of maximum freshwater density $\theta_m = 4°C$ into *polar* and *tropical* types characterized by $\theta_{max} \leq 4°C$, and $\theta_{min} \geq 4°C$, respectively, where θ_{max}, θ_{min} are extreme surface temperatures in a year cycle. An intermediate type in Forel's approach is *moderate* lakes with $\theta_{max} > 4°C$, and $\theta_{min} < 4°C$. Polar and tropical lakes are characterized by one full convective overturning during summer heating and winter cooling (in the northern hemisphere), respectively. In turn, two full overturning (in spring and fall) occur in moderate lakes. However, later the tropical lakes were found in the temperate zone as well as moderate ones in the polar latitudes. Similar inadequacy has been reported for high mountain lakes. Additional lake types, such as perennially ice-covered, polymictic, etc., not described by Forel's classification, have been documented.

2.2.1.2 Hutchinson–Loffler's classification

Hutchinson and Loffler (1956) considerably advanced Forel's approach. In their thermal classification (HLC) they indicated *moderate dimictic* ($\theta_{max} > 4°C$, $\theta_{min} < 4°C$), *warm* and *cold monomictic* characterized by $\theta_{min} \geq 4°C$, $\theta_{max} \leq 4°C$, respectively. Additional classes defined by the authors were *cold amictic* (perennially ice-covered lakes), *polymictic* and *oligomictic* lakes. Cold amictic lakes situated, e.g. in the high Arctic and Antarctic, are characterized by almost zero vertical mixing. The latter two types include equatorial lakes located in high mountains and close to sea level, respectively. Hutchinson (1957) indicated approximate qualitative boundaries of defined lake types in a (φ, Z)-space (latitude–above sea-level altitude), and Wetzel (1975) schematically arranged HLC spatially. The attempts to present HLC quantitatively in a (φ, Z)-space have been also made (Lewis, 1983; Håkanson and Peters, 1995). It should be noted that although thermal classifications of lakes based on other temperature characteristics have been developed (Birge, 1915; Tikhomirov, 1982; Kitaev, 1984, and others), HLC remains generally accepted in contemporary limnology. However, HLC, in spite of its clear physical background for nontropical lakes, gives neither type boundaries in a (φ, Z)-space nor detailed tropical classes.

Evident spatial boundaries of HLC in a (φ, Z)-space have been formulated (Ryanzhin, 1989, 1994) using the following approach. The boundaries of warm and cold monomictic lake distribution were found from the relationships $\theta_{min}(\varphi_w, Z_w) = 4°C$, $\theta_{max}(\varphi_c, Z_c) = 4°C$, respectively, where $Z_w(\varphi)$ and $Z_c(\varphi)$ are mountain boundaries of warm and cold monomictic lakes. Similarly, moderate dimictic lakes are located within the limits $\theta_{min}(\varphi, Z) < 4°C$, $\theta_{max}(\varphi, Z) > 4°C$.

Temperature conditions for the boundaries of freezing lakes and mictic lakes have to be $\theta_{min}(\varphi_{fr}, Z_{fr}) = 0°C$, and $\theta_{min}(\varphi_0, Z_0) = \theta_{max}(\varphi_0, Z_0) = \Delta\theta(\varphi_0, Z_0) = \theta_{max}(\varphi_0, Z_0) - \theta_{min}(\varphi_0, Z_0) = \bar{\theta}(\varphi_0, Z_0) = 0°C$ (where $\bar{\theta}$, $\Delta\theta(\varphi_0, Z_0)$ are annual mean and range of surface temperature variation), respectively. Note that the latter condition actually describes cold amictic lakes, which could be treated as intermediate boundary objects between mictic lakes and glacial ones. Values of latitude in the above relationships are attributed to sea level.

2.2.1.3　Straškraba's approach

Straškraba (1980, 1993) represents the annual course of lake surface temperature as $\theta(t) = \bar{\theta} + A_{12}$, where A_{12} is annual amplitude of surface temperature variation. Then, both variables were fitted by regressions *versus* latitude φ for the field data on more than 50 "medium size" lakes with a surface area $>1\,km^2$ situated close to sea level. In the northern hemisphere according to the linear regression derived, $\bar{\theta}$ decreases from 28.12°C at $\varphi = 0°$ to 0°C at $\varphi = 75.6°N$. Straškraba's model yields for the boundaries of freezing lakes (where $\theta_{min} = 0°C$), warm and cold monomictic lakes $\varphi_{fr} = 52°N$, $\varphi_w = 43°N$, $\varphi_c = 75°N$, respectively. Straškraba's model adequately describes on average the temperature trends and annual course of non-tropical non-freezing lakes located close to sea level, i.e. at $23.5° < \varphi < \varphi_{fr}$. However, this adequacy fails in this latitudinal frame for the following reasons (Ryanzhin, 1994). (i) In the tropics the surface temperature spectrum is bimodal and consists of annual and semi-annual harmonics with A_{12} and A_6 amplitudes, respectively both induced by analogous insolation behavior (see, e.g., Figures in Hutchinson, 1957; Wetzel, 1975; Straškraba, 1980, and others). A_{12} drops almost to zero at the Equator where, in turn, A_6 reaches its maximum ($\sim$1–3°C) beginning at $\sim$0°C at the tropical circle. (ii) In freezing lakes the longer the ice-cover period t_{fr} is the less adequate is Straškraba's approximation $\theta(t) = \bar{\theta} + A_{12}$. Therefore, the adequacy of $\theta(t)$ fitting $\bar{\theta}$ accompanied by one harmonic is considerably restricted. Thus, in spite of A_{12} and A_6 being less-biased statistical values, more general characteristics comparing amplitudes are required.

　　It should be underlined that in addition to lake-surface temperatures, Straškraba (1980, 1993) presents an excellent overview of many physical principles underlying latitudinal trends in freshwater ecosystems.

2.2.2　Global model for lake-surface temperature

The model for variation in freshwater lake-surface temperatures in a (φ, z)-space of the northern hemisphere has been developed (Ryanzhin, 1989, 1990, 1991, 1991a) and summarized (Ryanzhin, 1994). Annual average and extreme temperatures have been approached as principal characteristics.

Lake-surface temperature at sea level

To describe latitudinal variations in lake-surface temperature at sea level ($z < 200\,m$ asl) spline models were derived based on appropriate scaling, similarity hypothesis and appropriate boundary conditions imposed. First, cubic regressions have been calculated from the field temperature data on approximately 120 lakes. Integration of the regression for annual average temperature over the latitudes gives an important climatological estimate: 15.2°C, the mean annual surface temperature for lakes situated at $z < 200\,\mathrm{m}$ asl in the northern hemisphere. However, this value is close to 14.9°C, considerably less than 17.8°C for similarly averaged surface air and ocean temperatures, respectively, from North *et al.* (1981). Next, the values of annual mean temperature of the equatorial lakes at sea level $\bar{\theta}(0,0) = 30.0°\mathrm{C}$, the second derivative $\partial^2\tilde{\theta}_{max}/\partial\tilde{\varphi}^2 = -0.2$ at $\varphi = 0$, and $\varphi_0 = 80°\mathrm{N}$, $\varphi_{fr} = 53°\mathrm{N}$ were calculated from the regressions. Then $\bar{\theta}(0,0) = 30.0°\mathrm{C}$ and $\varphi_0 = 80°\mathrm{N}$ were approached as characteristic values for scaling. After the scaling and imposing of appropriate boundary conditions the following simple dimensionless spline model was derived to describe latitudinal variations close to sea level

$$\tilde{\theta}(\tilde{\varphi}) = 1 - 3\tilde{\varphi}^2 + 2\tilde{\varphi}^3 \qquad\qquad\qquad \text{at } 0 \le \tilde{\varphi} \le 1$$

$$\tilde{\theta}_{max}(\tilde{\varphi}) = 1 - 0.1\tilde{\varphi}^2 - 0.9\tilde{\varphi}^3, 1 - 3C_1^2\tilde{\varphi}^2 + 2C_1^3\tilde{\varphi}^3 \quad \text{at } 0 \le \tilde{\varphi} \le C_1^{-1}$$

$$\tilde{\theta}_{min}(\tilde{\varphi}) = \{0\} \qquad\qquad\qquad\qquad \text{at } C_1^{-1} \le \tilde{\varphi} \le 1 \qquad (2.30)$$

where $\tilde{\theta}(\tilde{\varphi})$ is the dimensionless annual mean temperature, and other dimensionless variables are denoted by "wave"; $C = \varphi_0/\varphi_{fr}$ is the parameter calculated from the field data.

The spline model (2.30) does not fit the data in a least-square-error sense. However, due to the appropriate scaling and the boundary conditions its general form is plausible (Fig. 2.12).

From (2.30) further thermal characteristics can easily be derived. Particularly, according to (2.30), the maximum range of annual temperature variations equal 22.5°C occurs at $\varphi_m = 46°\mathrm{N}$. The model (2.30) is governed by only four external parameters, $\bar{\theta}(0,0)$, $\partial^2\tilde{\theta}_{max}/\partial\tilde{\varphi}^2$ at $\varphi = 0$, φ_0, and φ_{fr}, to be calculated from field data. Therefore, the model can be easily applied to lakes located both in the southern hemisphere or separate meridian strip.

2.2.3　Modelling of altitude effects and thermal classification of lakes

2.2.3.1　*Altitudinal effects on lake-surface temperature*

Altitudinal effects on lake-surface temperatures were parameterized (Ryanzhin, 1989, 1990, 1991, 1991a, 1994) by analysis of field data on 45 mountain lakes, appropriate scaling, similarity hypothesis and imposed boundary conditions by the following dimensionless spline model (Fig. 2.13):

$$\tilde{\theta}(\tilde{Z}) = 1 - 0.5\tilde{Z} - 2\tilde{Z}^2 + 1.5\tilde{Z}^3 \quad \text{at } 0 \le \tilde{Z} \le 1$$

$$\Delta\tilde{\theta}(\tilde{Z}) = 1 - 0.5\tilde{Z} - 0.5\tilde{Z}^2 \qquad\qquad (2.31)$$

where $\tilde{\theta}(\tilde{Z}) = \theta(\varphi, Z)/\theta(\varphi, 0)$, $\Delta\tilde{\theta}(\tilde{Z}) = \Delta\theta(\varphi, Z)/\Delta\theta(\varphi, 0)$ are dimensionless annual mean and range of surface temperatures with characteristic values $\theta(\varphi, 0)$ and

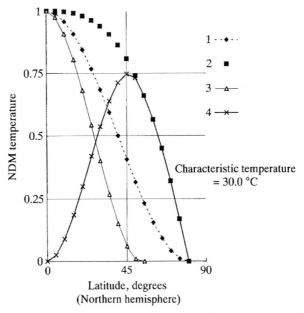

Fig. 2.12. Latitudinal dimensionless trends *versus* dimensionless latitude for the surface temperature of northern hemisphere freshwater lakes situated close to sea level ($Z < 200$ m asl). 1—ndm (dimensionless), average surface temperature; 2—maximum surface temperature; 3—minimum surface temperature; 4—range of surface temperature.

$\Delta\theta(\varphi,0)$ calculated from (2.30) for lakes situated at sea level; $\tilde{Z} = Z/Z_s(\varphi)$ is dimensionless lake elevation with characteristic value of snow-line altitude for a given latitude $Z_s(\varphi)$ taken from Kotlyakov (1968).

Discussing (2.30) and (2.31), the similarity should be pointed out in variations in annual mean temperature depending on latitude φ and altitude z (cf. Figs. 2.12 and 2.13). However, for $\Delta\theta$ such similarity was only found for lakes located at $\varphi > \varphi_m$. Dimensionless relationships have been derived from (2.30) and (2.31) for altitudinal boundaries of freezing $\tilde{Z}_{fr}(\tilde{\varphi})$ and cold monomictic $\tilde{Z}_c(\tilde{\varphi})$ lakes as:

$$\tilde{Z}_{fr}(\tilde{\varphi}) = \frac{(C_2^2 - 48C_2)^{1/2} - C_2}{12} \tag{2.32}$$

$$\tilde{Z}_c(\tilde{\varphi}) = \left[\frac{9}{4} - 2\frac{\theta_m}{\Delta\theta(\varphi,0)}\right]^{1/2} \tag{2.33}$$

where $C_2 = [\Delta\theta/\theta(\varphi,0) - 2]$, $\theta_m = 4°C$.

According to (2.32), dimensionless mountain boundary of freezing lakes increases monotonously from 0 (sea level) at $\varphi = \varphi_{fr} = 53°$ to 1 at the Equator. Thus, at the Equator freezing lakes occur in a single point $(0, Z_s)$ of (φ, z)-space. In turn, according to (2.33), dimensionless mountain boundary of cold monomictic lakes increases monotonously from sea level at $\varphi = \varphi_c = 76°$ with a decrease of latitude. However, it was shown (Ryanzhin, 1994) that the boundaries $\tilde{Z}_{fr}(\tilde{\varphi})$ and $\tilde{Z}_c(\tilde{\varphi})$ are

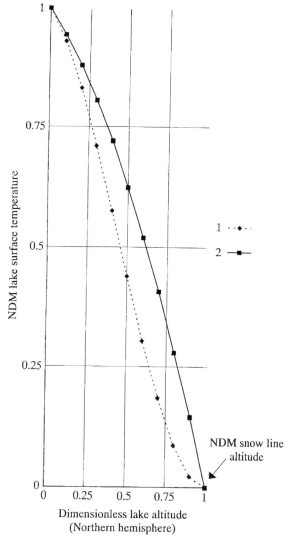

Fig. 2.13. Altitudinal dimensionless trends *versus* dimensionless elevation for the surface temperature of northern hemisphere freshwater mountain lakes ($Z < 200$ m asl). 1—ndm (dimensionless) average surface temperature; 2—ndm, range.

intersected in mountains at approximately $z/Z_s \sim 0.80$, $\varphi \sim 27°$. Thus, (2.33) is valid from $\varphi_c = 76°$ up to the latitude of intersection. In turn, it means that non-freezing lakes characterized by $\theta_{max} < 4°$C, $\theta_{min} > 0°$C, not previously described can occur high in the mountains close to the tropical cycle, e.g. in Andes.

2.2.3.2   *Thermal classification of lakes*

Thermal quantitative classification of lakes for (φ, z)-space has been suggested

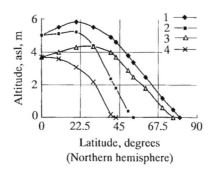

Fig. 2.14. Schematic arrangement of lake thermal classes with latitude and altitude in the northern hemisphere. 1—snow line; 2—ftreezing lakes; 3—cold lakes; 4—warm lake boundary.

(Ryanzhin, 1994) based on both models (2.30)–(2.33) and average behavior of snow-line altitude in the northern hemisphere adopted from Kotlyakov (1968). The principle point of classification, schematically shown in Fig. 2.14, is the relation of surface temperature variation to 4°C. The lakes were divided into general types: (i) *cold* ($\theta_{max} \leq 4°C$), (ii) *warm* ($\theta_{min} \geq 4°C$), and (iii) *moderate* ($\theta_{max} > 4°C$, $\theta_{min} < 4°C$). Cold amictic lakes fall into the cold type. Outside the Tropics lake types are similar to those suggested by Forel (1892) and Hutchinson and Loffler (1956). However, within the tropics new types such as *moderate*, *cold* and *warm* all *polymictic* are revealed instead of moderate dimictic, cold and warm monomictic, respectively. An additional new class, cold monomictic non-freezing lakes, presumably located in high mountains close to the tropical cycle have been indicated (Fig. 2.14).

2.3   THERMAL BAR IN A MODERATE LAKE

2.3.1   Discovery and study

Spring and fall convective overturning in a moderate dimictic freshwater lake, characterized by a minimum in a year cycle surface temperature <4°C, occurs in a form of *thermal bar* (TB). Thus, TB is an essential thermodynamic phenomenon associated with spring heating and fall cooling of lake water in the vicinity of maximum freshwater density at approximately 4°C. This phenomenon was dis-covered more than hundred years ago by Forel (1880) in moderate non-freezing Lake Geneva (Leman), 46.5°N, 8°E, 372 m asl, Switzerland/France. Forel presented the pattern of measured temperature distribution, supposed TB to be thermally convective in origin and first introduced the meaning *thermal bar* (*barre thermique littorale*). Then, TB was not studied for a long time until Tikhomirov (1959, 1963) and Rodgers (1965, 1966, 1968) independently "rediscovered it during their field measurements at Lake Ladoga and the Great Lawrentian Lakes, respectively. Since then vast field observations of TB mainly attributed to large lakes have been carried out at Lake Michigan (Noble and Anderson, 1968; Beer, 1971), Lake Superior (Hubbard and Spain, 1973; Spain *et al.*, 1976), Lake Ontario (Csanady, 1972, 1974;

Rodgers, 1971; Rodgers and Sato, 1970), Lake Ladoga (Tikhomirov, 1982; Popov and Kondratyev, 1982; Malm *et al.*, 1993, 1994; Malm, 1995; Naumenko, 1998; Naumenko and Karetnikov, 1998), Lake Vanern, Sweden (Ehlin, 1974; according to Malm, 1994), Lake Onega (Tikhomirov, 1982; Filatov *et al.*, 1990; Naumenko, 1994) and at some others lakes as well.

Laboratory experiments simulating the TB in a small tank have been done by Elliott and Elliott (1969, 1970), Kreiman (1989), Zilitinkevich and Kreiman (1989).

In order to model the TB, numerical hydrodynamic models based on 2- and 3-D nonlinear momentum and heat-transfer equations including Coriolis effects with different approximations of turbulent viscosity and heat conductivity have been developed (Csanady, 1971; Huang, 1971, 1972; Bennett, 1971; Brooks and Lick, 1972; Malm, 1995; Bocharov *et al.*, 1996). On the other hand, different versions of the 1-D heat conductivity equation have also been used to derive simple relationships for TB behavior (Elliott and Elliott, 1970; Elliott, 1971; Sundaran, 1974; Tikhomirov, 1982). Recently this approach has been considerably advanced (Zilitinkevich and Terzhevik, 1987, 1989; Zilitinkevich 1991, Zilitinkevich *et al.*, 1992, Zilitinkevich and Malm, 1993) by using the 2-D heat conductivity equation, appropriate scaling and self-similarity hypothesis.

2.3.2 General description of the phenomenon

In spring (at $<4°$C) heating at the air–water lake surface is distributed over a vertical water column mainly by gravity convection. The shallow nearshore area (*warm zone*, with $>4°$C) is heated rapidly compared with the offshore deep one (*cold zone*, $<4°$C). Thus, in the former temperature reaches and passes $4°$C faster and stable summer stratification with horizontal isotherms develops earlier than those in a cold zone (Fig. 2.15).

The cold zone remains at $<4°$C and vertically homogeneous for a longer time. The narrow TB zone characterized by strong horizontal temperature gradients of $\sim 10^{-2}$–10^{-1} °C m^{-1} is located between these zones and acts like a boundary convergence zone with sinking of surface $4°$C-water (Fig. 2.15). Field surveys show that horizontal temperature gradients are usually much higher at the onshore warm side compared with the offshore side of the $4°$C-isotherm. Estimates derived from field studies (Carmack and Farmer, 1982) and hydrodynamic models (Brooks and Lick, 1972; Csanady, 1972; Huang, 1972; Malm, 1995) give $\sim 10^{-1}$–10^{-4} cm s^{-1} for downwelling velocity in the spring TB. Rodgers (1966) reported that drogues placed on both sides of the $4°$C-isotherm converged and floated along with the TB. So, two circulating cells with more pronounced and strong circulation at the onshore side of the $4°$C-isotherm develop in a lake. Upwelling occurs in the center of the cold zone and near the lake shore in the warm zone. In the near-bottom TB layer the surface convergence changes to divergence (Fig. 2.15). In the course of annual warming and/ or cooling, the TB is displaced (progresses) with a characteristic speed of ~ 1 cm s^{-1} (e.g., Tikhomirov, 1959, 1963; Rodgers, 1965) towards the deepest sites of a lake until their temperature increases to $4°$C. Horizontal declination of the TB towards the cold zone increases with TB development (Fig. 2.15).

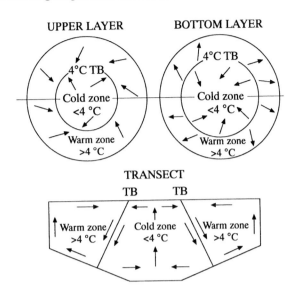

Fig. 2.15. Schematic general density-induced spring circulation (superposition of geostrophic and thermal bar (TB) circulation) in a moderate lake. Warm and cold zones with alternative upwelling and downwelling are seen. Note the inclination of TB caused by the larger area of the cold zone near the bottom compared with that near the lake surface.

Length of existence of the spring TB varies from days or even hours in a shallow small lake or ponds to 2–3 months in deep large lakes, such as Lake Superior and Lake Ladoga. For example, more than 20 years of observation at Lake Ladoga, maximal depth $H_m \cong 230$ m, gives long-term variation from 40 to 180 days (Tikhomirov, 1982). It should be noted that hypothetically the time of TB existence varies from 0 to infinity when a moderate dimictic lake transforms to a warm monomictic or cold monomictic lake, respectively (Ryanzhin, 1995).

Thus the TB exists as a density-induced circulation against a background of primary geostrophic circulation. General spring circulation in a moderate lake is by superposition of currents induced by both TB and geostrophic balance. However, it should be stressed that non-zero lake bottom slope μ is the principal condition for TB existence.

During fall cooling the TB occurs in the opposite direction to heat and buoyancy flux at the air–water boundary. Autumn TB originates at a shallow zone when surface temperature drops here below 4°C. Then it moves towards and vanishes in the deepest part of the lake. It should be noted that the autumn TB is considerably overshadowed and smoothed by strong wind mixing. On the other hand, TB is strongly smoothed in a riverine lake characterized by low-water residence time (Carmack, 1979; Carmack et al., 1979).

Since the TB insulates the warm stratified zone from the cold, well-mixed and oxygenized zone, it has important chemical and biological consequences (Menon et al., 1971; Scavia and Bennett, 1980) which have not so far been well studied.

2.3.3 Theoretical models and descriptions

The principal characteristic to be determined for the TB is its spatial–temporal position $l_b(t)$ with respect to lake shore.

Elliott–Elliott's approach

Simple relationships here called the Elliott–Elliott theory (EET) were derived from the 1-D heat conductivity equation (Elliott and Elliott, 1970; Elliott, 1971) and later independently (Tikhomirov, 1982). Consider heat conduction in a lake with the co-ordinate system assumed in §2.1. Suppose (i) no heat exchange at the bottom–sediment boundary ($F_H = 0$), (ii) the total kinematic heat flux at lake surface $F_0^* = F_0/c_w\rho_w$ and lateral (horizontal) heat flux to be time independent and negligibly small, respectively. Then, after integration of the 1-D heat conductivity equation from the lake surface $z = 0$ to the bottom $z = H$ with respect to lake depth and from the beginning of heating (cooling) $t = 0$ to some $t = t_1$ with respect to time t, one has:

$$\theta(t_1) - \theta_0 = \frac{F_0^* t_1}{H(x)} \tag{2.34}$$

where $\theta(t_1)$, θ_0 are vertically averaged temperatures at $t = t_1$ and $t = 0$; $H(x)$ is current lake depth at x distance from the shore.

At a constant tangent of lake-bottom inclination (bottom slope), $\mu = H(x)/x$ across the entire lake basin, (2.34) is rewritten:

$$\theta(t_1) - \theta_0 = \frac{F_0^* t_1}{\mu x} \tag{2.35}$$

According to (2.34) and (2.35), at invariable slope $\mu = H(x)/x$, the TB position $l_b(t)$ at $x = l_b$ corresponding to the temperature $\theta(t_1) = \theta_m = 4°C$ is expressed as:

$$l_b(t) = \frac{tF_0^*}{\mu(\theta_m - \theta_0)} \tag{2.36}$$

Note that (2.36) was incorrectly written in Zilitinkevich (1991), cf. his formula (6.2). For a freezing lake it is $\theta_0 = 0°C$, and (2.36) is reduced to:

$$l_b(t) = \frac{tF_0^*}{\mu\theta_m} \tag{2.37}$$

suggested by Elliott (1971). It was shown (Hubbard and Spain, 1973; Popov and Kondratyev, 1982; Malm, 1995, and others) that (2.34)–(2.37) adequately describe a spatial–temporal evolution of the TB on average. However, EET increasingly underestimates the rate of TB position at later stages of TB development.

Zilitinkevich–Terzhevik's approach

The next step to study was the Zilitinkevich and Terzhevik theory (ZTT) (1987, 1989). In some studies (e.g. Malm *et al.*, 1993) the theory is referred to as the Zilitinkevich, Kreiman and Terzhevik (ZKT) (1992) published later, however. ZTT takes into account the heating action of the warm zone on TB progress. Considering additionally to EET the horizontal advection of heat, assuming its analogy to a turbulent jet, after appropriate averaging and scaling the authors derived for "horizontal dynamic heat flux" due to horizontal entrainment of water from the warm zone into the TB:

$$F_l = C_t (ga\mu l)^{1/2} (\theta_w - \theta_m)^2 \tag{2.38}$$

where $a = 1.6509 \times 10^{-5}\,\mathrm{K}^{-3}$ is the coefficient in the quadratic equation of state for freshwater; C_t is the dimensionless coefficient empirically estimated as 0.008 (Zilitinkevich, 1991).

Then asymptotes for both dimensionless TB position $\lambda = l_b/l_c$ and temperature of warm zone $\vartheta = (\theta_w - \theta_m)/(\theta_m - \theta_0)$ depending on dimensionless time $\tau = t/t_c$ were derived as:

$$\lambda \approx \frac{2M - 1}{M^2} \tau + \frac{2C_t M}{(5M - 3)(2M - 1)^{1/2}} \tau^{3/2}$$

$$\vartheta \approx \frac{M^2}{2M - 1} - \frac{4C_t M^6}{(5M - 3)(2M - 1)^{5/2}} \tau^{1/2} \quad \text{at } \tau \ll C_t^{-2} \quad \text{and} \tag{2.39}$$

$$2\tau \approx M\lambda + C_t^{1/2} M^{-1} \lambda^{3/4}$$

$$\vartheta \approx C_t^{1/2} M^{-1} \lambda^{1/4} \quad\quad\quad \text{at } \tau \gg 10^{-1} C_t^{-2} \tag{2.40}$$

where $M = 1 + C_M \mu$ is the dimensionless parameter; $C_M < 1$ is a ratio of the width of the TB zone to the lake depth in that zone; and characteristic scales for time and length are taken as:

$$t_c = \frac{F_0^*}{ga\mu^3 (\theta_m - \theta_0)^3}, \quad\quad l_c = \frac{F_0^{*2}}{ga\mu^3 (\theta_m - \theta_0)^4} \tag{2.41}$$

Equations (2.39) and (2.40) state that with increasing τ the dimensionless rate of TB displacement $d\lambda/d\tau$ increases monotonously from $(2M - 1)/M^2$ to $2/M$, the dimensionless temperature of the warm zone ϑ monotonously decreases from $M^2/(2M - 1)$ to 0, and the dimensionless heat flux $\mu F_l/F_0$ increases monotonously from 0 to 1. Since $C_M < 1$ at not very high bottom slopes, say at $\mu < 0.1$, $M \sim 1$, so (2.39) and (2.40) are reduced to:

$$\lambda \approx \tau + C_t \tau^{3/2}, \quad\quad \vartheta \approx 1 - 2C_t \tau^{1/2} \quad \text{at } \tau \ll C_t^{-2} \quad \text{and} \tag{2.42}$$

$$2\tau \approx \lambda + C_t^{1/2} \lambda^{3/4}, \quad\quad \vartheta \approx C_t^{1/2} \lambda^{1/4} \quad \text{at } \tau \gg 10^{-1} C_t^{-2} \tag{2.43}$$

So, according to ZTT, only three external parameters μ, F_0, and θ_0 govern temporal–spatial development of the TB. Calculations made by Malm et al. (1993) using detailed field data from Lake Ladoga show that ZTT predicts more realistic values for the rate of TB progress than those given by EET (Table 2.3.).

ZTT was adapted for axisymmetric circular lakes by Zilitinkevich and Malm (1993). Particularly, the authors derived solutions for upper and lower limits of time

Table 2.3. Governing parameters for thermal bar displacement observed at Lake Ladoga and predicted by Elliott–Elliott (EET) and Zilitinkevich–Terzhevik (ZTT) theory (adopted from Malm *et al.* (1993).

Parameter	Survey			
	1–3	3–5	1–5	2–4
H (m)	24.5	33.5	24.5	23.0
l_b (km)	9.5	15.3	9.5	27.0
θ_w, °C	4.79	4.96	4.79	5.46
Rate of thermal bar displacement dl_b/dt, cm s^{-1}				
Observed	1.93	1.33	1.72	1.94
EET	1.29	0.90	1.30	1.61
ZTT	1.64	1.27	1.56	2.51

of TB existence in a lake characterized by a constant bottom slope μ, $H_{max} = r\mu$, $H_{avg} = r\mu/3$, where r is lake surface radius, as:

$$t_{upper} = \frac{H_{max}(\theta_m - \theta_0)}{F_0^*} \qquad (2.44)$$

$$t_{lower} = \frac{H_{avg}(\theta_m - \theta_0)}{F_0^*} \qquad (2.45)$$

The physical interpretation of time limits given by (2.44) and (2.45) is quite straightforward. The lower limit (maximum rate of TB displacement) corresponds to the highest intensity of heat entrainment to the TB zone when the warm zone has a temperature equal to that of maximum density, i.e. $\theta_w = \theta_m = 4°$, $\vartheta = 0$. The upper limit (minimum rate of TB displacement) characterizes the case of no heat entrainment at all to the TB zone when displacement of the TB is completely governed by cold zone heating. Note that this is the case of EET. Values predicted by (2.44) and (2.45) are not in conflict with observed field data (Table 2.4).

Certain limiting points should be underlined when discussing EET and ZTT. First, in both theories the total heat flux F_0 is assumed to be time independent. Whereas, during spring heating of moderate lakes F_0 monotonously increases reaching its annual maximum at the end of May–the beginning of June. Second, although horizontal temperature gradients are much higher at the onshore side than the offshore side of the 4°C-isotherm and geostrophic circulation seems to develop in a warm zone, the velocities of induced currents are too small to be considered a "turbulent jet. Therefore, the physical mechanism of heat entrainment suggested by ZTT remains unclear.

Other approaches

Naumenko (1998), and Naumenko and Karetnikov (1998) studying the long-term

Table 2.4. Governing parameters for time of thermal bar existence observed and predicted by (2.44) and (2.45) according to Zilitinkevich and Malm (1993).

	Lake Ladoga	Lake Onega	Lake Vanern (Varmlandssjon)
F_0^*, m °C s^{-1}	5×10^{-5}	5×10^{-5}	5×10^{-5}
θ_0, °C	0.5	0.5	2.0
H_{max}, m	230	115	106
H_{avg}, m	51	30	30
Time of thermal bar existence, days			
t_{upper}	186	93	49
t_{lower}	41	24	14
Observed	~60	~35	~20

field data on water temperature, total surface heat flux $F_0(t)$ and spatial–temporal position of the TB for Lake Ladoga, derived the following empirical regressions:

$$F_0(t) = 101.126 + 4.801t + 0.0282t^2 - 0.000\,5209t^3 \tag{2.46}$$

$$H(t, x) = 0.0766t + 0.0237t^2 \tag{2.47}$$

where $F_0(t)$ is in cal cm^{-2} day^{-1}, t is number of days after April 20.

Naumenko (1998) also indicated the increased rate of TB progress at late stages of development.

Remote sensing study of TB. The TB often manifests itself as a pronounced visible strip at the lake surface where foam, air bubbles, debris, algae, etc. collect. Relevant Figures are shown in § 2.7 where manifestation of the TB, revealed by remote sensing in the visible band, is discussed (e.g., air-borne photograph in Fig. 2.49). The TB due to strong horizontal temperature gradients is revealed in IR images (with ~0.1°C and ~1 km resolutions) obtained, for example, from NOAA-AVHRR (channels CH4, 10.5–11.3 μm, CH5, 11.5–12.5 μm) and KOSMOS satellites under cloudless conditions.

The vast studies of the spring TB using IR images accompanied by *in situ* temperature measurements have been carried out at Lake Ladoga (Karetnikov and Naumenko, 1993; Kondratyev *et al.*, 1996; Naumenko, 1998) and Lake Onega (Storm and Bengtsson, 1993; Naumenko, 1998) and others. The relationship for temporal decrease at the cold zone area of Lake Ladoga caused by May–July heating, derived from long-term sequences of IR images by NOAA satellite, was presented by Bichkova *et al.* (1989). Later Naumenko and Karetnikov (1998) using *in situ* long-term temperature data from Lake Ladoga calculated the relationships for variation of warm and cold zone areas during spring–summer heating and compared them with those derived in Bichkov *et al.* (1989). It was found that satellite data constantly overestimate the cold zone area by approximately 4000 km^2, i.e. by 20% of the total lake area. Inadequacy was blamed in Naumenko and Karetnikov (1998)

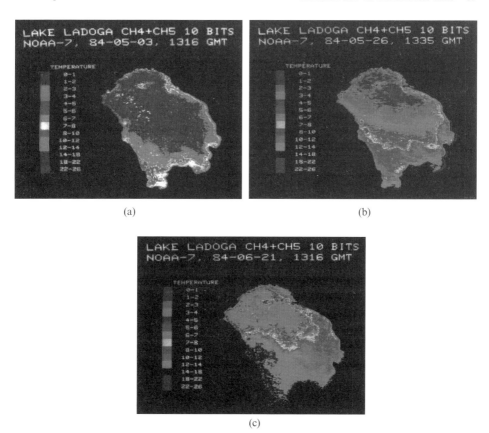

(a)   (b)

(c)

Fig. 2.16. Photos showing the sequence of surface IR-temperature distribution in Lake Ladoga's development of the thermal bar, an increase in the warm zone area and a decrease in the cold zone area (colored dark) are clearly seen (NOOA satellite, after Malm and Jonsson, 1993).

for incorrect estimation of TB shape by satellite data. Excellent temporal sequences of IR images of spring TB development at Lake Ladoga shown in Fig. 2.16, have been analysed (Malm and Jonsson, 1993, 1994). The authors compared TB locations calculated by (2.34) for 1984, 1986, 1988 and 1992 with those derived from corrected IR images obtained from NOAA-AVHRR. Agreement was in general good, except during the later stages of TB evolution. Storm and Bengtsson (1993) presented a similar kind of analysis applied for the spring TB at Lake Onega and also found adequate agreement. Important results on the TB at Lake Ladoga have been taken from IR images obtained from air-borne *Vulkan* IR-radiometer by Kondratyev *et al.* 1988 (cited in Naumenko, 1998). The authors found several cyclone- like vortices penetrated the cold zone from the TB. Naumenko (1998) estimated the vortices to be at 15–20 m depth, having 20–30 hours of dissipation and 0.05–1 km hour^{-1} speed of propagation. These observations help to explain the progressive rate of propagation of the spring TB.

2.4 LANGMUIR CIRCULATION

Langmuir circulation in natural water bodies

Langmuir circulation (LC) is a system of horizontal coherent counter-rotating vortices that develop in the ocean/sea/lake surface layer with moderate to strong winds blowing over the surface. Threshold wind speed V_{cr} is controlled by the initial hydrostatic stability of upper water and near surface air layers, state of the sea, etc. (Leibovich, 1983; Ryanzhin and Mironov, 1985). The axes of the vortices are aligned roughly parallel to wind direction and the water surface (Fig. 2.17).

Movement in LC, having longitudinal, lateral, and vertical current components, is substantially three-dimensional. Downward velocities in convergent zones of LC, reaching values of order $\sim 1\,\mathrm{cm\,s}^{-1}$, exceed the upward currents at divergent zones. Thus, the vortices are asymmetric. Two neighboring counter-rotating vortices form a circulating cell with a characteristic lateral and vertical scale (wavelength) L and, Z_L, respectively. The former is called LC *spacing* or streaks *spacing*. Field studies carried out in lakes (Myer, 1969; Scott *et al.*, 1969; Filatov *et al.*, 1981; Ryanzhin *et al.*, 1981; Ryanzhin, 1983, 1995, and others) show that LC, due to its coherent structure and high vertical velocities, is an effective mechanism of vertical mixing within the layer Z_L as compared with small-scale turbulence, gravity convection, internal wave breaking, etc. Besides, LC induces specific chemical, biological, and physical regimes within the upper layer that is revealed in redistribution of suspended matter, seaweed, plankton, fish eggs, oil spills, etc. The physical regime, for example, manifests itself in mixed-layer deepening and thermocline development (Faller and Caponi, 1978; Leibovich, 1983; Weller and Price, 1988), generation and entrainment of air-bubble inhomogeneities or clouds with a characteristic lateral scale L (Thorpe and Hall, 1982; Thorpe, 1984; Farmer and Li, 1995), etc.

The above qualitative description of LC was first given by the 1932 Nobel Prize winner Irving Langmuir in his pioneer article (Langmuir, 1938) based on the excellent field studies at small Lake George, 43.7°N, 73.6°W, New York. Since the scheme has been qualitatively and quantitatively confirmed and enhanced both theoretically and experimentally by many workers (see the above references and also Welander, 1963; Faller and Woodcock, 1964; McLeish, 1968; Dmitrieva and Ryanzhin, 1976; Kenney, 1977; Ryanzhin 1982, 1994; and others). It is commonly recognized that LC is wind-driven in origin. According to Craik–Leibovich theory (CLT) (see review Leibovich, 1983) and its further modifications, LC results from

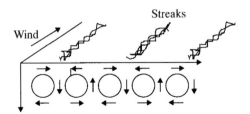

Fig. 2.17. Sketch of Langmuir circulation (LC).

nonlinear interaction between the Stokes drift gradient and velocity shear. There-fore, LC can be considered as a result of atmosphere–lake interaction in which wind–wave energy is transformed into the energy of small- scale turbulence. However, in spite of its evident importance, LC has not been incorporated in the large-scale hydrodynamic models so far. The reason is that CLT, partly based on K-theory of turbulence, is too complex and does not present simple parameterizations of LC effects to be directly included in models. Moreover, field data on LC remain too poor and occasionally obtained to be used for both CLT testing and incorporated in large-scale models in the form of empirical parameterizations. There have been no general experimental relationships between LC parameters (such as L and Z_L) and back-ground gross hydrometeorologic conditions (such as wind speed, wind-wave param-eters, surface heat and buoyancy flux, etc.). Furthermore, it is widely accepted in limnology and oceanography that LC is a rather rare phenomenon.

Probability of LC occurrence

At the surface LC is revealed and identified in a form of narrow *wind streaks* or *windrows* (convergent zones) where foam, bubbles, seaweed, plankton, etc. accumu-late. It seems the first streaks' description was given by Charles Darwin in 1839 during his voyage on *The Beagle*. Bubbles and foam directly make the streaks visible and compressed organic films reveal them by damping of capillary waves at the surface. Air photographs of LC streaks are presented in § 2.7 (e.g. Fig. 2.49). Streaks become visible when the wind speed reaches its critical value of $V_{cr} = 3$–$5\,\mathrm{m\,s}^{-1}$.

It has been shown (Welander, 1963; Scott *et al.*, 1969; Dmitrieva and Ryanzhin, 1976; Ryanzhin, 1980; Ryanzhin and Mironov, 1985) that LC is a commonly occurred phenomenon. Figure 2.18 shows the histograms of LC occurrence under different values of V. Corresponding statistics, such as probability of LC presence, biserial correlation between LC occurrence and wind speed R are given in Table 2.5. At least two conclusions follow from Fig. 2.18 and Table 2.5.

First, the total probability of LC occurrence, $p(+)$, varying from 0.32 at small Lake Punnus-Jarvi (Ozero Krasnoye), Karelia, Russia, to 0.67 at Lake Ladoga, NW Russia, reaches considerable values. It should be emphasized that values of $p(+)$ are close to each other under lake and sea conditions. Second, $p(+)$ increases with increase in V that is also seen from high positive correlation coefficients R. However, it should be stressed that at very strong winds ($V > 15$–$18\,\mathrm{m\,s}^{-1}$, storm conditions) LC seems not to be generated or to be destroyed.

Threshold conditions for LC

Vast field studies of LC have been carried out at Lake Ladoga during recent years (Filatov *et al.*, 1981; Ryanzhin, 1982, 1983, 1994, 1995; Ryanzhin *et al.*, 1981; Ryanzhin and Mironov, 1985). Unfortunately, these studies remain poorly known to western authors. Observations included the measurements of horizontal and vertical currents in CL cells, streak spacing, etc. under various gross meteorological con-ditions. Details can be found in Ryanzhin (1980); Ryanzhin *et al.*, 1981). Some sets of LC observations have been accompanied by measurements of tangential wind

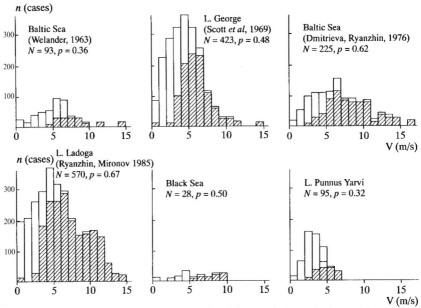

Fig. 2.18. Occurrences of Langmuir circulation under different wind speeds observed in water bodies. Height of the dashed bar gives the number of cases when LC has been observed; N, p is the total number of observations and probability of LC occurrence, respectively.

stress τ at the air–water boundary and vertical temperature profiling of the upper water layer. It was shown, based on these measurements, that LC does not appear under the smooth regime of surface wind friction. However, at the intermediate regime (from smooth to rough) LC is rapidly established (Fig. 2.19). The empirical relation for wind stress τ_{cr} critical for LC generation, dependent on initial Brunt–Vaisala frequency of the upper (thickness 1–2 m) lake layer N_i has been calculated as (Ryanzhin and Mironov, 1985):

$$\tau_{cr} = \tau_0 + CN_i^2 \qquad (2.48)$$

where $\tau_0 = 0.15\,\mathrm{dyn\,cm^{-2}}$, $C = 1.16 \times 10^3\,\mathrm{g\,cm^{-1}}$ are empirical coefficients.

According to (2.48), LC instability occurs at $\tau > \tau_{cr}$, and τ_{cr} increases with increase in original hydrostatic stability of the upper water layer. Theoretical criterion for LC stability in inviscid nonconducting fluid of infinite depth was earlier derived from CLT as (Leibovich, 1983):

$$R_i = \min\left\{ \frac{N_i^2(z)}{U'(z)U_s'(z)} \right\} \qquad (2.49)$$

where $U'(z)$ and $U_s(z)$ are the vertical gradients of wind shear current and Stokes drift, respectively.

According to (2.49), the upper layer is unstable with respect to LC at $R_i < 1$, when the minimum is taken over depth z. There are not any relevant field data to verify

Table 2.5. Statistics for Langmuir circulation occurrence in natural waters.

Location	Number of observations N	Probability of LC occurrence			Biserial correlation $R^{(1)}$	Source
		total $p(+)$	at $V > 5\,\text{m/s}$ $p_{>5}(+)$	at $V > 10\,\text{m/s}$ $p_{>10}(+)$		
Baltic Sea [2]	93	0.36	0.56	1	0.82 (±0.06)	Welander, 1963
Lake George, 43.7°N, 73.6°W, New York [2]	423	0.48	0.84	0.89	0.79 (±0.05)	Scott et al., 1969
Baltic Sea (Gulf of Finland)	225	0.62	0.84	0.84	0.78 (±0.07)	Dmitrieva and Ryanzhin, 1976
Lake Ladoga, 60.9°N, 31.4°E, NW Russia	570	0.67	0.93	1	0.86 (±0.04)	Ryanzhin and Mironov, 1985
Black Sea, Bulgarian shore	28	0.50	0.77	—	0.89 (±0.08)	This study
Lake Punnus-Jarvi 60.6°N, 29.7°E, Karelia, Russia	95	0.32	0.93	—	0.87 (±0.06)	This study

[1] Confidence limits of 0.95 are given.
[2] Statistics were calculated using the histograms published in cited articles.

(2.49). However, criterion (2.48) and (2.49) being close to each other, both state that some work against buoyancy is needed to establish LC instability.

LC spacing

L is considered as a lateral scale or wavelength of LC. L widely varies from meters to tens of meters even under similar gross hydrometeorologic and morphometric conditions, such as wind speed, wind–wave parameters, heat and buoyancy flux, water or thermocline depth, etc. (Faller and Woodcock, 1964; Ryanzhin, 1980, 1983, 1994; and others). Streaks length exceeds spacing between streaks and may reach several hundred meters or more (see, e.g., Fig. 2.49). It seems only Kenney (1977) has presented histograms of streaks length. In some cases the *secondary* less pronounced and developed streaks occur between the strong visible *main* streaks with spacing L_m and L_s, respectively (at $L_m > L_s$) (Ryanzhin et al. 1981). Thus, often many scales (modes) or cascade of LC exist simultaneously. Some cells grow and others disappear. In other situations only main streaks exist indicating the dominant

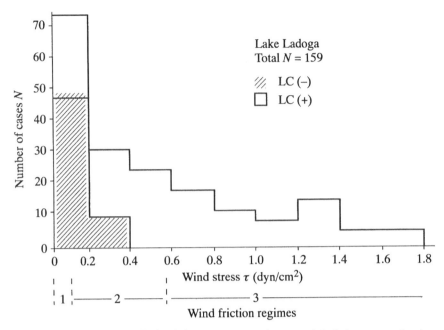

Fig. 2.19. Histograms of Langmuir circulation occurrence under tangential wind stress τ at the air–water surface (Lake Ladoga, 60.9°N, 31.4°E, NW Russia). Height of the dashed bar gives the number of cases when LC has not been observed ($-$LC); N is total number of observations; 1, 2, 3 are approximately the smooth, intermediate and rough regimes of wind friction at $\tau < 0.1$, 0.1–0.55, and, $>0.55\,\mathrm{dyn/cm}^2$, respectively.

mode of LC. Therefore, a statistical approach is valid for processing data on LC spacing. It should be emphasized that, in spite of its variability, L is the most easily and reliably measured LC parameter (using, e.g., airborne or remote sensing methods). Therefore, L should be approached as a main parameter to be analysed.

Field data on LC spacing obtained from lakes and seas are summarized in Table 2.6. Most of the original field data have been obtained using the measurement technique suggested by Faller and Woodcock (1964). In their study the original values of spacing for each set of measurements were obtained in a form of time intervals between streaks as the research vessel sailed across the streaks at a known speed and angle. Then, measured time values multiplied by the ship's crossing speed gave the spacing between streaks. Similar techniques have been used by (McLeish, 1968; Dmitrieva and Ryanzhin, 1976; Ryanzhin, 1980, 1982, 1994; Thorpe and Hall, 1982). Another approach was developed by Kenney (1977). He used aircraft photographs of the lake surface covered with LC streaks to calculate spacing. It seems B. Kenney first measured the LC spacings using remote sensing approach. Each set of data shown in Table 2.6 contains different amounts of measured spacing. In this sense the data obtained from Lake-of-the-Woods, Loch Ness and Lake Ladoga can be considered as the most complete and representative sets. About 150 original sets of L measurements have been collected for analysis. First, for each

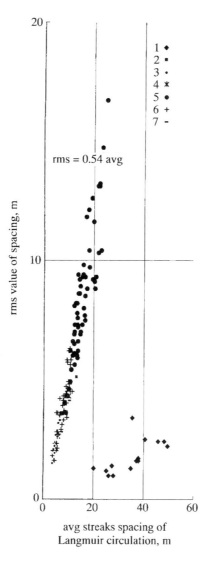

Fig. 2.20. Root mean square (rms) *versus* average (avg) spacing of Langmuir circulation. 1–3—Atlantic Ocean; 4—Lake-of-the-Woods, Canada; 5—Loch Ness, Scotland; 5–6—Lake Ladoga, Russia; 7—Baltic Sea.

original set, an average (avg) spacing $\bar{L}$ and corresponding root mean square (rms) δ were calculated or taken from the cited articles (Table 2.6). The plot $\bar{L}$ *versus* δ is represented in Fig. 2.20, where the data on main and secondary streaks obtained from Lake Ladoga are given separately. The variables vary a lot: from 3.8 to 25.1 m for average, and from 1.6 to 16.7 m for rms spacing (Table 2.6, Fig. 2.20).

Note that Faller and Woodcock (1964) documented considerably higher average spacings accompanied by extremely low rms values (Fig. 2.20) that could be

Table 2.6. Field measurements of LC spacing.

Location	Sets of observations		Range measured values			Reference
	Number of sets	Number of cells in each set	Average spacing, m	rms spacing, m	Wind speed, $m\,s^{-1}$	
Atlantic Ocean	14	Unknown	20.1–49.5	1–3.4	4.3–11.2	Faller and Woodcock, 1964
—	2	Unknown	5–6[1]	2.8–3.3[1]	5–10	McLeish, 1968
—	1	Unknown	13.3[1]	5.1[1]	Unknown	Myer, 1969
Baltic Sea, Gulf of Finland	3	4–12	6.5–10	2.2–4.1	9.4–12.0	Ryanzhin, 1994
Lake-of-the-Woods, 49.3°N, 94.6°W, Canada/USA	5	270–607	5.1–12.4	2.0–4.8	3.5–8.4	Kenney, 1977
Lake Ladoga, 60.9°N, 31.4°E, NW Russia	112	60–651	4.0–25.1	1.6–16.7	3.4–11.6	Ryanzhin, 1980, 1982
Loch Ness, 57.3°N, 4.5°W, Scotland	3	141–431	8.9–10.3	3.6–4.3[1]	7.1–9.8	Thorpe and Hall, 1982
Lake Punnus-Jarvi, 60.6°N, 29.7°E, Karelia, Russia	4	4–8	3.8–6.8	1.5–3.4	4.1–7.4	Ryanzhin, 1994
Black Sea, Bulgarian shore	5	4–10	4.3–7.3	2.5–3.9	6.7–9.4	Ryanzhin, 1994

[1] Values were estimated using the data published in cited articles.

attributed to (i) the small (secondary) cells have not been measured by Faller and Woodcock, and (ii) the authors reported that they transformed.

Their original data for some *"fundamental scale"* could presumably be used rather arbitrarily. On the other hand, their fundamental scale may be to a certain extent similar to spacing of the main LC streaks referred above. Thus, we did not include Faller and Woodcock's data in further analyses. There is strong positive correlation 0.85 ($\pm$0.05), between $\bar{L}$ *versus* δ accompanied by regression:

$$\delta = C_1 \bar{L} \tag{2.50}$$

here $C_1 = 0.54$ ($\pm$0.06) is the dimensionless regression coefficient.

It should be underlined that the above correlation becomes stronger at higher values of $\bar{L}$. Particularly, it was shown (Ryanzhin, 1994, 1995) that $C_1 \cong 0.47, 0.56$ at $\bar{L} < 12\,\text{m}$, $>12\,\text{m}$, respectively. It is also remarkable that regression (2.50) is almost identical for main and secondary streaks.

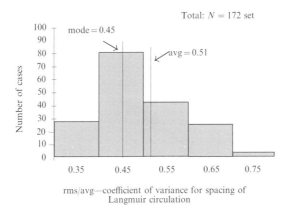

Fig. 2.21. Histogram of the coefficient of variance for spacing of Langmuir circulation (adopted from Ryanzhin (1994a). Original data from Atlantic Ocean (McLeish, 1968; Myer, 1969), Lake-of-the-Wood (Kenney, 1977), Lake Ladoga (Ryanzhin, 1980, 1982), Loch Ness (Thorpe and Hall, 1982), Black and Baltic Sea, Lake Punnus-Jarvi (Ryanzhin, 1994a, 1995). Total number of observations, mode 0.45 and global avg value 0.51 are shown.

The coefficient of variance $K_{var} = \delta/L$ for 172 LC sets of spacing was also calculated (Ryanzhin, 1994) from the field data listed in Table 2.6. As seen from Fig. 2.21 and Table 2.6, K_{var} vary from 0.34 at the Black Sea and Lake-of-the-Wood to 0.71 for main streaks at Lake Ladoga, giving the mode of 0.45 and total average value of 0.51. A slight positive skewness is also seen in the histogram (Fig. 2.21). However, no correlation between K_{var} and V was established in Ryanzhin (1994).

Nondimensional variables

As stressed above there have been no universal experimental relationships established between spacing and background physical conditions. Particularly, the attempts to calculate universal correlation between L and wind speed V from lake and ocean field data failed (Faller and Woodcock, 1964; Kenney, 1977; Ryanzhin *et al.*, 1981; Ryanzhin, 1980, 1982, 1983). Why are the earlier-derived dimensional relationships between L and V only locally adequate? The evident reason is that L correlates stronger with wind–wave parameters (such as, e.g., wind–wave length scale V^2/g, where g is gravity acceleration) than with local wind speed. Therefore, the original data from Fig. 2.20 were scaled with corresponding V^2/g values in Ryanzhin (1994). Dimensionless $L_n = Lg/V^2$ and $\delta_n = \delta g/V^2$ values (Fig. 2.22) vary within the limits 0.44–16.0 and 0.15–11.0, respectively.

The lowest values are attributed to Baltic Sea and the highest to Lake Ladoga data. However, dimensionless data reveal a lower scatter as compared with dimensional data and the general relationship for Fig. 2.22 can be given as:

$$\delta_n = C_2 L_n C_3 \qquad (2.51)$$

where $C_2 = 0.50$ (±0.03), $C_3 = 1.0$ (±0.02) are dimensionless regression coefficients.

Thus, LC spacing closely correlates with the characteristic length scale of wind

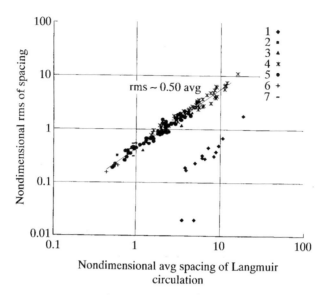

Fig. 2.22. Nondimensional rms ($\delta g/V^2$) *versus* avg ($\bar{L}g/V^2$) spacing of Langmuir circulation. Note that the slope of Faller and Woodcock's data is close to the slope of other field data shown with the curve: rms ~ 0.5 avg. 1–2—Atlantic Ocean; 3—Lake-of-the-Woods; 4–5—Ladoga; 6—Baltic Sea; 7—Black Sea.

surface gravity waves. It should be noted that Faller and Woodcock's (1964) Atlantic Ocean field data in their nondimensional form locate notably lower as on the graph than other processed data in Fig. 2.22. However, in dimensionless form, the general slope of their data is close to that given by (2.51).

2.5   ICE FORMATION AND ITS TEMPORAL/SPATIAL EVOLUTION, ICE DYNAMICS RELATION TO THE LAKE THERMAL REGIME AND WEATHER CONDITIONS

2.5.1   Study of freeze-up phase change on large temperate-zone inland water bodies using microwave observation

Radar and microwave radiometry have been widely used in recent years for remote diagnostics of various natural processes and phenomena taking place on the Earth's surface (Johannessen *et al.*, 1996; Kondratyev *et al.*, 1992, 1996b). This is connected to the well-known advantages of this kind of survey: possibility to obtain all-weather information at any time of the day, as well as relatively deep penetration of microwaves through the sounded surface. However, despite a large volume of accumulated global multispectral satellite information, the number of approaches to retrieve the characteristics of water and land surface features which can be quantitatively estimated is rather limited.

This section deals with the problems of using SAR techniques for studying ice formation and the ice regime of large lakes. For the first time the period of freezing-up has been divided into three subsequent phases and accompanying winter-time hydroecological phenomena are analyzed. The problem of diagnostics of ice-cover parameters of inland water bodies is studied using the ERS-1&2 synthetic-aperture radar (SAR) survey (wavelength $\lambda = 5.7$ cm).

Study of ice formation on lakes and reservoirs is known to be seriously hindered by specifically complicated applications of conventional methods of observations and information collection. It is for this reason that limnological literature still lacks the overview monographs dedicated to the ice regime of large lakes. Limitations of the techniques of remote diagnostics of fresh-water ice parameters are also substantial.

Probably, the number of factors affecting winter-time emissivity and reflectivity of the lake ecosystem are not less than in the open-water period. On the one hand, the ice cover, being more inert than the open water surface, responds more weakly to some external forcings, and on the other hand, it can hold and "remember" the forcings, keeping, accumulating and conserving some of them, which is fixed on the autumn–winter radar "portrait" of the lake.

To reveal the processes of ice-cover formation, meteorological information has been used, as well as data from aircraft survey, field snow and ice survey, and also morphological and hydrological characteristics of the water bodies under study. The freshwater ice terms and definitions are given here in accordance with recommenda-tions (*Recommendations*, 1989).

The results of long-term multi-year observations have been used (Tikhomirov, 1959, 1963, 1982), and features of the annual thermal cycle for each concrete water body have been taken into account. Thermal structures in the annual course of their development and especially in the autumn–winter hydrological seasons, as well as information on heat supply and temperature of lake-water masses serve the basis for the thematic decoding and the interpretation of microwave and radar images of large temperate-zone lakes and reservoirs. This is the novelty of the proposed approach, which substantially broadens the possibility of thematic decoding of microwave information.

2.5.2 Basic laws of the process of ice-cover formation on large inland water bodies

Accumulation and distribution of heat supplies in lake water masses in spring–summer heating are determined by their size and depth. In the cooling period this explains the time difference between freeze-up terminal point and duration in different lakes under the same meteorological conditions.

The main factors governing the process of formation of ice cover of inland fresh-water bodies (and their parts) are: (a) intensity of heat released from the water surface to the atmosphere; (b) wind; (c) heat supply in the lake; (d) morphological and morphometric features of the lake hollow and its parts. Most characteristic manifestations of the freeze-up phase change connected to changing thermal structures of water masses are considered below with Lake Ladoga and Lake Onega, the largest lakes in Europe, as an example (Fig. 2.23).

The magnitude of the heat release by the lake per unit time is strongly governed by air temperature and efficient radiation, sharply increasing when cloud amount decreases, the role of winds in the process of ice formation being important and

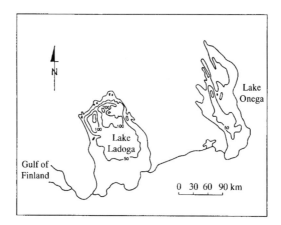

Fig. 2.23. Bathymetric map of Lakes Ladoga and Onega.

diverse. A light breeze somewhat intensifies the heat release off the water surface, but does not cause any heat influx from deep layers of the lake. At negative air temperatures the breeze favors formation of a smooth ice crust on the water surface.

On the contrary, in frosty weather, high winds disturbing thick water layers can prevent ice formation until the temperature through water thickness reaches a certain critical value.

According to Tikhomirov (1963, 1982), the formation of ice on lakes takes place only in regions where the vertical profile of water temperature is below 2°C. It is an important condition discovered from observational data on Lakes Ladoga and Onega and confirmed by observations on Lakes Baikal and Michigan (Rossolimo, 1957; Church, 1943). For this reason, to study the processes of freezing-up on large lakes, special attention has been paid to studies of spatial distribution and temporal variability of the location of the 2°C isotherm. Note that in the analysis of hydrological processes this important regularity has been so far treated lightly.

It was known long ago that lakes start freezing along the shore; the ice has the form of concentric isolines, gradually moving from shallow waters to the center of the water body. However, no convincing explanation of this phenomenon has been given so far which is, apparently, connected to the inadequately studied thermal regime of large water bodies of the temperate climatic zone, especially the thermal conditions in the late hydrological autumn and early hydrological winter.

The first freeze-up phase on large lakes starts from the second half of the hydrological autumn, when the autumn-time thermal bar appears (Tikhomirov, 1963, 1982). At this time the thermal conditions of water masses are characterized by a vertical isothermal-cold area (inverse to spring-time conditions with its spatial deep-water inhomogeneity); that is, when water temperature in the lake at equal depths from the surface to the bottom is similar, increasing with growing depth of the hollow. As mentioned above, the time of the terminal point varies strongly depending on the characteristics of the state of each concrete water body and

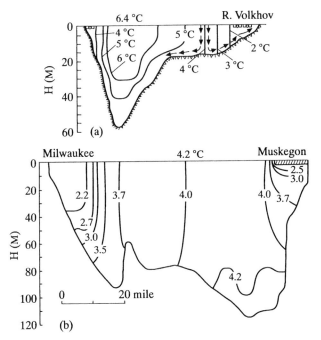

Fig. 2.24. (a) Water temperature distribution and location of thermobar of the first freeze- up phase on Lake Ladoga (Melentyev *et al.*, 1997); (b) water temperature distribution and location of thermobar of the first freeze-up phase on Lake Michigan (January 27, 1942) (Church, 1943).

weather conditions. So, for example, for Lake Ladoga in mean climatic conditions the first phase begins in late October–early November. In this time period the water temperature in the pelagic area is 6°C (and sometimes higher). Gradually decreasing towards the shore, it reaches 0°C by the water edge. Therefore it is over shallow water that new ice is formed. Here the fast ice forms, gradually growing in area— Fig. 2.24.

Between depths of 12–15 m along the lake coastline the water at this time of the year has the temperature of maximum dense water 4°C, and the thermal bar is located in the form of a stable narrow strip (on Lake Ladoga this area is at a distance of 10–12 km from the southern shore). The thermal bar is not distracted by wind or by roughness, it is the frontal zone where the waters with temperature above and below maximum density, flowing from the shore and from the lake, get mixed. The mixing of these waters produces water of maximum density which, immersing, creates a vertical dynamic "blind" continually maintaining convergence and mixing of waters. This zone is marked in Fig. 2.24a with short vertical arrows.

Due to this convergence of surface waters and, respectively, divergence of deep waters, the autumn-time thermal bar breaks the closed circulation in the lake, it divides the lake into the thermo-active area (TAA) and thermo-inert area (TIA), favouring a more intensive cooling of shallow waters near the shore. The thermal bar itself, with cooling of shallow and off-shore waters, gradually shifts towards deeper

Fig. 2.25. Fast ice on the 20-m isobath, south of Bay of Andrusov (north-eastern shore of Lake Ladoga), December 13, 1957.

waters. Together with the thermal bar, the 2°C isotherm moves in the same direction being followed by new ice and by the still-fast ice edge.

During the first phase the water temperature in the TIA of large lakes exceeds 4°C, the fast ice floes are brought here and snow falling on the water melts, and therefore only open water is observed here. On Lake Ladoga, by the end of the first phase of freezing, the still-fast ice edge borders the south-western, southern and north-eastern shores along the 20-m isobath—Fig. 2.25.

From calculations of mean climatic conditions on 1 December, the 2°C isotherm is located over the 25-m depth, the TIA water temperature is 4.4°C, and that of TAA is 0.7°C. By this time the fast ice gets covered by a snow layer 20–30 cm thick. This snow on the ice surface is a multi-layered repeatedly frozen porous structure, originating from multiple moistening, freezing and melting of snow. This is connected with special climatic conditions in the north-western region of Russia: alternating frosty weather and thawing with an intrusion of warm air masses from the Atlantic. As shown below, this feature of the snow state can be used in decoding satellite radar images.

The onset of the second phase of the lake freezing-up has been attributed to the moment when the deep-water temperature reaches 4°C and conditions for the thermal bar disappear; that is, the hydrological winter sets in. On Lake Ladoga this happens around 15 December when, resulting from the transformation of thermal conditions, the spatial deep-water thermal inhomogeneity of the late autumn-time cooling (Fig. 2.24) gradually transforms into an inverse thermal stratification (Fig. 2.26).

In the deep-water area the water temperature decreases below 4°C, forming the winter-time dome of dense water (in Fig. 2.26 it is marked with horizontal shading), which, as in summer-time warming, tries to flatten out. The denser and "warmer"

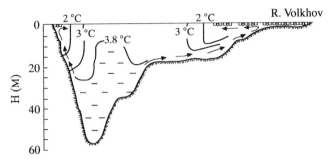

Fig. 2.26. Water temperature distribution and location of the fast ice edge of the early second freeze-up phase on Lake Ladoga (Melentyev *et al.*, 1997).

water of the dome starts moving in the bottom layers towards the shore, ousting the less dense cold water beneath the fast ice cover near the shore in the gulfs and bays of the lake. This water exchange favours the inflow of polluted near-shore waters from the gulfs and bays to the lake and the transport of deep waters to the regions near the shore. The conclusion has been drawn that this kind of circulation in lake waters favors, on the whole, the ecological state of inland water bodies, since due to ousting of polluted coastal waters, it promotes self-cleaning of the areas of large lakes and reservoirs under maximum anthropogenic stress. The compensating inflow of deep waters from the center of the lake to the littoral region can have a dual effect. If these waters have excess oxygen, their effect on coastal biota is, no doubt, favorable. But with input of poorly aerated deep waters the temporal change of the quality of coastal waters becomes more complicated. In winter these waters can negatively affect the life of the water body. However, in the spring, in view of the special melting of the ice of inland water basins (the disintegration and melting of ice start from the shore) these waters can make up for deficient oxygen at the expense of intensive wind-driven mixing even at the first stage of lake ice disintegration.

Heat advection, occurring in the beginning of the second phase near the bottom with the water temperature at 3°C, slows down or (depending on weather, wind, and air temperature) delays the formation of fast ice and the growth of the fast ice floes floating near the edge. But then, when the effect of advection starts to cease, frosty weather again leads to formation of floating ice and growth of fast ice. The fast ice floes (often 10 km and more in width) may be wind-scattered over the water area; in the deep parts of lakes they melt, and around the 2°C isotherm they become centers of formation of new ice. Such floes are marked by repeatedly re-frozen multi-layered snow cover (Fig. 2.27).

At this time there is still a clear interface between fast ice and floating ice, as well as between floating ice and open water, since the latter in deep parts of the lake (from depths 80–100 m and deeper) still retains a temperature above 2°C, and ice drifting here melts. For the same reason, the slush from solid precipitation in this period cannot form on open water. The time of completion of the second phase of freezing is determined by when the water temperature over deep parts of the lake drops to

Fig. 2.27. Fast ice, floating ice cake, fast ice and open water—the scene of the ice cover distribution of the second freeze-up phase on Lake Ladoga.

2°C; that is, when in hard frost (down to −10°C and lower) conditions the large lakes get frozen-up.

Thus, owing to the formation of new ice as the 2°C isotherm approaches the center of the lake, the freeze-up on large lakes during its first and second phases has the form of concentric isolines following the outline of the shore.

The third phase of the freeze-up starts when the deep-water temperature drops below 2°C. In mild winters, usually characterized by high winds, the under-dome water, first rushing towards the shore, then gets involved in the cycle and, together with the heat preserved while freezing, is transported to the deep part of the lake, owing to which the latter remains open until spring. In severe winters, when a spell of calm frosty weather may last for several days, ice rind is formed over large areas of the open water of the lake.

The ice rind, binding the floating ice over the whole surface of the lake, forms the fields of ice breccia, connecting the edges of fast ice from one shore to another. The complete freeze-up sets in. However, a new fall of snow cannot be firmly held by the smooth surface of the ice rind. Subsequently, with changing wind directions, thin patches of snow fancifully change their position on the even ice of the lake— Fig. 2.28.

Snow cover of this kind does not further affect water cooling through the ice thickness. The ice rind grows rapidly, and often by the end of the winter it can be even thicker than the first-phase fast ice covered by autumnal multi-layered snowfall and growing only at the expense of solid precipitation in winter.

The duration of each freeze-up phase is strictly individual for each of the lakes: changing as a function of heat supply and morphology of the water body, it can vary from days (and even hours) to several months. On Lake Ladoga the second and third

Fig. 2.28. Crystalline ice—ice rind covered with thin snow patches, 4 February 1958—the third freeze-up phase on Lake Ladoga.

phases of winter cooling last, on average, from 15 December to 15 March; that is, for about 3 months. After the last date the ice thickness on Lake Ladoga stops growing (Tikhomirov, 1982).

The end of winter cooling and the beginning of spring warming of the lake are characterized by a clearly manifested inverse thermal stratification and are followed by horizontal isothermy, when water temperature at equal depths over the entire lake remains practically the same (varying within $0.1°C$), with the exception of the deep-water bottom layer. By this time, the heat supply of a large lake reaches a minimum with water temperature $0.3–0.6°C$.

2.5.3 Radar signatures of the ice cover of large inland water bodies ($\lambda = 5.7$ cm)

First experience of the use of microwave information to study ice cover features was the successful launch in 1983 of the Soviet satellite "Kosmos-1500" carrying side-looking radar (SLR) ($\lambda = 3.15$ cm). The first satellite carrying synthetic-aperture radar (SAR) was "Seasat" launched in 1978. The working wavelength was 23.5 cm, the spatial resolution 25 m, the bandswath 100 km. Later on, SAR were installed on Soviet space vehicles "Kosmos-1870" and "Almaz" ($\lambda = 9.6$ cm). The ERS-1 and ERS-2 satellites launched, respectively, in July 1991 and August 1995, provided data on the state of sea and fresh-water ice, as well as continental ice and that of frozen soils at $\lambda = 5.7$ cm.

From the data of satellite survey, extensive factual material has been accumulated on various parameters of Arctic sea ice conditions: ice cover and concentration of drifting sea ice, distribution and age characteristics, fast ice state, dynamics of ice drifting and deformation. The satellite data, obtained in this monitoring regime, revealed the succession of the processes of ice formation (as well as ice melting and disintegration) over the entire Arctic water basin (Johannessen *et al.*, 1996).

Nevertheless, the literature practically lacks information on satellite and aircraft radar investigations of fresh-water ice. There are only isolated data on the signatures of ice cover of inland water bodies, which are very difficult to use because they were obtained at different frequencies. But since microwave sounding is special, emitting and reflecting properties change greatly across the spectrum.

Consider fresh-water ice radar signatures at $\lambda = 5.7$ cm needed for thematic decoding of the ERS-1&2 microwave images. This summary of data is based on our multi-year experience of working with aircraft and satellite microwave images of different types of water bodies (Kondratyev *et al.*, 1992, 1996b). The signatures were revealed from data of aircraft survey and field snow and ice observations.

The variability in revealed signatures strictly corresponds to the succession of the freeze-up phases considered above. The emphasis is on the description of the condition of the surface–air interface, since it is the water surface and ice condition in the autumn–winter period that is responsible for special features of reflecting properties.

At the start of the first freeze-up phase, in calm frosty weather, young coastal ice and fast ice form near the open shallow-water sloping shore. Their crystalline structure has no visible edges, they lack air bubbles, mineral and organic particles. This ice is transparent (Fig. 2.29), and its radar signature at $\lambda = 5.7$ cm looks like a deep dark homogeneous tone.

The outlines of this ice from the shore and from the open water are usually fixed on the "radio-portrait" of the lake as brighter signals appearing, respectively, due to compressing and deformation of the off-shore fast ice edge as well as possible fractures at its base.

Fig. 2.29. Transverse ice boundary in the Yakimvar Bay of Lake Ladoga, November 1958—fast ice formed in calm frosty weather.

Fig. 2.30. Ice formations near the southern shore of Lake Ladoga—fast ice formed in high winds and snowfall.

Fast ice forming at negative air temperatures and high winds is characterized by a lower density, small-grain structure, chaotic orientation of the optical axes of ice crystals. It looks whitish, turbid and opaque (Fig. 2.30). This ice often forms during a snowfall followed by the formation of slush, freezing together of shuga and transformed in-water ice.

The radar signature of ice formed in windy and rough conditions appears as a bright white exposure.

Let us describe the development of snow on the "radio-portrait" of the first-phase fast ice accumulated in a layer 20–30 cm thick. If this snow is dry and clean, at $\lambda = 5.7$ cm it is transparent. The smooth fast ice covered by this snow develops on the image as a deep black tone. If the snow layer is humid or overfrozen, the snow-covered ice signature has a bright white tone due to anisotropy of the mechanical and thermal characteristics of the snow cover, causing changes in the electric and microphysical parameters of ice and snow.

Ice splashes near the boulder and granite steep-dipping shore (Fig. 2.31) can have a brighter white exposure on the image.

Ice splashes on the hummocking strips and on the conglomerations of broken ice formed during freezing on shallow-water banks and shoals give a similar signal. These ice formations can be observed in different parts of Lake Ladoga and other large lakes of north-western Russia; their height often reaches 8–10 m, and they stretch sometimes for several kilometers.

As mentioned above, the outside edge of the fast ice clearly seen on the ERS-1&2 images marks the boundary of ice propagation and dynamics of the $+2°$C isotherm, revealing the respective depths.

An increase in ice area on the lake after the onset of freezing can be calculated from the sum of daily negative air temperatures $\sum T_{air}$ using data from lakeside

Fig. 2.31. Ice bank lapped over the shore of Mantinsari Island, 9 February 1957—the first freeze-up phase on Lake Ladoga; on the right of the frame is still fast ice.

hydrometeostations (HMS) and observation points. To calculate the Ladoga and Onega ice areas, the following empirical formula has been proposed:

$$S = K(\sum T_{air}) \quad (\text{km}^2)$$

where $K = 22$ is the dimensionless coefficient.

In this formula for Lake Ladoga the data on air temperature should be used from the HMS of Vidlitsa (east coast of Lake Ladoga), and for Lake Onega from the HMS of Petrozavodsk. In addition, regional peculiarities of the formation of ice on large lakes should be borne in mind. So, for example, the ice areas calculated for the first phase of freezing on Lake Ladoga refer mainly to its southern and eastern shallow- water shores, whereas the areas of new ice forming at this time near the western and northern steep-to shores are rather restricted. On Lake Onega the pattern of the initial phase of freezing is somewhat different: here the ice at this time forms, mainly, along the south-eastern shallows and in the northern skerries. Near the western shore it almost does not form.

Multi-year observations and analysis show that for inland water bodies the time of fast ice formation and its progress towards the center of the lake calculated for each 10-m isobath is strictly determined. For example, the formation of fast ice on the 10-m isobath takes place when the sum of daily negative air temperatures for the given lakeside HMS approach $-166°C$, and further increase in this sum by every $-166°C$ determines the moving of the fast ice into the next 10-m isobath. As calculations show, for complete freezing-up of Lake Ladoga it is necessary that $\sum T_{air} = -800°C$ for the HMS in Vidlitsa and of Lake Onega approximately

$-500°$C for the HMS in Petrozavodsk. In this connection, it is clear how to solve the respective inverse problem of satellite meteorology: determination from the data of microwave survey of the freeze- up terminal point unequivocally signifies that by this time the sum of negative air temperatures in this region had reached the values given above. Also, satellite radar data can be used for more detailed winter-time climatological studies. Possibilities of remote sounding of the surface air temperature and of the quantitative estimate of its regional variability are very important in studies of variability in regional and global climate.

The results of calculations suggest one more important conclusion with respect to remote control of limnic processes. It turns out that even without the bathymetric map of the water body one can state that in November–December, even in the severest winter conditions in north-western Russia one cannot expect any ice formation in lake regions >50 m deep. Note that experience of working with satellite IR survey data of lake-surface temperatures in the autumn period testifies to water temperatures systematically underestimated by $1.0–1.5°$C compared with *in situ* observations (Kondratyev *et al.*, 1992, 1996b; Melentyev *et al.*, 1997). This fact must be taken into account for remote monitoring of ice-thermal processes in large lakes in the case of complex use of microwave IR survey data.

The second phase of the freeze-up on large lakes connected with formation of ice floating along the still fast-ice edge is characterized by specific features with respective signatures. This form of ice, not bound with the shore or with the bottom, can either stand still or be wind- or current-driven (drifting ice). Ice of this kind as well as fast ice floes do not melt in shallows. The signatures of new ice, the age of which varies from several hours to 30 days, are rather diverse, and fast ice floes retain the radar indications they had at the moment of their formation; that is, either dark or, on the contrary, bright white tones.

New ice—frazil ice, sludge, slush, shuga—covers the lake surface in isolated patches. Their thickness does not exceed 5 cm. The signature of frazil ice and sludge shows itself as dark patches against signals of different exposure from the water surface. The signature of slush (the freezing layer of sticky snow mass) and shuga (porous lumps of ice formed of sludge, slush and emerging in-water ice in conditions of rough lake surface) is a signal of bright white tone.

Young thin ice is nilas, grey and pancake ice which form a solid layer on the water surface. Nilas is a plastic thin ice crust on the water up to 10 cm thick. Dark nilas forming in calm or breezy weather gives a signal of dark tone. Light nilas formed of sludge, slush and shuga in windy weather, has a brighter signature than that of dark nilas. Under compression, nilas forms layers, which, respectively, brighten the tone of the signal on the "radio-portrait" of the lakes.

Grey ice forms layers and hummocks, the height of the hummocks averaging 15 cm—Fig. 2.32.

In places of stationary fractures, the so-called "backbone" cracks, usually coincident with changes in the bottom profile, the height of hummocks can reach 40–60 cm. Note that the total thickness of lake ice at the expense of tucked-up ice and hummocks may reach 1.5 m. The grey ice signature is a light-grey tone of the

Fig. 2.32. Hummocks on grey ice—second freeze-up phase on Lake Ladoga.

signal, the hummocking areas are distinguished by ridging strips of additional brightness.

The size distribution of floating ice during the second phase of formation is not diverse owing to relatively thin ice (not thicker than 15–20 cm). This ice consists of ice cake (10–25 m in diameter) and ice lumps as well as brash ice (<2 m in diameter). At this stage floating ice is usually little covered with snow, drifting ice floes can be covered by a thin layer of water. On radar images this ice is well decoded: on the ERS-1&2 images it is clearly seen as sufficiently homogeneous patches of grey tone, the boundaries of their propagation into the ice formations of different age as well as their borders on open water are well outlined. Note also that the thickness of snow and its area on Lake Onega exceed the respective parameters for Lake Ladoga owing to special atmospheric processes, earlier freezing-up, fewer warming returns.

The signature of the third phase of freeze-up on large lakes, characterized by formation of fields of floating and still ice breccia, has more contrasting indications compared with the preceding phase. These differences in radiosignal appear due to special features of the physico-mechanical and microphysical properties of ice rind playing, at this stage of ice formation, the main connecting and "cementing" role.

Ice rind is a transparent ice crust up to 5 cm thick, of increased hardness (ice-cutter). It forms of sludge in calm weather or directly on the calm water surface when it freezes. Ice rind has a smooth, sometimes specular surface; this crystalline ice structure can break into small plates when deformed. Ice rind binds the polynyas both between the still fast-ice edge and drifting floes and between the floating blocks of ice, it connects the formations of small ice cake and shuga. Eliminating everywhere areas of open water, it can cover areas of tens and hundreds of square kilometers at the center of the lake.

Fig. 2.33. Wintertime ice core—ice formed through accretion of young ice and reaching its maximum thickness at the end of the hydrological winter (Siberia, Limbeyakha Lake, March 1991).

Fig. 2.34. (a, b) Wintertime ice—mapping of the ice using the Georadar system (Siberia, Limbeyakha Lake, March 1991).

Since ice rind is characterized by low reflecting properties in the direction of the source of sounding, it is fixed at $\lambda = 5.7\,cm$ as a deep homogeneous black tone on the image. At the same time it should be borne in mind that when the thickness of ice rind increases, the resulting microcracks of thermal and dynamical origin cause a brightening and reduction in the dark tone of the radiosignal—Fig. 2.33.

However, in spite of this, during the whole hydrological winter the signature of the growing ice rind retains a sufficiently dark tone, strongly contrasting with the light white tone of the signature of the combined fields of the winter ice breccia— Fig. 2.34a, b.

Fig. 2.35. Polynya in a totally frozen lake due to seiche (Yakimvar Bay of Lake Ladoga, 4 February 1958).

Open water on large inland lakes of the temperate zone is present during both partial and complete freeze-up (Fig. 2.35). In frosty winters the open water areas appear in the form of chains of polynyas concentrated usually near the "backbone" cracks on the 20-m isobath along the shore and around some islands and banks. The signature of the open water surface is a signal of both dark and bright white tone, in contrast to ice rind. This is connected with the near-water wind speed and with the effect of small-scale capillary formations, always present at the water–air interface.

2.5.4   Freeze-up phase classification and perspectives of SAR thematic decoding development

The studied succession of the freeze-up phase change on large inland water bodies of the temperate zone makes it possible to broaden the possibilities of thematic decoding of the aerospace microwave information. The results discussed show that ice cover is a tracer of various natural processes and phenomena, including those that are ecologically important. As will be demonstrated later the summary of signatures for each phase of the freeze-up opens up possibilities to study hydro-dynamic and hydrological processes and phenomena accompanying ice formation such as regulated structures and rings, directed drifting of ice masses, drawing polluted coastal waters towards the center of the lake, winter-time seiches and manifestations of inner waves.

Quantitative estimation of climatically important meteorological parameters as well as referencing and indexing by time and location of the distribution of ice is important, the change and transformation of sums of negative temperatures in the region over a given period of the hydrological winter.

Without dwelling upon the results of analysis of the ERS-1&2 survey of the lakes Ladoga, Onega and Pskovo-Chudskoye and the data on their ice regime over the period 1994-1998, which will be discussed later in Chapter 4, it should be noted that complex decoding of microwave images of ice cover of large lakes in north-western

Russia makes it possible not only to diagnose ice conditions at the moment of satellite survey but also to retrieve the history of the lake-ice formation.

2.6 LAKE HYDRODYNAMICS AND HYDROLOGY: WEATHER AND CLIMATE

2.6.1 Spatial and temporal scales of water motions in lakes

The first long-term multi-year observations of currents and water temperature on Lakes Ladoga and Onega were made at the end of the 1960s–early 1970s. Between 10 and 15 autonomous buoy stations (ABS) were evenly distributed around the lake. The maximum length of the time series was 5 months with the time step ranging from 10 to 30 minutes. These data allowed variability of currents and water temperature with characteristic time scales from several hours to 15 days to be studied. These data were averaged for all periods of observations (May–November) and they reveal the tendency in the lakes to anti-clockwise circulation (Fig. 2.36). This kind of circulation characterized the so-called "climate" of currents. The spectra of currents were calculated using the tensor analysis technique (Filatov, 1991).

In particular, the variability in currents was investigated using four invariant and two non-invariant characteristics. The tensor of spectral density of currents was determined using Fourier transformations of the corresponding correlation tensor. The frequency-temporal spectrum (linear invariant) and bispectra of currents in Lake Ladoga illustrates the non-stationary character of the current variability. Analysis of currents showed spectral transformations from nearshore to offshore regions. In Lakes Ladoga and Onega (Fig. 2.37), spectral constituents with frequencies of 0.02 rad/h (corresponding to a time scale about 2 weeks), 0.06–0.12 rad/h (2–4 days), 0.24 rad/h (I day) and frequencies close to the inertial frequency corresponding to latitude 61°N (0.45 rad/h) (13.5 hour) stand out more clearly.

There are several peaks corresponding to frequencies which are lower than the local inertial oscillations (about 13.5 hours). In offshore regions, low-frequency oscillations from days to weeks are usually caused by large-scale synoptic wind variation. In the coastal zone some low-frequency waves should also be taken into consideration, e.g. it may be the manifestation of internal Kelvin waves or topographic waves which can be generated by wind and can persist long after the decrease in initial wind impulse, causing the characteristic increase in current velocity. According to the probabilistic analysis of currents, water temperature and winds over Lakes Ladoga and Onega, the low-frequency oscillations in nearshore currents with periods up to 4 days are indeed connected with the internal Kelvin waves. There is a pronounced peak in the current spectrum, which corresponds to the frequency of local inertial oscillations. This peak may be caused by pure inertial oscillations provoked by spatially inhomogeneous winds and also by the internal Poincaré waves which dominate in large lakes at distances exceeding the internal Rossby radius of deformation (about 3–5 km from shore). To analyze the nature of the oscillations we used scalar-vector characteristics. Functions of cross-correlations of water

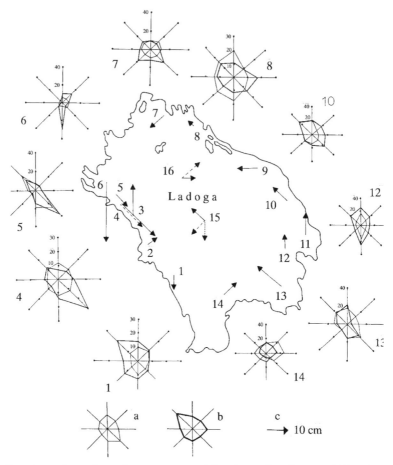

Fig. 2.36. Long-term observations of current averaged for observations between 1969 and 1975. The schema shows number of ABS, distributions of directions (a), speed of currents (b) and vector of currents in cm/s.

temperature and currents were calculated (Filatov, 1991). These characteristics reveal that inertial oscillations inherent in Poincaré waves show up very clearly. Inertial motions in the lakes have an alternating character. In Lakes Ladoga and Onega the life-time of these motions does not exceed two or three inertial periods. Bispectral analysis of currents indicates the occurrence of phase shifts between low-frequency oscillations and inertial oscillations which indicates the possibility of non-linear interactions between spectral constituents and energy flows from one spectral band to another (Filatov, 1991). The data reveal large-scale current and temperature variability in both lakes (see § 2.6.2) with several energy peaks reflecting the lake's response to atmospheric forcing and the lake's own system movements. The distribution of typical climatic temperature and currents on Lake Onega and so-called "weather" follows the atmospheric weather changes. The wind-induced coastal

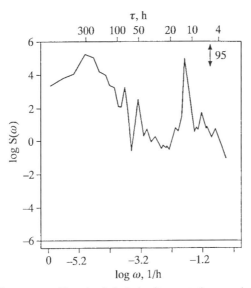

Fig. 2.37. Spectrum of currents at 10-m depth in Lake Onega at the near-shore station, summer 1989.

upwelling is among the main components of hydrological weather in the lake. In response to the wind, coastal upwelling continually changes their strength and locations, which can most easily be traced in the surface temperature field. Features of hydrological climate and weather will be shown in § 2.6.2.

2.6.2 Hydrodynamics of the largest lakes in Europe: Ladoga and Onega*

While the first measurements of currents and temperature in Lakes Ladoga and Onega were made at the end of the last century, the systematic study of thermal structure and circulation in both lakes began only in the late 1950s. Tikhomirov (1963) was the first to reveal the thermal bar in Lake Ladoga; afterwards intensive lakewide water-temperature observations from spring to autumn were undertaken in the two lakes (Tikhomirov, 1982). The experimental programs included shipborne current observations which were compared with geostrophic currents calculated by Okhlopkova (1961, 1972) based on observed temperature data. The first mooring stations were deployed in the middle of 1960s on Lake Onega and the results of numerous observations of currents and temperature variability were summarized by Bojarinov *et al.* (1994); Filatov *et al.* (1990). Complex hydrological and meteorological data have been collected in Lakes Ladoga and Onega during the last two decades and created a database for physical analysis and modelling experiments. Synoptic ship surveys have been conducted during the period of spring and autumn thermal-bar formation and full stratification period too. Long-term current and temperature data have been recorded using a network of buoys (ABS). Primary

---

* This section is based on a paper of Beletsky F., Filatov N., and Ibraev R., 1994.

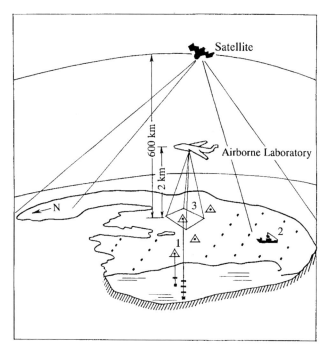

Fig. 2.38. Schema of the multi-level hydrophysical "Onega experiment". Locations of mooring stations (triangles), temperature observations (dots) in summer 1987–1989 experiments.

emphasis is given to temporal and spatial variability in currents and temperature, numerical modeling of hydrophysical fields and model verification. The data obtained during the "Onega" experiment (Filatov *et al.*, 1991) have the best temporal and space resolution compared with data collected on other large lakes of the world (Fig. 2.38).

For example, ship surveys of water temperature for Lake Baikal, Ontario, Erie or Ladoga over 3–7 days shows temperatures equal to or more than the synoptic period. Real synchronic ship surveys for Lake Onega have been conducted over the period of one day and were repeated three times by three research vessels of the Soviet Union Academy of Sciences. At the same time data, using a network of buoys stations with currents and temperature measurements on several horizons (in epi-, meta- and hypolimnions), were recorded together with remote sensing IK airborne and satellite observations. These data were used for calibration, parametrization and verification of 3-D models for European large lakes. The most developed three-dimensional diagnostic and prognostic circulation models for description of meso- and macro-scale processes in Lakes Ladoga and Onega have resolution in the horizontal plane of approximately 2–10 km (Beletsky *et al.*, 1994). The 3-D circulation model using the finite-element method was designed by Jakovlev (1991) who described some meso-scale processes in the coastal zone. These are upwelling, coastal jet and other water motions. The above-mentioned data were combined with the mathematical simulation model of the transformation of nutrients, organic matter and plankton community in Lake Ladoga (Leonov *et al.*, 1995). Water exchange between typical boxes in this model was determined from a 3-D diagnostic model run for all the seasons. The model also included data on evaporation, precipitation and river inflows.

Three-dimensional mathematical modelling of the evolution and modern condition of Lake Ladoga's ecosystem have been created by Menshutkin *et al.* (1998). The model adequately reproduces eutrophication processes and includes 3-D hydrodynamical processes. It describes the annual climatic circulation and shows that climatic annual changes are rather noticeable in ecological parameters. But these models do not reproduce meso- and small-scale processes especially in the near-shore coastal zone. Large attention to description of these features in the coastal zone has been made for the Lawrentian Lakes (see, e.g., Schertzer, Murthy, 1994) and two-dimensional depth-averaged finite-element models of circulation were developed by Podsetchine *et al.* (1995) and applied to the northern part of Lake Ladoga. Malm, Jonsson (1993) have designed a model for the spring thermal bar utilizing AVHRR data from NOAA satellites. A numerical particle tracking method has been used to obtain the maximum number of trajectories of dissolved contaminants from the main pollution sources on the northern shore of the lake. Modern 3-D forecasting models for hydrophysical processes for the Great American Lakes have been developed by Schwab, Bedford (1994). Despite computer capacity increasing strongly in 3-D models, this model uses some simplifications. For example, the vertical coordinate component is replaced by the special vertical coordinate. In the former case the initial system transformed and simplified calculation equations (Schwab, Bedford, 1994). But in the latter case problems arise with pressure gradients in bottom relief. Combination of traditional and special bathygraphical curves are used as a morphometric submodel of bottom relief by creating a simulation model of seasonal and inter-annual variability in the lake's ecosystem (Kolodochka, 1997). Thus, transition from 3-D physical space description of simulation model to 2-D is carried out by integral reflection of its plan metric. The origin of large-scale internal boundaries in deep cold-water lakes of the temperate climatic zone is bound up first of all with such thermohydrodynamic phenomena as formation of spring and autumn thermal bars and summer thermocline. Positions and speed of distribution of this frontal boundary is regulated to a considerable extent by depths field. Moreover, plan length and interface area are related to its basic morphometric parameters. Integral characteristics of bottom relief, such as bathygraphic curves, are sufficiently convenient to reflect peculiarities of dynamics and internal structure of lake-water masses by creation of simulation models of seasonal and inter-annual variability in the lake ecosystem. First then, the opportunity arises for model parameters to take into consideration by not only displays of principal lake-bottom relief peculiarities but also generalized influence of smaller relief forms on limnological processes. Second, essentially for lowering the measurement of grid approximation of the physical model space by solving boundary-value problems of lake thermohydrodynamics. Third, for creation of a sufficiently economical algorithm of spatial and temporal variability in the box-type structure of the lake ecological model. This coordinate approach fits the ecological model of Lakes Ladoga and Onega (Kolodochka, 1997).

2.6.3   Hydrodynamics of Lake Ladoga. Large-scale circulation: model results

New techniques developed over the last two decades have allowed measurements of long-term time series of currents and water temperature. Advances in modelling and temperature observations have made it possible to calculate spring and summer currents in the lakes more precisely. Special observations on smaller-scale hydro-dynamical processes such as turbulence, Langmuir circulations, internal waves and thermocline deepening were also made during the "LADEX" physical experiment in 1980–1983 (Filatov, 1983). Large-scale multi-level hydrodynamic experiments took place on Lake Onega (Filatov et al., 1990). For numerical calculations of currents in Lake Ladoga, over the thermal bar and full stratification period, a 3-D nonlinear diagnostic model (Filatov, 1983) was used, which was previously applied for ocean dynamics studies (Demin and Sarkisyan, 1977). The spatial step for numerical calculations was 10 km, the vertical resolution was 8 levels and the time step was 600 s. Though the coefficients of turbulent viscosity can vary over a wide range, especially in the vicinity of the springtime thermal bar front (Filatov 1983), it should be noted that the results of diagnostic calculations usually depend rather weakly on the choice of horizontal and vertical turbulent viscosity. In numerical experiments the ranges of these coefficients were: $\mu = 10^4$–$10^6\,cm^2/s$; $\nu = 1$–$10\,cm^2/s$ with average values of $\mu = 10^6\,cm^2/s$ and $\nu = 10^6\,cm^2/s$, respectively. Calculations showed that in spring when the thermal bar exists the circulation pattern is rather regular. Lake-wide cyclonic circulation induced generally by density gradients occupies a narrow zone between the shore and the front of the thermal bar. Detailed analysis shows that cyclonic circulation is enhanced due to the joint effect of baroclinicity and relief (JEBAR) (Demin and Sarkisyan, 1977). In the central part of Lake Ladoga, which is practically homogeneous at that time, the circulation patterns respond more readily to wind forcing. Wind observations from meteoro-logical stations around the lake reveal that at the early stages of the thermal bar a local atmospheric cyclone appears due to sharp temperature gradients between the lake and the land surface. This cyclonic vorticity of the wind serves as an additional source of cyclonic circulation in the lake. Indirectly, the persistence of cyclonic circulation at the early stages of the thermal bar may be confirmed by analysis of the surface temperature pattern. For this purpose we used remote sensing observations of Lake Ladoga and Lake Onega from an airborne laboratory AN-28 and satellites equipped with infrared radiometers. These infrared images have been made regularly by the hydrometeorological service over 20 years (1968–1991) on Lakes Onega and Ladoga. As the observations indicate (Filatov, 1991), wind-induced upwellings are practically absent until the stratified zone occupies less than approximately 40% of the surface area. Presumably cyclonic circulation can be persistent at that time in both lakes even in the presence of strong winds (Fig. 2.39).

Progressive warming of the lake during the summer leads to the advance of the thermal bar front into the deep part of Lake Ladoga and to the increased spatial

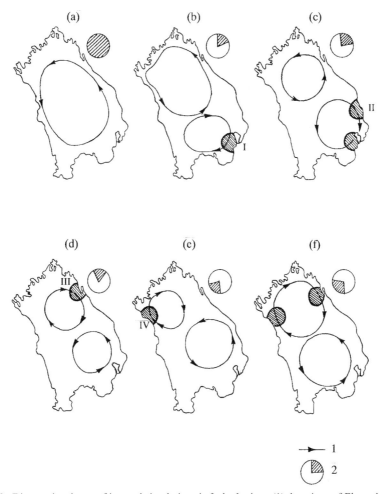

Fig. 2.39. Diagnostic schema of integral circulations in Lake Ladoga (1), locations of Ekman's nearshore upwelling zones (I–IV) for early spring with thermal bar (a), summer with thermal bar (b and c), period of full stratifications (d and e) and autumn (f) under specific wind directions (2).

extent of the stratified zone. Thermal gradients between land and water become weaker and the local atmospheric cyclone vanishes—this makes lakewide cyclonic circulation more and more responsive to wind fluctuations. In this period wind-induced upwellings arise in various parts of the lake (Fig. 2.39). Wind-induced upwellings are also frequent during the full stratification period. Though cyclonic circulation remains typical especially for the deep central and northern basins, local anticyclonic circulation connected to upwellings in the southern shallow part of Lake Ladoga are frequently generated. Some features of coastal jets in the nearshore zone are seen in spite of the rather poor spatial resolution of the model, and the whole pattern of currents varies rather weakly with depth. The spatial resolution is restricted mainly by limited computer capacity but it should be noted that model density gradients do not completely reflect the complicated picture of the lake density field. For example, the zones with sharp temperature gradients in the areas of upwelling usually demonstrate mesoscale eddies which were not recorded during the field surveys in Lake Ladoga at the scale of a $10 \times 10\,\mathrm{km}$ grid. From a practical point of view it was rather difficult to attain more precise results during the experiments on a large lake, such as Lake Ladoga, because of the spatial vastness of the water area drawn into synchronous field observations. This goal has since been achieved during experiments on the smaller Lake Onega.

2.6.4   Hydrodynamics of Lake Onega

An important step in the investigation of the hydrodynamics of Lake Onega was made at the end of the 1980s during the hydrophysical experiment "ONEGA". The main focus was to measure the spatial and temporal variability of currents and temperature fields in spring and summer with characteristic time scales from hours to weeks. The most intense work was done during the summer of 1986–1989 when the number of hydrological stations was increased from 100 in 1987 to 250 in 1989 (Filatov *et al.*, 1990). In 1989 a very fine grid of observations, up to $4\,\mathrm{km}$ in the central basin was used, in which the surveys were carried out in 1.5 days by three research vessels. The study provided perhaps the finest temperature measurements ever obtained in a large lake. While current meters show currents only in certain locations, a three-dimensional numerical model covering the whole lake was used with emphasis on diagnostic and adaptation versions (Beletsky *et al.*, 1991).

Model formulation

The model is based on the system of primitive equations written in Cartesian coordinates x, y, z, with axes directed eastward, northward and downward respectively. The domain of interest is limited by boundary conditions, which includes the lateral surface (on upper boundary $z = 0$ and bottom relief $H(x, y)$. Using the traditional approximations of Boussinesq, hydrostatics, incompressibility and Simons' approximation for the fresh-water equation of state (Simons, 1973) model equations can be written as follows:

$$du/dt - fv = -(1/\rho_0)\,\partial p/\partial x + \mu\Delta u + \partial(\nu\,\partial u/\partial z)/\partial z$$

$$dv/dt + fu = -(1/\rho_0)\,\partial p/\partial y + \mu\Delta v + \partial(\nu\,\partial v/\partial z)/\partial z$$

$$dp/dz = gp$$

$$\partial u/\partial x + \partial v/\partial y + \partial w/\partial z = 0$$

$$dT/dt = \mu_T\Delta T + \partial(\nu_T\delta^{-1}\,\partial T/\partial z)/\partial z$$

$$\rho = \rho_0(1 - 0.68T \times 10^{-5}(T - 4)^2)$$

where u, v, w are components of velocity, p is pressure anomaly, ρ is density anomaly, T is temperature, t is time, ρ_0 is mean density, f is Coriolis parameter, g is the Earth's gravitational acceleration; μ, ν, μ_T, and ν_T are the coefficients of vertical and horizontal turbulent diffusion of momentum and heat, $d/dt = \partial/\partial t + u(\partial/\partial x) + v(\partial/\partial y) + w(\partial/\partial z)$.

Coefficient δ parameterizes mixing in the case of unstable vertical stratification:

$$\delta = 1 \quad \text{if } \partial p/\partial z \geq 0$$

$$\delta = 0 \quad \text{if } \partial p/\partial z < 0$$

The following boundary conditions are imposed on the system:

$p_0 \nu(\partial v/\partial z) = -\tau_s$	$w = 0$	$\partial T/\partial z = 0$ at $z = 0$
$p_0 \nu(\partial v/\partial z) = -\tau_b$	$w = u(\partial H/\partial x) + v(\partial H/\partial y)$	$\partial T/\partial n = 0$ at $z = H(x, y)$
$\nu_n = 0$	$\partial v n/\partial n = 0$	$\partial T/\partial n = T$ at σ

where $v = (u, v)$ is the vector of horizontal velocity, $\tau_s = (\tau_{sx}, \tau_{sy})$ is wind stress, $\tau_b = (\tau_{BX}, \tau_{BY})$ is bottom stress, n and τ are the external normal and tangent to the boundary. The wind and bottom stress is related to surface winds and bottom currents by conventional quadratic stress laws.

The finite-difference scheme of the model is constructed using Bryan's conservative box method and Arakawa's grid (see Demin and Sarkisyan, 1977). The lake surface level is used as the integral function of the model (Demin and Ibraev, 1989). The basic scheme of integration by time is the "leap-frog" scheme with respect to pressure gradient and advective terms and Euler scheme for the diffusion terms. A horizontal mesh size of 4 km and 8 levels in vertical were used. The wind observations on two islands located in the open part of Lake Onega were used to define the wind stress. The coefficients of vertical and horizontal turbulent diffusion of momentum were chosen as 10 and 10^5 cm^2/s respectively (Filatov, 1983), the turbulent heat diffusion coefficients were by one order less. The time step of integration was 1200 s.

2.6.5   Water circulation connected with topographic effects

While the Coriolis parameter only slightly changes with latitude, free Rossby waves are absent from the lakes. However, in a closed water body with complex bottom relief wind fluctuations cause topographic vortices, eddies, circulations or Rossby waves. Topographic waves have been considered by Csanady (1977) for water bodies with ideal shape and bottom. In a real water body with a complex shape and bottom relief variations in the shore and bottom slope form a wide spectrum of wave motions, Kelvin waves, topographic waves (including "hybrid" waves), as a result of interactions between baroclinic Kelvin waves and barotropic topographic waves. When the scales of motions in lakes are large due to topographic effects, gravitational forces lose their dominant role in water dynamics to effects related to rotation. Barotropic waves become "planetary", manifested as slowly spreading macroscale currents. They are identified not by relative equilibrium level fluctuations, but by variability in the horizontal velocity field. The effect of the lake topography

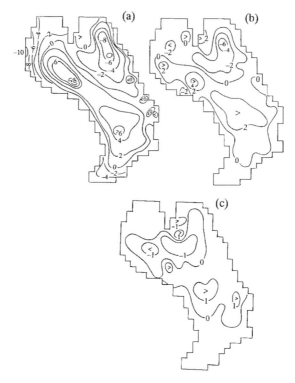

Fig. 2.40. Barotropic topography motions (free surface level in mm) in Lake Onega calculated by diagnostic model: a—5 hours, b—24 h, and c—48 h after cessation of the action of wind.

(discontinuities of the bottom relief, shore slope and the water body closure) is the most important factor in the formation of the whole-lake circulation. Water circulation in the lake is of the nature typical for "boundary" layers. Besides, there are particular features characteristic of the nearshore zone limited by the distance of the order of baroclinal radius of Rossby deformation and manifested as a coastal jet, trapped waves, Kelvin waves; and there are specific features related to topographic effects manifested also in the offshore part of the lake as topographic circulation waves. Allender, Saylor (1979) have described the generation of topographic waves in a lake of simple elliptic shape with a parabolic bottom. Calculations of topographic waves for several phases show that topographic motions normally have "double-gyre" circulation. We have studied the features of barotropic motions by controlling the relief shape of the free surface elevations at different time stages of calculations using the diagnostic model for Lakes Ladoga and Onega (Fig. 2.40).

Five hours after the cessation of the action of wind, water circulation with three major circulations was observed in the lake, and the level gradient in Onega proper reached 10–16 mm. In 24 hours the level gradient was only 8–10 mm, in two days 2 mm. Current velocity in the upper layer dropped from 15–20 cm/s to 1 cm/s. Further on, after the effect of the wind there appeared signs of the development of motions with a tendency towards anticlockwise rotation. The main gyres fall apart and relatively small eddies with the horizontal size of the order of several kilometers are formed. Difference in bottom and shore slope in the western and eastern parts of the lake caused a disagreement in the frequency of topographic waves and baroclinal Kelvin waves. Hence, topographic waves by the eastern shore left Kelvin waves "behind", and the spectra of current fluctuations contain two energy maxima corresponding to the mentioned waves in the low frequency region.

2.6.6   Currents and dynamics of wind-induced upwelling

The data reveal large-scale current and temperature variability in both lakes with several energy peaks (Fig. 2.37), reflecting the lake's response to atmospheric forcing and their strong variability in a hydrological "weather scale". Bearing in mind Mortimer's (1979) remarks about "underwater" weather, which follows atmospheric weather changes, we can say that wind-induced coastal upwellings are among the main components of the above-mentioned hydrological weather in Lake Onega. In response to the wind, coastal upwellings continually change strength and location, which can be most easily traced in the surface temperature field. In cases of especially strong winds, the area occupied by the cold upwelled water, can reach up to 30% of Lake Onega's surface. In the evolution cycle of coastal upwelling, three main phases can be discerned: generation, steady-state and relaxation (Beletsky *et al.*, 1994). The generation of coastal upwelling follows the occurrence of steady winds over the lake. Usually after a few hours of moderate wind forcing a double-gyre circulation becomes dominant in the lake. The prominent features of double-gyre circulation are narrow coastal currents directed by the wind and the broad counter-current in the deep part of the lake (Bennett, 1974). A case of wind-induced upwelling close to the steady- state one was observed in Lake Onega during one of the ship surveys in 1987. The wind was from a southeastern storm of 2 days duration and the ship cruise was done immediately after the winds decreased. The thermocline can reach the lake surface in cases of such strong winds. This fact is vital for the lake ecosystem because of the enhancement of water exchange between epilimnion and hypolimnion and, as a consequence, the ventilation of the hypolimnion and increased supply of nutrients to the epilimnion. Shipborne and airborne observations have shown that if calm and warm weather follows the storm, abundant growth of phytoplankton in upwelling zones occurs (Filatov *et al.*, 1990). The diagnostic model calculations have a pronounced double-gyre circulation (Fig. 2.41b).

After cessation of the wind, upwelling relaxation begins. Following Csanady (1977), we distinguish two types of relaxation events. Type-1 relaxation is characterized by simultaneous anticlockwise propagation of cold and warm temperature fronts and by the relevant coastal jet reversals around the lake. These Kelvin wave-like patterns were studied intensively in Lake Ontario during IFYGL (International Field Year Great Lakes) (Csanady, Scott, 1974). In contrast to type 1, only one front moves in the case of type-2 relaxation. Mortimer (1963) was the first to discover such a warm front propagation along the southern coast of Lake Michigan; later another case of type-2 relaxation was described for Lake Ontario (Simons and Schertzer,

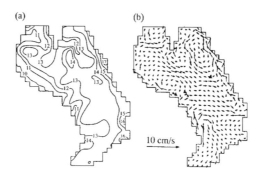

Fig. 2.41. Currents (A) and distribution of water temperature on horizon 10 m (B) in Onega Lake. Diagnostic experiment.

1987). It is interesting also to note that type-2 relaxation events have ocean counterparts: Somalia's upwelling relaxation in the Indian Ocean (Zalesny, 1990) and to some extent the famous El Niño in the Pacific Ocean (Gill, 1982). The reason for the existence of type-2 relaxation is not clear. Bennett (1973) shows that in the cases of non-linear Kelvin waves a difference in the speed of warm and cold water propagation occurs and warm wave movement becomes dominant. Another reason may lie in the irregular lake morphometry, preventing wave propagation around the lake. It is also typical for the northern coast of Lake Onega and may be responsible for the observed type-2 relaxation in 1989. The summer 1989 observations were made after the cessation of strong northern winds which dominated over the lake for 3 days before the first ship survey. Observations showed that the upwelling zone near the eastern coast with a sharp thermal front still occupied a large part of the lake (Fig. 2.41). Diagnostic calculations have shown that lake circulation consists of the main cyclonic gyre and some smaller gyres. Near the eastern shore the coastal jet structure is clearly seen, intense currents are also observed along the thermal front zone. During the next 4 days no significant wind events occurred. The observations showed that the upwelling front moved along the eastern coast to the north, in the direction coincident with the direction of Kelvin wave propagation (Fig. 2.42). The result of upwelling front propagation was the rapid restoration of the thermocline near the coast and also the restoration of a cold-water dome in the lake. Diagnostic model calculations have demonstrated that type-2 relaxation leads to enhancement of cyclonic circulation in Lake Onega (Fig. 2.42a). The field data obtained have shown that the combined effect of stratification, Earth's rotation and the presence of lateral boundaries makes large lake dynamics very complicated during the summer. It will be noted that intense dynamical phenomena such as strong mesoscale eddies can be found not only in the coastal waters (Rao and Doughty, 1981) but offshore as well. For example, the most intense eddy revealed during 1989 observations was located in the center of the lake, stretching through the thermocline from the surface to 20 m depth. It should be noted that even such strong thermocline eddies may show weak surface signatures and indeed, we had rather weak thermal gradients in that area on the infrared images. Another interesting detail concerning remote sensing

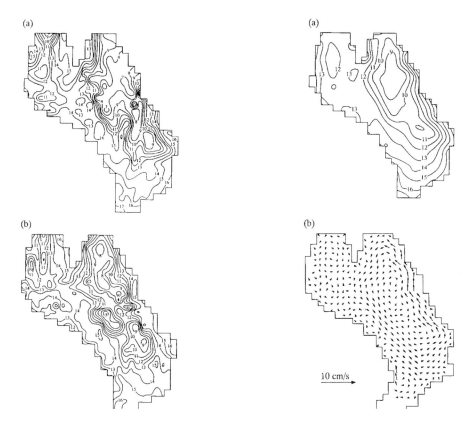

Fig. 2.42. Temperature at 10 m depth in Lake Onega, (a) 10 July 1989, and (b) 14 July 1989.

Fig. 2.43. Lake Onega prognostic model results for 14 July 1989 for (a) temperature at 10 m depth, and (b) currents at 10 m depth.

data was that the strong frontal zones near the surface and at depth did not correspond completely. The horizontal scale of the upwelling front meandering and eddies was about several kilometers which coincides with the characteristic Rossby baroclinic radius deformation (3–5 km). This suggests that their origin is connected to the baroclinic instability of the frontal current. To confirm a suggestion of restoration of cyclonic circulation during type-2 relaxation events a prognostic model of Lake Onega was used. Initial temperature distribution and currents have been obtained from diagnostic model results for the first ship survey in 1989. Wind has been neglected and time integration was held up by 4 days, until the second survey. The prognostic model was able to reproduce the main relaxation mechanism—the upwelling front propagation along the coast; enhancement of the cyclonic circulation was also observed (Fig. 2.43).

Consequently, the baroclinic relaxation mechanism which takes into account internal pressure gradients may supplement the "surface" mechanism of Emery and Csanady (1973) in producing cyclonic summer circulation in large lakes. however, It should be noted that the speed of thermal front propagation was reduced because of insufficient horizontal resolution and high damping. The thermal diffusivity coefficients were also too high which caused strong smoothing of the initial temperature gradients. There was no eddy generation as well. Consequently, at least some of the intense mesoscale phenomena could not be properly treated in the numerical models with a coarse grid resolution.

Lakewide circulation patterns typical for spring and summer conditions in two of the largest European lakes were discussed, based on field observations and numerical modelling. It has been shown that the lake's circulation depends strongly on two important hydrodynamical processes such as thermal bar and wind-induced upwellings. Evolution of wind-induced upwellings has been described with focus on thermal front dynamics and circulation pattern changes. One type of upwelling relaxation was investigated with particular attention to fine grid observations, though further investigations are needed to explain the existence of various types of relaxation. Though much attention was drawn to the short-term variability of currents and temperature, some conclusions relating to the cyclonic character of mean circulation in large lakes were also made. There are still few appropriate data concerning this important area. More work is required to investigate mesoscale dynamics. It is obvious that intense mesoscale eddies observed in large lakes have to be taken into account for an accurate description of pollutant transport in the coastal and offshore zones. New models with finer grids (at least with 0.5-1.0 km resolution) are needed for proper simulation of Kelvin and topographical trapped waves interactions, upwelling front meandering, mesoscale eddy generation and their interactions with lake-wide circulation.

2.6.7   Technology of lake observations by remote sensing and assimilation of these data in a models

Remote sensing techniques have been applied to investigations of large lakes of Europe. But it is important it to provide that invaluable information for data gathering which is useful for assimilation in prediction models.

More often NOAA AVHRR temperature data have been applied for analysis of spring thermal bar for large lakes (Malm, Jonsson, 1993). Cloudiness was introduced in forecasting hydrological "weather" 3-D models (Schwab, Bedford, 1994).

An unsolved problem is how to restore vertical distribution of temperature and apply this data in models. Many years' observations on Lakes Ladoga and Onega have provided us with knowledge about manifestation of upwelling, and frontal zones in lakes showed ways for restoration of 3-D water temperature in lakes. It is possible to calculate currents in a lakes by measurements of water level. The altimetric sensor satellite data now apply in ocean models but accuracy of satellite sensor is rather low for lakes. Experiments with altimetric data have been done on Lake Ladoga. In the near future more precise altimetric data from satellite will be available and it will be possible to assimilate it for monitoring and hydrodynamical models too. Some useful information from Russian and others satellites with SAR, AVHRR, TM techniques has been obtained for spatial and temporal characteristics of patchiness, gyres, eddies, current and waves, pollutants spills, meteorological features (cloudiness, temperature), ice cover, surface roughness, chlorophyll plumes and macrophytes with time intervals from days to months. Remote sensing measurements need to development of calibrations with data obtained from field observations.

2.6.8   Lake hydrodynamics, water ecosystem and climate changes

An important question is how to get knowledge about scenarios of climatic and anthropogenic change for application to ecological models? For example, over the last 10 years fertilizer use in the watershed territory of Lakes Ladoga and Onega has decreased more than five times. In this case of anthropogenic stress on the water system has strongly declined. In this it is easier to show the role of climatic factors in the current state of Lakes Ladoga and Onega. From this point of view it is very important to use the Global Climate Model (GSM) for watershed territory of Lake Ladoga and Onega and incorporate it in ecological models. Menshutkin *et al.* (1998) showed noticeable climatic variability in the ecological system of Lake Ladoga. Hudon (1997) has shown variability of biomass in water ecosystems of the Great Lakes related to climatic conditions in the basin.

It is important to understand mechanisms of water transport mixing and how they vary from year to year in response to climate change. Of course, these properties cannot be understood in isolation, without consideration of their catchment area. We need to determine mechanisms and rates of transfer of nutrient, tracer exchange features between water and biotic and abiotic processes and to estimate the role of global warming on complex processes in the near shore and pelagic zone of the lake. Special attention should be given to coastal zone problems. During climate change, fluctuating water level and stronger changes are found in the littoral zone of the lake. This region of the lake is the first to show signs of climatic and anthropogenic factors. But for long-term changes we must understand the scenario of climate change. In this case, modelling can be fundamental to effective management of lakes. The expected consequences of global climate change illustrate the need for accurate hydrological models. It is very important to jointly organize monitoring and sampling collection data, and to manage the water ecosystem on an international scale because the watershed territory of the lakes overlaps national boundaries. In this case it is important:

—to create a database for identification of changes;
—to choose classes of the model used to imitate the up-to-date status of the ecosystem and chemical–biological situation;
—to forecast the behavior of the system under anthropogenic trends and climate changes in the region.

It is also important to identify long-term natural cycles in order to differentiate them from changes caused by natural climatic impacts. Only in this case can we predict the future condition of the lake.

2.7   CHARACTERISTICS OF WATER MOTION AT VARIOUS SPATIAL/ TEMPORAL SCALES

The whole variety of thermohydrodynamic phenomena in lakes: waves, circulations, gyres, eddies, fronts including thermal bars, diverse streaks and patches (spills, spots) can be conventionally divided into macro- and microscale ones. The appearance of a number of macroscale phenomena on the lake surface can be caused by their interactions with microscale processes. For example, the appearance of internal waves on the lake surface is due to their interaction with wind-driven, microscale waves. Below we will demonstrate some opportunities of identifying a number of dynamic phenomena in lakes using remote-sensing satellite measurements. Table 2.7 shows some requirements for the accuracy and replicability of remote-sensing measurements of dynamic processes and phenomena in lakes made from satellites.

2.7.1   Macroscale phenomena

2.7.1.1   Gyres, coherent structures

The description of lake-water dynamics using data from a ramified system of observations at autonomous buoy stations (ABS) or from research vessels by measurements on sections at various complexes of hydrometeorological conditions is quite promising but so far impracticable. Especially difficult is description of formations which move in space such as gyres 20–30 km in diameter with horizontal axis, coherent structures, mono- and dipoles. To study these phenomena essentially novel observation tools are needed: first of all, remote-sensing aerospace facilities are the most adequate means for this task. Target-oriented multilevel experiments using air- and space facilities started on Lakes Onega and Ladoga in the mid-1990s. Observations within the "Onega" experiment, which started in 1986, were carried out using all available means: ABS, research vessels, hydrometeorological standard

Table 2.7. Requirements for accuracy and replicability of satellite remote-sensing imaging.

Process, phenomenon	Space scales, km	Duration of the survey	Accuracy of space resolution, km	Possibility
Macroscale circulations	10–200	1 day	1	Yes
Topographic waves	20–50	2–12 h	10	No
Mesoscale circulations, gyres	5–20	1 day	1	Yes
Upwellings	5–20	0.5 day	0.5	Yes
Fronts	1	1	0.5	Yes
Internal waves	1	3–4 h	0.5	Yes
River plumes	1	1	0.5	Yes
Ice fields	1	1	0.5	Yes
Surface waves	0.01–0.001	0.1	0.001	No

stations, on-board laboratory of AN-28 aircraft, satellites of the "NOAA", "Meteor", "Kosmos-1939", "Kosmos-F1", "Almaz" and ERS types. These observations of thermohydrodynamic processes and phenomena have no analogues in the magnitude, complex and synchronous nature (see scheme of the experiment in Fig. 2.38). Measurements during the "Onega" experiment were made for a broad spectrum of thermohydrodynamic phenomena with spatial scales from centimeters to hundreds of kilometers both in the thin surface layer and for the whole depth of the waterbody. In this research we used medium and high-resolution data (MSU-SK and MSU-E with ground resolution 170 m up to 45 m in the ranges 0.5–0.6, 0.6–0.7, 0.7–0.8, 0.8–1.1 µm from a Resurs-01 type satellite ("Kosmos-1939"), Almaz, as well as results of space imaging made from satellite ERTS-1 with a multispectral scanner, and measurements made by an infrared radiometer, thermal scanner "Vulkan", and by lidar on board a flying laboratory.

The water temperature in the surface layer of a large lake can be effectively measured by remote-sensing instruments: by IR radiometers set to aircraft, research vessel and satellites. These measurements for survey of Lakes Ladoga or Onega from an aircraft require several hours, from a NOAA satellite several minutes. Synchronous aircraft, research vessel and satellite observations enable us to study the problems of remote-sensing measurements of water surface temperature (WST) with regard to the specific features of hydrometeorological conditions. Let us demonstrate the opportunities of estimating physical processes in the water body relying on data from observations of the surface layer of the lake. These data reveal considerable "patchiness" of the phenomena occurring in the lake, which is essential from the point of view of standard hydrological measurements. Disregarding or underestimating the "patchiness" of a phenomenon may result in inaccurate calculations of the fields of thermohydrodynamic properties and erroneous interpretation of the lake water-quality indices.

A survey made using side-scan radar from the all-weather satellite "Almaz" simultaneously with airborne and shipborne photography indicated that according to the shade and texture of its surface layer Lake Onega was divided into two parts with patches sized from dozens to hundreds of meters (Fig. 2.44).

These patches are correlated with breaks in the thin film on the lake surface, water turbulence caused by ships, interactions of surface and internal gravity waves. The division of the lake in this period is caused by southern point windblasts, which cause surface roughness in the western half of the lake. When

Fig. 2.44. SAR image of Lake Onega from Russian satellite "Almaz" (14 July 1989).

the southern point winds grow stronger nearshore upwelling is formed at the western shore, and a zone of maximum WST—at the eastern shore. In these conditions chlorophyll values at the eastern shore near the River Vodla in Lake Onega reach 14 mg/l (this value in the central part is 4 mg/l). As this happens micro- and mesoscale discontinuities in WST and chlorophyll fields of the size of several dozen meters disappear, and macroscale ones (several dozen kilometers) are formed instead. As the wind grows weaker WST regains evenness, no nearshore upwellings are observed, the distribution of water temperature becomes typical for the stratification period (Fig. 2.39). Thus, the action of wind breaks the "typical" (climatic kind) pattern of the density and velocity field, and mesoscale discontinuities of the velocity and density fields are formed as demonstrated in § 2.6.2. Quasi-steady anticlockwise (cyclonic) water circulation is rarely observed during summer stratification when a number of mesoscale circulations exist whose properties depend on the effect of baroclinicy, morphometry and wind. WST can be used to estimate the kinematics of physical processes going on in the water body when the wind is absent and daily thermocline does not produce the sealing effect of the typical dome structure of isotherms.

We will now consider the specific features of motions with spatial scales from meters to several dozen kilometers found in lakes using space imagery data. The image of Lake Ladoga in the visible spectrum region obtained from the ERTS satellite on 10.09.79 was preceded by aerophotography on 03.09.79 (Fig. 2.45a) and on-site (synchronously with space imaging) measurements of water temperature and currents as well as observations of meteorological parameters (wind velocity and direction, pressure, cloudiness) and wind-driven waves. In some verticals internal waves and currents were measured as well.

Computations using a diagnostic model (Demin, Ibraev, 1989) with a grid size of 10 km, and computations of wind-driven and gravity-driven currents from rivers with grid size 5–7 km were made for the studied hydrometeorological conditions (the period of stable thermal stratification, wind velocity about 4 m/s). All available information was used to reveal the nature of the discontinuities in Lake Ladoga water dynamics including eddies and internal waves in offshore and nearshore zones demonstrated by satellite image. Zero cloudiness (clear sky) during the space survey enabled easier interpretation of recorded specific features of water dynamics. Eddies and internal waves were indicated by suspended matter carried by the Rivers Volkhov and Svir to the south part of Lake Ladoga and by plankton.

Figure 2.45b shows interpreted characteristics of water dynamics in the southern and central parts of Lake Ladoga and specific features of water dynamics in the nearshore zone (Kondratyev *et al.*, 1989). Gyre A with the diameter of about 25 km and anticlockwise circulation, and anticyclonal gyre C can be singled out. The lake depth in the A-gyre periphery is 45 m. A mushroom-like structure B can be observed. Streaks E at a 10 km distance from each other are found between gyres A and C. These streaks are related to the appearance of internal waves on the lake surface.

Let us consider lake-water dynamics relying on the data of observations and calculations based on models. Analysis of the results of numerical modelling using a

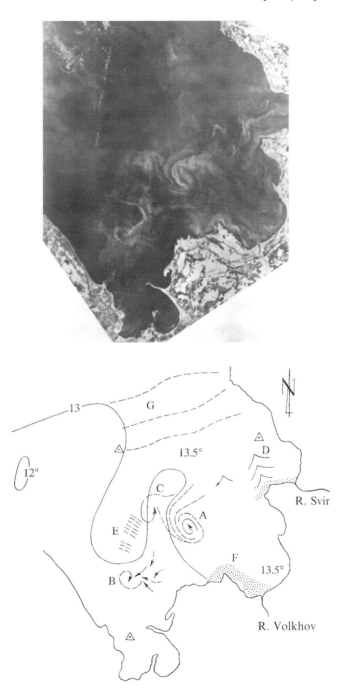

Fig. 2.45. (a) ERTS satellite image of Lake Ladoga, 3 September 1979. (b) Interpretation of the satellite image. A and C—Gyres, B—mushroom-like formation, E and G, D—surface manifestations of internal waves, F—River Volkhov flow, triangles—locations of mooring stations with current measurements.

nonlinear diagnostic model indicates that the lake is governed by cyclonic circulation with the upper 10-m layer dominated by wind-driven currents, and velocity on the surface of 10–15 cm/s. An anticlockwise vortex is recorded in the southern part of the lake. Calculated thickness of epilimnion at wind velocity 3–4 m/s was 10–15 m. Several days before the survey the speed of the wind from the same direction reached 12 m/s resulting in considerable development of wind-driven currents. Currents in the nearshore zone of the lake become unstable; gyres and trapped waves by the shore are formed. The nearshore zone of the lake can be determined by the inner radius of Rossby deformation, which makes up 3–5 km. Thus, the gyres, eddies and waves recorded 20–40 km from the shore belong to the offshore part of the lake.

The major traits of total and local circulation and eddies can be explained by the conservation of absolute potential vorticity: $f + \zeta/H = \text{const.}$, where H is depth; ζ is vertical component of relative vorticity. The dominant contribution to the changes in eddy characteristics in relatively small basins, which is the case with lakes (large lakes as well), is provided by variations in depth H. In a simple case integral water circulation in the central and southern parts of a lake can be represented as two circulations separated by a nodal line with anticlockwise precession (Fig. 2.45b). The frequency of these currents at latitude 61° in Lake Ladoga approximated by an ellipse with the parabolic bottom is nearly $\omega = f/7 = 0.08$ rad/h; velocity in the middle of the basin and near the shore is of approximately the same order, whereas maximum current velocity for Kelvin waves is noted in the nearshore zone. On bottom relief discontinuities macroscale circulation may evoke local circulations with scale of several dozen kilometers. The nature of the vortex depends on the sign of the value $\partial(f/H)/\partial y$. If the value is positive, the vortex has anticlockwise circulation. In the A-gyre area (Fig. 2.45) this value is negative, and hence the circulation is anticlockwise. The derivative in the B-gyre area is positive, and water circulation is clockwise. $\partial(f/H)/\partial y$ along the critical line is equal to zero and an open jet is formed. The main factor influencing these local eddies is the tendency inherent in eddies to preserve their state. The water column moves so that the height and relative vorticity change whereas the value of potential eddy remains unchanged. Gyres A and B have the same ratios of vertical and horizontal (H/L) dimensions— 0.001. Spectra of currents and water temperature were calculated from the data of observations at autonomous buoy stations. An important feature of the spectra of the nearshore and offshore parts of the lake is that the nearshore zone spectrum has no maximum at the local frequency of inertia f. Inertial fluctuations of currents in the lake appear ellipsoid at the distance of several radii of inertia This radius at average velocity $u = 7$ cm/s is 1 km. Thus, there exists a maximum at the inertia frequency $\omega = 0.45$ rad/h in the offshore zone several km from the shore. These may be Poincaré inertial-gravity waves; or in some cases they may be expressed as purely wind-driven inertial oscillations. According to the image typical spatial scales of the motions about 1 km long are related to internal Poincaré waves (E and G) (Fig. 2.45).

The spectra of Lake Ladoga hydrophysical characteristics have a maximum in the low-frequency region of the spectrum $\omega > 0.10$ rad/h. In the nearshore zone these movements may be caused by baroclinic Kelvin waves which are restricted to the distance equal to the internal radius of Rossby deformation R_b. The data of limnograph measurements reveal a slight surge to the southwestern part of the lake at the moment of space survey, which then was 30–40 cm. Knowing that the distance between the level monitoring stations is about 100 km the estimated current velocity is then 5–10 cm/s. The diagnostic model yields the same values of velocity. Measurements of currents carried out in several points simultaneously with the space survey corroborate the estimates and calculations presented above. For example, the values averaged for 1 day were: of the current in the 5 m horizon 5 cm/s, in the 10 m horizon 4 cm/sec. However, density streaks showing the coverage of waters with different characteristics in the A and B gyres found in the image suggest that the current velocity should be about 20–30 cm/s. Cold water containing little suspended matter which points to water rising up was recorded in the center of the anticlockwise gyre A. The streaks observed in the northwestern part of the lake may be the manifestation of internal waves about 10 km long on the surface.

The effect of rivers on the formation of total circulation in the summer period is negligible. Streamflows become visible and are recorded by satellite surveys at a distance from several hundred meters to several kilometers depending on the river discharge and bottom slope (Fig. 2.45A–F).

Coastal streams observed in lakes are divided into jets and plumes. The jets are flows which do not possess neutral buoyancy in relation to the lake water, but do have momentum; the plumes—a jet with positive buoyancy but zero momentum. Jets and plumes can also be recorded by space imagery data. Jets in lakes may fall under the effect of being trapped by the shore. This happens when the ratio of the flow height h and the site depth H is less than 0.1. Figures 2.45 and 2.46 show an example of river flow in Volkhov Bay with mild bottom slope in Lake Ladoga. River flow may not occur under certain seasonal distributions of water temperature.

The trapped flow, the presence of thermal bar and coastal jet do not enhance the mixing of the runoff with the main water body of the lake. When estimating the distribution and transformation of sewage from enterprises located on the shore of water bodies using mathematical models it is important to give due consideration to the above-listed characteristic features of nearshore zone hydrodynamics. Waste water has a distinct boundary–front. Their distribution and intensity of dissipation depend on currents, bottom relief, stratification and other reasons.

River jets on the steep bottom slope near the west coast of Lake Ladoga $h/H > 0.1$ are trapped by alongshore jets. These circulations spread several kilometers into the lake with anticlockwise direction of motion under slight action of eddy diffusion, which is demonstrated by the appearance of the "plume" in the space image. During most intensive upwellings the dilution of sewage in this area is accompanied by the movement of sewage from the hypolimnion to the surface. Anticlockwise water circulation in lakes Ladoga and Onega in the spring period promotes the spreading of little-diluted sewage all over the lake.

Characteristic features of nearshore upwelling in Lake Ladoga have been estimated using space imagery data (Scanner MSU-E) from the 0.5–0.6 and 0.6–0.7 µm ranges. The upwelling is visible in the image (in the optical range) due to the existing difference in the amount of suspended matter in the water of the Volkhov

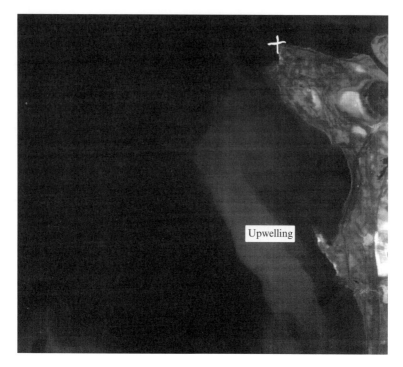

Fig. 2.46. Satellite image of Volkhov Bay of Ladoga Lake from Russian satellite "Kosmos-1939", Scanner MSU-E, ch. 0.5–0.6 μm, 7 July 1989. Inside the frame is shown the mushroom-like formation on Ekman's coastal upwelling.

River itself (the content of suspended matter is about $20\,mg/m^3$) (Fig. 2.46) and in the nearshore upwelling water. The water of the latter comes from the lake hypolimnion and contains about $2\,mg/m^3$ of suspended matter. The difference between WST in the upwelling and the central part of the bay is 4°C. The characteristic dimensions of the upwelling are $25 \times 10\,km$; a fungiform structure sized $5 \times 5\,km$ with a tendency towards clockwise circulation is observed in the northern part of the upwelling.

Transverse structures in the coastal upwelling zone

Sun-glint can sometimes by used to reveal the patchiness of processes and phenomena on the lake surface and diagnose the state of the water body. In multi-zonal aerial photography sun-glint provides a view of the cellular structure of water with cells sized from several dozen centimeters to several meters, which are similar to those of Benard cells (Bukata and McColl, 1973).

Spectro-zonal (SZ) space survey of the lake surface revealed streaks perpendicular to the shore. The frontal zone on the upwelling boundary is easily marked in the infrared range. Transverse streaks and jets in the sea appear as a result of desalinization of sea water by river waters. These streaks in lakes are due to the

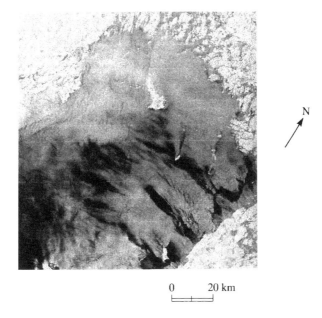

0   20 km

Fig. 2.47. SAR satellite image of Lake Ladoga in summer 1995. In the east part of the lake manifestations of internal waves and other mesoscale structures are shown.

appearance of internal gravity-wave packets on the surface. In a zone limited by upwelling there appear multinodal standing waves, which interact with surface waves. These waves are visible in space images and can be marked when lit up by a sun flash. Internal waves show up best on the lake surface with the thermocline located close to the surface where the upper thermocline–metalimnion boundary is found at the 5–15 m depth. In this case the currents evoked by internal waves are particularly strong in the epilimnion. Internal waves appear on the surface in the form of low-frequency surface waves with the periodicity of internal waves. They are visualized on the surface when wind-driven waves are transformed on currents by means of nonlinear mechanisms of interactions between surface and internal waves. SAR image of Lake Ladoga shows the manifestations of internal waves on the surface with wavelength of few tens kilometers (Fig. 2.47).

2.7.1.2 Some specific features of hydrodynamic processes in a lake in the winter period

Analysis of the space images of Lakes Ladoga and Onega reveals the regularly repeated features of ice cover crevassing. There is a relatively stationary lead fringing Lake Ladoga at some distance from the shore, approximately above the 20 m isobath (Fig. 2.48).

In the north the lead is found to the south of Valaam Island in Lake Ladoga. When winds are stronger than 5 m/s the central massif which is covered in cracks is observed to drift. After the wind attenuates the lead regenerates. Formation of

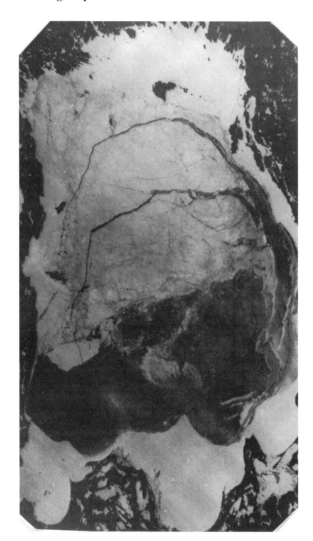

Fig. 2.48. Satellite image of Lake Ladoga by high resolution "Fragment" technique in the winter period.

cracks in the central lake massif is most probably due to the effect of seiches (standing waves). In this period seiches may emerge in the lake. The distance between longitudinal and transverse seiches in Lake Ladoga ranges between several to 20–30 km. Cracks appear mainly in seiche oscillation loops. The distance between the cracks corresponds to the length of the seiche. The period and velocity of currents under ice can be calculated by certain variance equations of Merianne's type (Hatchinson, 1969). The presence of a stationary lead in Lake Ladoga may also be related to dynamic processes, currents of cyclonic nature, which show the speed maximum at the bottom slope.

2.7.1.3   *Characteristic features of water dynamics according to satellite*
altimetry data

Let us estimate the dynamics of Lake Ladoga waters using data from the "Seasat" satellite for summer 1978 (Verner, 1983). There were altogether three noncollinear series of measurements along the routes crossing Lake Ladoga in the period from July to September. It is difficult to investigate mesoscale phenomena such as seiches or storm surges using the data which do not cover the whole lake because the time scale of these phenomena are less than 3 days while the time step of satellite measuments is 3 days. These measurements can help to evaluate the characteristic features of macroscale lake dynamics. The observations available for Lake Ladoga for the period in question were instrumental ones of the currents and lake level from several points obtained with the 3 h discreteness and 1 cm precision. Data on the topography of the free water surface were used in the diagnostic model to calculate currents.

It should be stressed that the described estimating method cannot be applied for shallow large lakes such as Balkhash and Balaton, where water circulation depends on wind rather than combined effect of baroclinicity and bottom relief. This method is unsuitable for estimating current in the coastal jet zone due to low spatial resolution while the flow width is usually less than 7 km. We will consider the feasibility of estimating water circulation with the time scale of the order of 6–8 days, i.e. features of circulation involving the whole lake. The advantage of satellite altimetry in studying the features of currents is its independence of cloud cover and the spatial step of measurements (7 km). As shown by calculations using nonlinear diagnostic models the change in lake level across the water area is 30–40 mm for a distance of about 100 km. The lake surface slope is thus of the order of 10^{-6}–10^{-7}, i.e. comparable with oceanic slope. Modelling data indicate a level reduction in the lake center, the difference between the level in the center and near the shore reaching 30 mm. The obtained level field shows that currents in the water body are mainly anticlockwise. Current velocity in the upper 10 m layer is 10–15 cm/s. In the 10–20 m horizon it decreases to 2–5 cm/s. The surge effect is seen in the nearshore zone of the lake (several kilometers), which is not covered by the resolution capacity of the satellite.

Local clockwise circulation is observed in the southern, relatively shallow part of the lake. It is several dozen kilometers in size. Current generation here is mostly related to wind field with a typical scale of several days. Thus, knowing the relatively unstable nature of baroclinal summer water circulation, current velocities can be calculated using satellite altimetry data in the period of hydrological summer after the thermal bar disappears from the lake. A dense network of limnograph observations and more frequent (every other 24 hours) observations along the same routes, which can be simultaneously obtained by several satellites, are necessary to estimate the currents with spatial scales of about several days and less. In this case, a two-dimensional field of lake level can be obtained, and mesoscale discontinuities in lake-water dynamics can be estimated. It has to be mentioned that in the future satellite altimetry data will only be applicable for estimating total water circulation in large lakes.

2.7.1.4   Upwelling

Upwelling, or the movement of water from hypolimnion up to epilimnion, was studied on Lake Onega by Bojarinov *et al.* (1994); Filatov (1991), on the Great Lakes by Csanady (1977); Simons and Schertzer (1987). The basis for study of upwelling in Lake Onega was mainly data of long-term observations at ABS. Several types of upwelling were singled out: Ekman nearshore upwelling, flotation, offshore upwelling in the center of anticlockwise circulations. Among the most common remote sensing techniques we have used to study upwelling were the measurements of water temperature fields by IR sensors from ships and satellites. Additional information on the characteristic features of hydrophysical fields in the upwelling zone was obtained by photography and videotape recording from on board an aircraft.

Observations from on board a ship during the "Onega" experiment show that in six of the ten recordings of the water temperature fields intensive Ekman nearshore upwelling was found at the middle part of the western shore of Lake Onega. In one instance at the eastern shore three cases of calm condition upwelling were not recorded. According to the estimates of the vertical compound of current velocity using diagnostic models the speed of upwelling in the upper 15-meter layer was 10^{-3}– 10^{-5} cm/s, which is in agreement with the calculations of the upwelling speed made earlier by Bojarinov *et al.* (1994). The average period of upwelling relaxation is less than the synoptic period. Therefore, simultaneous manifestations of upwelling by the western and eastern shores are quite rare. However, for Lake Ladoga such cases have been recorded in the data of long-term remote sensing IR-observations from airborne laboratories of the Hydrometeorological Service for 1968–1992. These and satellite-based observations of both lakes indicate that there simultaneously existed an "old" and a "new" nearshore upwelling. This is due to the large size of Lake Ladoga as compared with Onega, and hence a somewhat longer "life" (about 24 hours) of Ekman nearshore upwelling. Aerial and space remote sensing combined with ship-based observations and results of modelling at upwellings in different lake zones adjusts integral water circulation schemes obtained earlier under various hydrometeorological conditions. In the hydrological spring and autumn in the presence of the thermal bar, Ekman nearshore upwellings do not appear in the lake (Fig. 2.39). However, water circulations are quite changeable even when winds are the same. Proceeding water circulation is important for the formation of circulation at specific conditions, although sufficiently prolonged (up to 0.5 of the synoptic period) winds of certain directions quickly change water circulation. For the formation of Ekman nearshore upwellings these are northern and southern point winds. The period of upwelling relaxation is about two times shorter than the period of upwelling formation. Our estimates suggest that the speed of water rising and sinking was 10^{-2} and 5×10^{-2} cm/s respectively. Figure 2.39 shows integral water circulation schemes for Lake Ladoga at various hydrometeorological conditions, and the zones of Ekman nearshore upwelling. Noteworthy are some distinctions between the thermo-hydrodynamic state of the lake water at Ekman nearshore and flotation upwelling. Flotation upwelling causes changes in water temperature fields at the western shore, which chiefly involve the epilimnion. However, south winds of the same velocity (4–8 m/s) result in Ekman nearshore upwelling with considerably changed velocity and water temperature fields. In addition to the known features identified by the photographic images of streaks in ENU related to seiches, we have found meanders with diameters up to several kilometers and eddies which had separated from the ENU front (Naumenko, 1994; Naumenko and Karetnikov, 1994). Moreover, the formation of new upwelling which forces the old water masses away from the shore was recorded. The old ENU is succeeded by a new one with a period less than synoptic period. Data of long-term observations (1965–1992) in Lake Onega were used to describe more than 50 cases of ENU in different zones of the lake. Sometimes they are recorded simultaneously. The observations and modelling results show that no ENU occurs in the lake in the spring and autumn thermal bar period, and there is usually no total anticlockwise water circulation in the period of complete thermal stratification. The role of wind and JEBAW (joint effect of baroclinity and wind) in the formation of circulations is more significant than the role of JEBAR (joint effect of baroclinity and bottom relief). It is much easier to diagnose the state of water body on the basis of temperature and velocity using observations of these parameters on the surface. Still, the fields of water temperature obtained by remote sensing techniques alone cannot provide definite conclusions. This requires data on wind velocity and direction both during measurements and in the proceeding synoptic situation. With sufficiently strong north and south point winds more than 5 m/s the identified ENU zones can also be used to estimate integral water circulation, whereas the correlation of the fields of water temperature on the surface and in the water masses is not observed at low winds or under calm condition.

2.7.1.5   Frontal zones, thermal bar in lakes

A typical characteristic of lakes is uneven distribution of water masses, the presence of numerous frontal zones with high water temperature and electrical conductivity gradients, large amounts of various debris, foam, slicks. These result in considerable effects related to the presence of jets in the front zone, cellular circulations, vertical connective overturns. According of Fedorov's (1983) classification describing several types of fronts: those of runoff, thermal bar, upwelling, microscale circulations, convective cells are mainly defined as microscale fronts. The features of runoff fronts depend on river discharge, bottom slope and differences in the characteristics of water in the river and lake. In large lakes this front spreads several kilometers into the lake. Observations show that its width in Lake Onega is 100–200 m. This front can be identified using both contact measurements and data of remote IR imaging as well as the visible band of the spectrum made by MSU-SK and MSU-E scanners from a "Kosmos-1939" satellite. Observations carried out within the "Onega" experiment indicate prolonged (more than the synoptic period) life of runoff fronts. Csanady (1977) hypothesized a mechanism of discontinuity mixing in a lake including runoff discontinuities. We have demonstrated the possible changes in discontinuities of three scales: macroscale—related to the inflow of water from tributaries; mesoscale—of the order of several hundred meters to kilometers; and microscale—meters. Microscale discontinuities are transported practically unmodified by mesoscale discontinuities, which in their turn only deform the boundary of the runoff front. Changes in meteorological situation in the lake cause gradual migration of the front boundary, upwelling, thermal bar, river water plume, resulting in the appearance of meanders with horizontal dimensions from several hundred meters to several kilometers, and eddies which had separated from the front (Naumenko, 1994). At the spring thermal bar mesoscale eddies with clockwise rotation are formed in the stratified zone, eddies with anticlockwise rotation form in the homogenous lake zone. In the past it was impossible to identify and describe such eddies in lakes using the data of relatively sparse contact surveys. Up to now we have managed to identify the eddies with the help of spatial thermal aerial photography by the thermal mapper "Vulkan". Thermal fronts of upwellings are generally formed as a result of the thermocline rising to the surface. The distance of the upwelling zone from the shore is several kilometers; the frontal zone dimensions are several dozen meters. Jets with a velocity of 10–50 cm/sec are observed in the vicinity of the front boundary. The front boundary in this case can be identified by remote sensing using MSU-E scanner in the visible region of the spectrum, and from SAR data. The thermal bar front usually consists of several local fronts with a distinct foamy boundary on the lake surface, with corresponding color and brightness characteristics (Fig. 2.49).

Langmuir streaks (circulations) 30–50 m long and 5–10 m wide are observed in the stratified zone between the shore and the front. Ripples are not observed in this zone even if there is wind. In the stratified zone wind-driven waves (Beaufort scale 1–2) and ripples are seen on the surface. Foam is not always present on the lake surface in the front–thermal bar zone, though IR radiometer records a sharp temperature gradient between cold and warm zones. A foamy boundary emerges at certain ratios of air and lake temperature, which results in adiabatic processes in the front zone, intensive bubbling along the front boundary and formation of foam as the bubbles collapse (looks like a kind of cold boiling). Circular cells sized 20–50 m distinguished by color and ripple are observed in the thermo-inert (homogeneous or non-stratified) zone. These cells are probably convective heating cells. It is also possible that the zone may contain small eddies penetrating from the warm to the cold zone. Where the foamy boundary disappears the circular cells elongate acquiring the shape of curves, integrals, and represent the intrusion of warm water into the cold, relatively homogenous zone of the lake. Fronts in the stratified lake part are more distinctly visualized due to more vivid expression of wind streaks, Langmuir circulations on the lake surface. The stripes may also run perpendicular to the front. Secondary streaks which are practically never found in the cold zone appear in the stratified region during transition through a thermal bar (see Fig. 2.49).

Latitudinal differences in sunshine duration over the Lake Onega water area only start to appear in April, while in December–March the index is almost equal for the

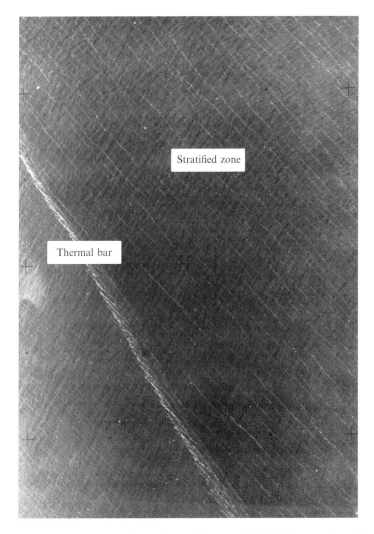

Fig. 2.49. Air photo of Lake Ladoga from airborne laboratory AN-28 during the thermal bar.

northern and southern parts of the lake. The amount of solar radiation reaching the underlying surface depends both on geographical latitude and on the specific features of atmospheric circulation processes. A large waterbody like Lake Onega transforms these processes creating local circulation which manifests itself near the shore causing frequent changes in wind direction. We have used the data from meteorological posts as well as satellite, ship and actinometric observations to characterize the specific features of this local atmospheric circulation. Quite clearly seen is the local circulation demonstrated in the scanned images from the Resource-01 (MSU-SK) satellite for Lake Onega (Fig. 2.50).

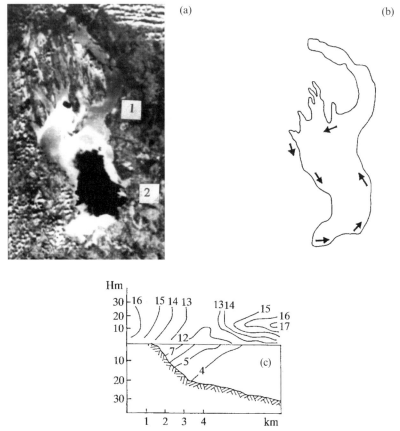

Fig. 2.50. Manifestation of thermal bar and frontal zone in Lake Onega. Satellite image from "Kosmos-1939", 12 June 1987. Ch. 0.6–0.7 μm MSU-SK (a), local atmospheric circulation calculated from data of hydrometeorological stations (b), distribution of water and atmospheric temperatures (c).

2.7.2   Microscale phenomena

2.7.2.1   *Wind-driven waves*

Wind-driven waves on the lake surface—the motions most open to direct observations and measurements out of all other types—are nevertheless insufficiently studied. They are characterized by irregular, stochastic nature. The parameters of the waves largely depend on the configuration, morphometry of the lake, stability of the over-lake wind field. Waves obey the main hydromechanical laws, and the description of the major regularities of waves should be made within the limits of the theory of statistical hydromechanics (Monin and Jaglom, 1968). As for the process of wave generation and development a common theory has not yet been developed. This process is rather complicated in lakes, where the effect of islands, shores, shallow water is added to limited space for acceleration length and its vagueness. Therefore, waves on the surface of even the deepest lakes are never well developed.

Initially, capillary waves appear. As the wind grows stronger gravity waves are generated, which in their turn change their irregular three-dimensional appearance to a two-dimensional one under sufficiently prolonged action of wind. As the wind attenuates ripple-free waves are generated in larger lakes. If the action of wind at a given stable speed is long enough the growth of waves ceases, the "saturation" of waves is observed. Experiments on the lakes have shown that this phenomenon takes place in lakes when the ratio of velocities of waves and wind is 0.6. Further strengthening of the wind will cause part of the energy to be transmitted to promote the increase in progressive motion of the waves. Waves in the nearshore zone are far more complicated and varied than in the deep-water zone. Waves are divided into several sections according to the nature of disturbance. At some critical depth ($H_{cr} = 2h$, where h is average height of the group of waves) waves begin to transform. In the breaker zone with the depth $H_{ovt} < H_{cr}(H_{ovt}0.65H_{cr})$ the crests of the largest waves "overturn" forming breakers, and in the overwash littoral zone the waves ultimately break down. Wave period in the shallow water up to the breaker zone reduces when compared with the deep-water zone. The indices of wind-driven waves change notably under the effect of nonstationary and non-uniform air flow. The theory of waves in the nearshore zone is not yet fully developed. The pool of instrumental observations of waves in the nearshore zone is also insufficient. The theory of linear waves (or finite amplitude waves) in deep water is nearly completed, whereas the nonlinear theory of finite amplitude waves still has questions to answer, especially in what concerns the study of interactions between finite amplitude waves. There still exists no orderly theory of wave transformations in the nearshore zone, and linear approximation is not applicable in this case. Finite amplitude waves on a sloping bottom are nonstationary, complex due to the accompanying currents. The most typical feature of wind-driven waves in lakes is their irregular, seemingly chaotic nature. Wave generation under relatively stable conditions consists of a sequence of waves with what seems to be irregularities. We will further give an example of wind-driven waves observed on Lake Onega. During the "Onega" experiment wave records were made and simultaneous remote sensing by a side-scan radar "Nit" from an Il-18 aircraft carried out (for a description of the aircraft laboratory see Matishov *et al.*, 1993). The interpreted data revealed that the wind disturbance was a nonstationary, non-uniform probability process. To make the analysis for short-time spans easier we will accept that the process is ergodic, stationary. We show section of statistical characteristics along the direction of wind. The moments of $x(t)$ process distribution are then expressed through the correlation function and the function of spectral density (Fig. 2.51).

Moreover, wavy surface is defined as a set of visible waves; wave as a geometric object described by a system of random values (height h, length L, period T, crest length l), and the multivariate functions of the distribution of these values are considered. The spectrum of wind-driven waves $S(f)$ is well approximated by a number of known dependencies also typical for inland seas with limited space of acceleration distance. As the wind attenuates damping of wind-driven waves and formation of free waves, ripple begins. Total energy of waves decreases, its attenuation proceeding proportional to frequency. The spectrum maximum shifts

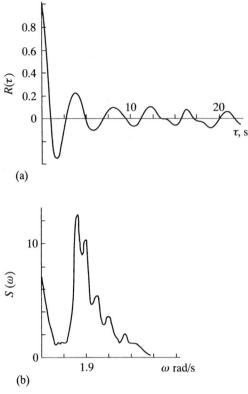

Fig. 2.51. Auto-correlation function (a) and spectra (b) of surface wave in offshore zone of Lake Onega. Calculated data from airborne observations.

into low-frequency range. Equilibrium and transitional regions gradually disappear from the frequency spectrum with only the maximum remaining. Swell in lakes has been studied rather thoroughly on Lakes Onega and Ladoga (Rogkov and Trapeznikov, 1990). Swell is typical of the weather shore of lakes. Both local swell, appearing when wind velocity decreases in the space of acceleration distance, and swell which had migrated from other lake areas are observed. The indices of swell waves depend on proceeding wind-driven waves. The distribution of wave height and especially period deviates from Weibull's asymptotic law. The regularities of these deviations are yet to be studied. Distribution functions of wave elements experience the effect of background hydrometeorological conditions within one season and across years. The properties of wave regime in lakes are so far poorly studied. Roughness intensity in lake gulfs and bays is lower, the probability of waves larger than 2 m is 6%. The calmest months are May to July. The probability of waves larger than 2 m is then only 2.5%, of waves 0.5 m high and lower 58%. The probability of waves with height 4 m in this season is less than 0.3%. In November the probability of waves with the height 7 m is as low as 0.3%. Stratification of the lowest atmospheric layer above the lake surface in this period is highly unstable, and the growth of waves due to wind strengthening requires less time than in the spring–summer period. Reliable approximations of the functions of multiannual distribution of individual waves have not been obtained for lakes due to the lack of long-term instrumental observations.

Spatial spectrum of waves in Lake Onega has been estimated using remote-sensing data from a radar (SAR). In this technique the scattering of electromagnetic waves is determined by irregularities resembling radiation waves, i.e. by capillary and high- frequency waves. The radio signal is much influenced by foam on the surface, a–disperse mixture of air and water bubbles differing from the foamless surface in dielectric properties. Bubbles and foam form due to air trapping at high turbulence level on the surface. The mechanisms of bubble formation and retention are different in marine and fresh water, and play an important ecological role in the surface layer of a waterbody aerating it. Experimental data show that bubbles are 1.5 more quickly destroyed in freshwater than in salt water. Hence, foam and whitecaps also disappear more quickly in lakes than in the sea. With regard to the formation speed and life period of whitecaps Monin and Krasitsky (1985) calculated that given equal external conditions: wind velocity, space of acceleration distance, the state of the lowest atmospheric layer above the water surface, the area occupied by foam will be about 1.5 times smaller in the lake than in the sea; in areas rich in phytoplankton the area is larger than in plankton-poor oligotrophic areas. It is important to be aware of these factors when studying the characteristics of wind-driven waves in lakes using standard means. An interesting feature of wind-induced agitation intrinsic to lakes is observed in the period of thermal bar in the lake. At north winds during the spring thermal bar period in Lake Onega the direction of currents is observed to coincide with the direction of agitation they have evolved in the stratified zone, whereas in the homogenous zone this coincidence is not recorded. In addition to this, roughness contrasts recorded both by the SAR data, and in the visible region of the spectrum obtained from an aircraft and satellite are observed. When the directions of wind-driven waves and currents coincide the slope of capillary and gravity waves reduces, while in the homogenous lake area the surface roughness is different. This situation enables identification of the front zone–thermal bar. Moreover, the lake surface in the front zone becomes even smoother due to the accumulation of surface-active film, allowing recording of meanders, eddies and other phenomena in the zone. The least studied region of the spectrum of surface waves in lakes is the high-frequency zone of wind-induced agitation and capillary waves. It is these motions that are of considerable interest for proper application of remote sensing techniques. Unfortunately, thorough investigation of surface waves in lakes are at present conducted only on the Great Lakes under the supervision of M. Donelan (private communication).

Wind streaks on the lake surface

Multi-zonal survey data show that a great variety of wind streaks with distances between them from centimeters to several meters appear on Lake Onega's surface at different hydrometeorological conditions. They were shown to appear in the open lake in the cool zone (homogeneous in vertical) behind the thermal bar, on the thermal bar and in the warm (stratified) lake zone, in the impact zone of the river flow at different depths and different ratios of water and air temperature. It was not our intention in this book to find out the genesis of the phenomena the streaks have revealed on lake surface. Our task is to demonstrate their diversity under various hydrometeorological conditions. More information is needed to determine the appurtenance of the streaks to certain phenomena, and to study the characteristic features of other phenomena occurring in the lake. Monin and Krasitsky (1985) have proved that not all streaks on the lake or sea surface are connected with Langmuir circulations. Our data relate streaks either to Langmuir circulations, convection, or other phenomena. Filatov *et al.* (1981) have shown that the distance between the main Langmuir streaks roughly corresponds to the depth of the upper mixed layer or epilimnion. Observations on Lake Ladoga recorded no such correspondence under the considered hydro-meteorological conditions. First, the distance between the streaks reduces considerably when the wind is blowing on-shore. However, aerial photography data from a small lake (less than 1 km) indicate that the streaks observed on the lake surface and heading for the shore tend to "meet" the shore without changing the distance between them. The streaks on the surface of a small lake show the following regularity. Wind streaks appear on just one part of the lake, the weather half. A distinct boundary dividing the lake into two parts runs across the middle of the lake. One of the parts has wind streaks at several-meter distance from each other, the other part only shows ripple. The reason why the streaks appear in just one part of the lake though the wind velocity over the lake is invariable is that the cold surface water layer is driven to the southern part of the lake, and that the difference between water and air temperatures varies over the lake water area. This situation is characteristic of Lakes Onega and Ladoga. The distance between the streaks in the nearshore stratified zone is of the order of several meters, which corresponds to the thickness of the epilimnion. The streaks are clearly seen at both "onshore" and "offshore" blowing winds, and are practically indiscernible in photographs of the cool zone (Fig. 2.49). Lake surface in this zone shows ripple and less expressed streaks, while secondary streaks are absent from the zone (Fig. 2.49). A small lake with an island in the middle has streaks both in the weather and lee parts. The streaks in such a lake run into the island bank in one part, and reappear immediately behind the island. The width of the streaks in front of and behind the island is the same. Streaks "in front of " Sukho Island situated in the open offshore zone of Lake Ladoga (Fig. 2.52a) are order of 50–100 m long with about 0–20 m between the main streaks.

When approaching the island the streaks "disarrange", the distance between them reduces abruptly to several dozen centimeters. By the island bank the streaks disappear leaving a homogenous zone. The distance between the streaks "behind" the island grows reaching its former values of several dozen meters. Curiously, the streaks in front of and behind the island run under different angles of deflection in relation to wind. The island apparently changes the direction of currents, the angle between the streaks and wind changing as well. In the nearshore zone with the wind blowing "onshore" the streaks are parallel to the shore, the distance between the

Fig. 2.52. Wind streaks (Langmuir circulations): near Sukho Island in Lake Ladoga.

main streaks decreases from 5–20 to 1 m, the length of individual streaks from 50–100 to 5–10 m. When the streaks pass the reedy zone in the shallow water area their typical dimensions do not change or change direction in relation to wind. Wind streaks are also observed in rivers, though only in some stretches rather than throughout the course. The current direction in the river does not coincide, but corresponds to wind direction (Fig. 2.52). Streaks in the open lake deflect 10–20° to the right of the wind direction. At certain viewing angles visually cellular structure is observed between the streaks. Cells in the form of Benard cells seen in lakes often occur in freshwater, as well as the streaks denoting Langmuir motions, which are elongated along the wind. The major formation mechanism of such streaks is probably connected with the thermocapillary effect in lakes. The cells are usually hexagonal in shape.

2.7.2.2   Slicks, patches, spots, ship trails

Slicks are defined as smoothed surfaces generated by passing ships, internal waves, connective motions, eddies of various size, direct action of wind or oil film. Slicks generated by internal waves in the lake as well as in the ocean are shaped as streaks; those generated by convection have the cellular structure of streaks, integrals. Patches on lake surface may be of various origins. Thermal scanner and IR radiometer have recorded non-uniform temperature of the surface of Lake Onega. The patches were caused by the degradation of the cold film on the surface in the summer time. The thin film, or skin layer is rather stable and cannot be broken in the ocean even at high wind up to 10 m/sec. The skin layer on lakes should be stable just like foam on lake surfaces. When the skin layer is broken cellular structures with dimensions up to several meters, which are the manifestations of thermics on the surface, can be seen from an aircraft or ship. In addition to these patches there are also patches caused by the destruction of surfactants, oil film. Simultaneously with visual observations of Lake Onega from an aircraft the structure of waves behind ships was photographed. These trails had the same characteristics as in the ocean,

Fig. 2.53. The structure of waves behind a ship on Lake Onega.

and were shaped as a system of longitudinal and transverse waves. The angle of longitudinal waves did not depend on the ship speed and size, making up 39° (Fig. 2.53).

Ship trails are quite easily discernible when observations are conducted from the "Almaz" satellite equipped with synthetic aperture radar (SAR). The lake surface is smoothed by turbulence, but the trail exists for several more hours. Moreover, ship trails are rectilinear, and do not meander unlike oil spills on the lake surface.

2.8   LAKE HYDROCHEMISTRY AND HYDROBIOTA (LAKE LADOGA AS AN ESAMPLE)

2.8.1   Seasonal and spatial distributions in the lake hydrochemical and hydrobiotic constituents. Consequences of anthropogenic influences.

2.8.1.1   Total phosphorus and nitrogen

In spring, the southern rivers (Volkhov, Svir, and Syas) account for more than 70–80% of the total phoshorus and ~60% of the total nitrogen input into Lake Ladoga

(Lake Ladoga, 1992). Consequently, the southern portions of the lake proper are more nutrient-rich than other portions of the lake. For instance, in the early 1980s, vernal concentrations of total phosphorus varied from 30 µg/l to 40 µg/l, occasionally reaching maximum values as high as 70 µg/l, whereas concentrations of total nitrogen averaged 0.7–0.9 mg/l. Further distribution of total phosphorus and nitrogen in Lake Ladoga is controlled by the development of the thermal bar. Gradually, waters rich in both total phosphorus ($C_{P_{tot}}$) and total nitrogen ($C_{N_{tot}}$) being entrained by a strong anticylonic transport begin spreading northwards, attaining first central and then northern regions of Lake Ladoga. The $C_{P_{tot}}$ and $C_{N_{tot}}$ levels therein acquire values of ~20–30 µg/l and 0.6 mg/l, respectively.

During the summer cyclonic circulation-driven currents effectively transport nutrients from north to south resulting in lake-wide reductions in the amounts of both $C_{P_{tot}}$ and $C_{N_{tot}}$. Average values of $C_{P_{tot}}$ and $C_{N_{tot}}$ remain at ~30–40 µg/l and ~0.5–0.6 mg/l, respectively.

Strong autumnal convective and wind-driven mixing generates a further reduction and a further homogeneity in the concentrations of both total phosphorus and total nitrogen. Analyses of multi-year data (Raspletina, 1992) illustrate a slow decrease in $C_{P_{tot}}$ and $C_{N_{tot}}$ in near-surface waters:

—1980–1985 average annual $C_{P_{tot}}$ decreased from ~32 µg/l to ~24 µg/l; average annual $C_{N_{tot}}$ decreased from ~0.86 mg/l to ~0.73 mg/l,

—1987–1989 average annual $C_{P_{tot}}$ decreased from ~23 µg/l to 20–21 µg/l; average annual $C_{N_{tot}}$ remained about the same level;

—1981–1984 average annual $C_{P_{tot}}$ in the Neva River outfall diminished slightly from 28 µg/l to 26 µg/l; average annual $C_{N_{tot}}$ in the Neva River outfall increased from 0.66 mg/l to 0.74 mg/l.

In 1987–1989, the mean phosphorus content in Lake Ladoga waters was ~21 µg/l in the pelagic zone and 32 µg/l in the littoral zone. In 1992–1993, the $C_{P_{tot}}$ values ranged from 15 to 29 µg/l. The $C_{N_{tot}}$ values remained over these years at the level of the mid-1980s.

2.8.1.2 *Dissolved organic matter (doc)*

Spring runoffs and floods enrich coastal waters, particularly in the southern regions of Lake Ladoga. The thermal bar inhibits the offshore spread of carbon-rich water. However, as the lake approaches stratification and the thermal bar retreats to the pelagic region, the onshore/offshore *doc* distribution contrast becomes much less severe. It should be noted that during the summer the pelagic central waters of Lake Ladoga display both elevated epilimnetic and hypolimnetic values of C_{doc}. Currents entrain and transport organic-rich waters northward along the eastern coastline.

In autumn, the spatial distribution of dissolved organic matter becomes more uniform. From Tregoubova and Koulish (1992), the historical evolution of inter-annual variations in mean Lake Ladoga concentrations of *doc* was as follows:

(a) early 1960s to late 1970s: mean concentrations 8.2–8.5 mg C/l; range of seasonal and interannual variations <10–12%;

(b) early 1980s: seasonal and interanual variations increased, spring C_{doc} ~4 mg C/l, summer near-surface values ~11 mg C/l;

(c) 1983–1984: mean concentrations rose to 9.5–9.8 mg C/l; further increases in interseasonal and interannual variations;

(d) 1986–present: mean concentrations gradually reduced to 8.0 mg C/l, a consequence of reduced summer values of C_{doc}. Interseasonal and interannual variations are still <10–12%.

2.8.1.3   Trace metal ions

Chernykh (1987) reported the trace element ion concentrations in Lake Ladoga water averaged over the years 1982–1985 to be:

Iron: $30\,\mu g/l \leq C_{Fe} \leq CFe \leq 1150\,\mu g/l$ with $\langle C_{Fe} \rangle = 156\,\mu g/l$
Aluminum: $10\,\mu g/ll \leq C_{Al} \leq 470\,\mu g/l$ with $\langle C_{Al} \rangle = 66\,\mu g/l$
Magnesium: $0.5\,\mu g/ll \leq C_{Mg} \leq 97.5\,\mu g/l$ with $\langle C_{Mg} \rangle = 6.8\,\mu g/l$
Copper: $1\,\mu g/ll \leq C_{Cu} \leq 33\,\mu g/l$ with $\langle C_{Cu} \rangle = 6.6\,\mu g/l$
Lead: $0.5\,\mu g/ll \leq C_{Pb} \leq 6\,\mu g/l$ with $\langle C_{Pb} \rangle = 1.4\,\mu g/l$
Cobalt: $C_{Co} < 0.5\,\mu g/l$ throughout the time interval.

Water column Fe, Al, and Mn ions are most abundant in the littoral zones of Lake Ladoga, zones that are most vulnerable to riverine inputs. Fe, Cu, and Pb ions are most abundant in the central and northern regions of the lake. The seasonal dynamics in metal ion concentrations in the lake waters exhibit distinct decreases during the vegetation growth season due to enhanced uptake by aquatic organisms. However, generally, planktonic Fe is most abundant in spring rather than summer. Aquatic organisms indigenous to Lake Ladoga can accumulate trace-element concentrations exceeding water-column concentrations by 5 to 6 orders of magnitude (Zaytseva, 1987).

The highest concentrations of trace metals are found in bottom sediments where C_{Fe}, C_{Cu}, and C_{Pb} exceed their columnar counterparts by factors of several thousands.

As well as the above trace elements being absorbed by benthic and planktonic organisms, various species of mollusks uptake additional trace elements such as tin, silver, zirconium, cobalt, gallium, and vanadium. Their shells are rich in manganese, copper, nickel, and chromium. Strontium concentrations are consistently high in the tissues of mollusks indigenous to all regions of Lake Ladoga.

2.8.1.4   Suspended matter

From longitudinal transects conducted between 1979 and 1984, Yudin (1987) reported that mean annual concentrations of suspended matter (*sm*) varied from $0.8\,g/m^3$ to $1.6\,g/m^3$. Highest values of C_{sm} are found in the shallow southern and near-central regions of the lake as well as in some northern fjordic regions. In spring, the shallow southern regions, receptacles for Volkhov River discharge, contain values of CSM as high as $6–8\,g/m^3$.

In spring, 50–60% of total *sm* is organic in all parts of Lake Ladoga with the exception of the southern region where 55–70% of the total *sm* is inorganic.

In summer, the southern regions do not undergo significant changes in surficial C_{sm}. However, the rest of the lake displays enhanced surficial C_{sm} values up to $\sim 2 \, g/m^3$.

In autumn, C_{sm} decreases down to $\sim 5 \, g/m^3$ in the southern and $< 1 \, g/m^3$ in central regions.

2.8.1.5   Algal fungi

Being a natural component of the aquatic ecosystem, algal fungi, along with bacteria, take part in mineralization of organic matter which constitutes the exclusive nutrient resource for these colourless chlorophyll-free lower plants consisting of assemblages of cellular filaments called hyphae. Depending on the nutrient substrate (living or dead) algal fungi fall into two groups: parasites and saprotrophs.

The heterotrophic nature of algal fungi makes them efficient indicators of water-body ecological state: the bulk water-spore concentration level proves to be an indicator of the lake trophic status, and an intense growth of mycoflora along with priority changes in fungal community composition point unmistakably to the steadily developing anthropogenic eutrophication of the water body.

Thirty-two species have been identified in the algal fungi community indigenous to Lake Ladoga (Iofina, 1992). These represent five classes: Hyphochytridiomycetes, Chytridiomycetes, Oomycetes, Zygomycetes, Dentoromycetes (24 genera in sum). Species of taxonomic groups of Oomycetes and imperfect fungi prove to be dominant in the lake algal fungi community. Most numerous are imperfect fungi, specifically, the Penicillium genus. They are rich in yeasty organisms ($\sim 9\%$). A large fraction of algal fungi in the lake is constituted by saprophytes except for three types of fungi which are parasites: *Olpidiopsis* sp., *Rhizophidium planktonicum*, and *Woronina polycistis*.

The number counts of algal fungi, highly variable throughout the lake in all seasons except the summer, are almost independent of abiotic factors. However, in summer there is a distinct dependence of fungi content on water temperature and organic matter availability. While in deep oligotrophic lakes, the spatial distribution of fungi exhibits elevations in the coastal and near estuarial areas, in Lake Ladoga large concentrations (e.g. the number counts were registered at 6–7 thousand diaspores per liter in the summer of 1986 and 32.5 thousand diaspores per liter in the spring of 1988) of dominant genera of algal fungi occurring in deep parts of the basin are believed to be unambiguously indicative of at least sustaining or developing anthropogenic eutrophication.

2.8.1.6   Bacterioplankton

In the early 1960s, the Lake Ladoga bacterioplankton were almost uniformly distributed over the entire lake area, the only obvious enhanced microbial densities being recorded in the vicinity of the Priozersk pulp-mill (in the northwestern coastal zone of the lake). However, by the mid-1980s the microbial densities had increased substantially (Kapoustina, 1992), the most significant occurring along the southern coastal zone where the bacterial counts increased by a factor of three.

The springtime distribution of Lake Ladoga bacterioplankton almost invariably follows the three-dimensional field of phytoplankton distributions in the near-surface waters: the onset of algal growth and the maximum bacterial count ($\sim$1.5–2.0 million cells/ml) are restricted to shallow coastal zones of Lake Ladoga where the water temperature is relatively high and the nutrient inputs from surface runoff and riverine discharge are closely confined to the shore by developing lake stratification and thermal bar. The pelagic bacterial counts are, of course, considerably lower ($\sim$0.01–0.1 million cells/ml), being controlled by water temperature (generally, around 1.3–1.5°C during early spring).

In summer, the spatial distributions of bacterioplankton are basically configured by hydrophysical patterns and lake morphology. In June–July, the bacterioplankton counts within the coastal zone drop down to about 0.65 million cells per ml. In late summer, the bacterioplankton distribution returns to the spring pattern with high concentrations (10–20 million cells/ml) in the southern shallow regions and considerably lower concentration values throughout the central (deep) areas (1–5 million cells/ml).

In autumn, the cell counts usually decrease and acquire a more or less homogeneous spatial distribution over the entire lake: in the epilimnetic layers, the microbial concentration could be about 0.6–0.7 million cells per ml, and in the hypolimnion, it is as low as *ca.* 0.4 million cells per ml.

Interannual variations in the bacterioplankton concentrations averaged over the epilimnion indicate a substantial rise from 0.4 million cells/ml in 1977 to 0.85 million cells/ml in 1987. A comparable hypolimnetic bacterioplankton increase was observed, *viz.* from 0.2 million cells/ml in 1977 to 0.5 million cells/ml in 1987. These bacterial population increases are taken to be a direct consequence of lake eutrophication.

In general, bacterioplankton production in lakes is lower than phytoplankton production by a factor of 3 to 5. However, in Lake Ladoga both productions tend to be numerically very close. The proportion of organic matter that is decomposed by bacterioplankton in surficial Lake Ladoga waters varies from 25% to 94% with a mean value of about 70%, characteristic of mesotrophic water bodies. This strongly suggests that aquatic micro-organisms mineralize more organic matter than is photosynthetically produced. During the period 1981–1986, the bacterial decomposition rates in Lake Ladoga exceeded primary production rates by a factor of 1.5 to 3.7. It is interesting to note that a similar condition is exhibited even more dramatically by the other great lake of northern Europe—Lake Onega. In Lake Onega, bacterial decomposition rates are 15 times higher than primary production rates (Sorokin and Fedorov, 1989).

2.8.1.7   Phytoplankton

The phytoplankton population of Lake Ladoga in the late-1960s was not typical of other large, cold, oligotrophic temperate lakes. While the number of phytoplanktonic species indigenous to such lakes is very large, the number of most abundant species (concentrations exceeding 100 000 cells per litre) is comparatively small and its composition fairly stable. For example, pre-1970 data presented by Petrova (1990) reveal the presence of 380 planktonic species in Lake Ladoga, the most abundant and diverse being, not unexpectedly, diatoms, greens, and blue-greens.

During the last two decades, however, shifts in the phytoplankton populations have resulted from anthropogenic eutrophication of Lake Ladoga. Again from Petrova (1990), groups of algae indigenous to Lake Ladoga are:

(i) dominant and subdominant species which, irrespective of anthropogenic eutrophication, retain their role in the phytoplanktonic community: *Anabaena circinnalis*, *Aphanizomenon flos-aquae*, *Woronichinia naegeliana* (Cyanophyta), *Aulacosira clistans* var. *alpigena*, *Aulacosira islandica* ssp. *helvetica*, *Stephanodiscus niagare* (Bacillariophyta), *Tribonema affine* (Chrysophyta), *Sphaerocystis schroeteri* (Chlorophyta);

(ii) species rapidly becoming dominant following the onset of anthropogenic eutrophication: *Anabaena flos-aquae*, *Microcystis aerugenosa*, *M. pulverea* ff., *Oscillatoria bornetti* f. *tenuis*, *Oscillatoria rubescens*, *Oscillatoria tenuis* (Cyanophyta), *Aulacosira granulatum* var. *granulata*, *Cyclotella mineghiniana*, *Diatoma elongatum* var. *elongatum*, *Stephanodiscus binderanus*, *Stephanodiscus hantzschii*, *Stephanodiscus tenuis* (Bacillariophyta);

(iii) species receding from the group of dominants following the onset of anthropogenic eutrophication: *Dinobryon* spp. (Chrysophyta), *Atthea zachariasii*, *Rhizosolenia* spp., *Cyclotella bodanica*, *Cyclotella stelligera* (Bacillariophyta).

Lake Ladoga displays rather complex seasonal dynamics in taxonomic phytoplankton composition within a restricted spectrum of dominant phyta, *viz.* Cyanophyta (blue–greens), Chrysophyta (yellow–greens), and Chlorophyta (dark- greens). Diatoms, which generally are associated with waters of relatively high clarity (and impart a yellowish-green hue to large oligotrophic lakes), are present in Lake Ladoga throughout the year. However, their concentrations are overwhelmed in summer by Cyanophyta, Chrysophyta, Chlorophyta, and dinoflagellates (in particular the reddish-hued Pyrrophyta). Thus, relative clarity in the pelagic regions of Lake Ladoga is confined to spring and fall. Table 2.8 lists the principal seasonal plankton algal communities indigenous to Lake Ladoga throughout the time interval 1975–1990 (culled from Petrova, 1990).

As a consequence of (a) the closing of a principal polluting pulp mill and (b) the imposition of stringent phosphorus-limiting legislation regulating agricultural practices, the phosphorus input to Lake Ladoga was dramatically curtailed in the early 1990s.

The ecological status of Lake Ladoga could, perhaps optimistically, be considered as exhibiting tentative signs of improvement, and that a new wave of phytoplankton succession may have been triggered.

The current number counts of Lake Ladoga algal cells in the top two meters of the

Table 2.8. Principal seasonal phytoplankton community compositions in Lake Ladoga, 1975–1993.

Spring	Summer	Autumn
Aulacosira (Bacillariophyta)	*Asterionella* (Diatoms)	*Aulacosira/Tribonema*
Aulacosira/Asterionella	*Asterionella*	*Aulacosira/ Gloethila*
Aulacosira/Diatoma	*Asterionella/Tribonema*	*Stephanodiscus* (Binderanus)
	Tribonema/Uroglenopsis	*Oscillatoria/Aulacosira*
	Tribonema/Stephanodiscus (Binderanus)	*Woronichina/Tribonema*
	Tribonema/Peridinea, Fragilaria, Oscillatoria	*Aphanizomenon, Asterionella/Oscillatoria*
	Aphanizomenon, Mycrocystis	

water column vary widely throughout the seasonal cycle, averaging from several hundred thousands (central deep-water region of the lake) to several millions per litre (coastal zone).

Apart from species seasonal succession and number counts dynamics, Lake Ladoga algae display both lateral and vertical distribution profiles (dynamics of the upper layer understandably dictated by water temperature, nutrient availability, shoreline location, water transparency, and atmospheric conditions). Figure 2.54 illustrates spring, summer, and autumnal spatial distributions of chlorophyll-*a* concentrations as recorded in May 1981 (Fig. 2.54a), in May, 1982 (Fig. 2.54b), in June 1982 (Fig. 2.54c), in July 1982 (Fig. 2.54d), in August 1982 (Fig. 2.54e), and in September 1982 (Fig. 2.54f).

The spring pattern reflects the division of lake area by the thermal bar (areas encapsulated by the thermal bar and the shoreline are rich in chlorophyll due to higher temperatures and nutritional inputs). In summer, the spatial distribution heterogeneity becomes even more pronounced. However, due to ambient phosphorus concentrations being substantial throughout the lake, nutrient-limiting chlorophyll concentrations are not a factor in the generation of this inhomogeneity. In autumn, algae growth is restricted to middle depth regions of the lake (Fig. 2.54f).

The lake mean phytoplankton production (in units of gC/m^2) over the growth season has displayed substantial variability: 44.9 in 1982; 54.5 in 1983; 20.6 in 1984; 139.5 in 1985; 75.8 in 1986; 38.1 in 1987; 65.3 in 1988; 96.5 in 1989. The total lake seasonal phytoplankton production escalated to 799.9 thousand tons in 1982 and 969.8 thousand tons in 1983. The strong heterogeneity in phytoplankton production, the point sources of nearshore nutrient injections, and the dynamic/thermal properties of the lake have generated severe hypolimnetic anoxia in the coastal zones and moderately severe hypolimnetic anoxia in the pelagic regions.

2.8.1.8   Zooplankton

Zooplankton species diversity is thought to be one of the numerous indicators of water body trophic status (Andronikova, 1996). Table 2.9 unambiguously indicates

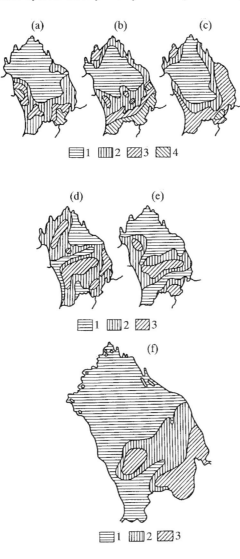

Fig. 2.54. Spatial distributions of chlorophyll-*a* (μg/l) over Lake Ladoga in (a) May 1981, (b) May 1982, (c) June 1982, (d) July 1982, (e) August 1982, (f) September 1982. For (a), (b), (c) hatching 1 indicates *chl-a* <0.5, hatching 2 indicates 0.5–2, hatching 3 indicates 2–5, and hatching 4 indicates >5 μg/l. For (d) and (e) hatching 1 indicates *chl-a* <2, hatching 2 indicates 2–5 , and hatching 3 indicates >5 μg/l. For (f) hatching 1 indicates *chl-a* <0.5, hatching 2 indicates 0.5–2, and hatching 3 indicates >2 μg/l.

that the number of species systematically increases from oligotrophic and acidic dystrophic lakes to mesotrophic and eutrophic lakes. As to the ratios of taxons higher than species or rank (class Rotatoria, order Copepoda, suborder Cladocera), it is well known that quantitatively, Rotatoria and Cladocera become more important whereas the count numbers and biomass of Copepoda significantly recede with increasing trophic level of the water body.

Table 2.9. Number of zooplankton species indigenous to lakes of different trophy (Andronikova, 1996).

Lake trophic status	Number of species	Mean values
Oligotrophic	17–120	46
Mesotrophic	22–128	51
Eutrophic	22–165	58
Dystrophic (acidic)	4–47	15

Spatial inhomogeneity in Lake Ladoga zooplankton distribution is particularly pronounced during the spring warming period. Maximum counts (about one thousand organisms per m^3) and biomass values are generally associated with areas of interfacing/merging pelagic and coastal zone water masses.

Due to strong influences from major input rivers Volkhov, Svir, as well as numerous other eastern and western tributaries, zooplankton residing within the shallow to medium deep regions of southern Lake Ladoga display a high degree of taxonomic diversity. Copepoda, Diatoma, and Cladocera are in abundance, as are numerous other species: *Karatella quadrata*, *Synchaeta* spp., *Kellicotia longispina* are present in concentrations of 30, 200, and 5000 organisms per m^3. The biomass of these zooplanktons in southern Lake Ladoga vary in springtime values from 123 to 390 mg/m^3, whereas eastern coastline values are at 13 to 36 mg/m^3 and central lake values at only 10 mg/m^3.

In summer, the areas of enhanced zooplankton development gradually shift to central regions of Lake Ladoga. For example, peak 1983 mid-lake zooplankton production was ~40 gC/m^2, with ~80% of the central and northern zooplankton populations comprising *Asplanchna priodonta*. Lake Ladoga has displayed substantial zooplankton growth, particularly in the central regions. The main August zooplankton pelagic biomass recorded in 1993 was in the range 6.1–19.4 gC/m^2 (Telesh, 1996).

Attributing all the changes in the recorded ecological status of Lake Ladoga exclusively to anthropogenic eutrophication may not be completely defensible, since these changes have been in concert with the climatic changes of the past decades (although one could suggest the not unreasonable linkage between the climate change and anthropogenic invasion of the atmosphere, see Chapter 1).

2.8.1.9 Ichthyofauna

As is typical of the worldwide plight of ichthyofauna, the combination of anthropogenic invasions of habitat and unreasonably zealous fishing practices has had an impact on fish populations in Lake Ladoga. Spawning locales have been drastically reduced due to floating wood debris within the river systems and the construction of dams. Introduction of modern trawling facilities has resulted in rapid decreases in noble fish populations (e.g. salmon, pike, sig, white fish, trout) (Kudersky *et al.*, 1996). Sig has been declared an endangered and disappearing species. Recently, urgent legislation has been adopted to prohibit the trawling and to return to sweep-net fishing. However, less noble species, such as smelt, are still abundant within Lake Ladoga, and retain their important role in the fishing industry.

As eutrophication progresses, hypolimnetic anoxia becomes more severe. The direct impact of excessive nutrients, toxic chemicals, and oxygen depletion on salmon, sig and other noble species has resulted in the ascendancy of less economically (and possibly environmentally) valuable species that are either less sensitive to oxygen depletion or their habitats are in upper, still well-aerated layers. Detrimental consequences of anoxia have also been felt by the benthic populations, impacts that reverberate throughout the Lake Ladoga food chain.

REFERENCES

Adamenko V.N. 1985. *Climate and Lakes*. Leningrad: Gidrometeoizdat., 264 pp. (in Russian with summary in English).

Adamenko V.N., Kondratyev K.Ya., Pozdnyakov D.V., Chekhin L.V. 1991. *Solar Radiation Regime and Optical Properties of Lakes*. Leningrad: Gidrometeoizdat., 300 pp. (in Russian with summary in English).

Allender J., Saylor J. 1979. Model and observed circulation throughout the annual temperature cycle of lake Michigan. *J. Phys. Ocean.*, **1**, 573–583.

Andronikova I.N. (ed.). 1980. *Eutrophication of Mesotrophic Lake*. Leningrad: Nauka Press, 247 pp. (in Russian).

Andronikova I.N. 1996. *Zooplankton Structural/Functional Organization in Lake Ecosystems*. St. Petersburg: Nauka Press, 190 pp.

Andronikova I.N., Mokiyevskii K.A. (eds.). 1984. *Peculiarities of Water Quality Formation in Different Lakes of Karelian Isthmus*. Leningrad: Nauka Press, 21 pp. (in Russian).

Beer L.P. 1971. Natural and unnatural water temperatures in Zion-Waukegan, Illinois area of southwest Lake Michigan. *Proceedings 14th Conference Great Lakes Research*. Milwaukee: Intern. Assoc. Great Lakes Res., pp. 507–521.

Beletsky D., N. Filatov and R. lbraev. 1994. Hydrodynamics of Lakes Ladoga and Onega. In special issue, *Water Pollution Res. J. Canada*, **29**, No. 2–3, 365–385.

Beletsky, D.V., Yu.L. Demin and N.N. Filatov. 1991. Complex investigation of hydrophysical fields in Lake Onego as an imitating ocean model. *Bull. (lzv.) Acad. Sci. USSR. Atmos. Ocean. Phys.*, **27**, 1172–1182 (in Russian).

Bennett J.R. 1971. Thermally driven lake currents during the spring and fall transition periods. *Proceedings 14th Conference Great Lakes Research*. Milwaukee: International Association of Great Lakes Research, pp. 535–544.

Bennett, J.R. 1973. A theory of large amplitude Kelvin waves. *J. Phys. Oceanogr.*, **3.**, 57–60.

Bennett, J.R. 1974. On the dynamics of wind-driven lake currents. *J. Phys. Oceanogr.*, **4**, 400-414.

Berlyand T.G. 1961. *Distribution of Solar Radiation over the Continents.* Leningrad: Gidrometeoizdat., 227 pp. (in Russian).

Bichkov I.A., Viktorov S.V., Demina M.D. 1989. Using of regular satellite IR images for thermal bar and upwelling studies. *Oceanologia (Russia)*, **29**, No. 5, 759–766 (in Russian).

Birge E.A. 1915. The heat budgets of American and European lakes. *Trans. Wiscon. Acad. Sci. Arts Lett.*, **18**, 341–391.

Birge E.A., Juday C., March H.W. 1927. The temperature of the bottom deposits of Lake Mendota. *Trans. Wiscon. Acad. Sci.*, **23**, 187–231.

Birkett C.M. 1995. The contribution of TOPEX/POSEIDON radar altimeter to the global monitoring of large rivers and wetlands. *Water Resource Res.*, **34**, No. 5, 1223–1239.

Birkett C.M. 1998. The contribution of TOPEX/POSEIDON to the global monitoring of climatically sensitive lakes. *J. Geophys. Res.*, **100**, No. C12, 25179–25204.

Birkett C.M., Mason I.M. 1995. A new global lakes database for a remote sensing program studying climatically sensitive large lakes. *J. Great Lakes Res.*, **21**, No. 3, 307–318.

Bocharov O.B., Vasilyev O.F., Kvon V.I., Ovchinnikova T.E. 1996. Mathematical modelling thermal bar in a deep lake. *Dokl. (Trans.) Acad. Sci. (USSR)*, **349**, No. 4, 530–532 (in Russian; English transl. *Scripta Technica*).

Bogorodskii V.V., Gavrilo V.P. 1980. *Ice. Physical Properties. Modern Methods of Glaciology. Handbook.* Leningrad: Gidrometeoizdat., 384 pp. (in Russian).

Bojarinov P., Palshin N., Petrov M. 1994. Thermal processes in Lake Onega. In special issue, *Water Pollution Res. J. Canada.* **29**, No. 2–3, 403–422.

Brooks J., Lick W. 1972. Lake currents associated with the thermal bar. *J. Geophys. Res.*, **77**, No. 30, 6000–6013.

Budyko M.I. 1956. *Heat Balance of Earth Surface.* Leningrad: Gidrometeoizdat., 255 pp. (in Russian).

Bukata R., McColl No. 1973. The utilization of sun-glint in a study of lake dynamics. *J. Remote Sensing Water Res. Management*, **17**, 351–367.

Carmack E.C., Farmer D.M. 1982. Cooling processes in deep temperate lakes: A review with examples from two lakes in British Columbia. *J. Marine Res.*, **40** (suppl.), 85–111.

Carmack E.K. 1979. Combined influence of inflow and lake temperature on spring circulation in a riverine lake. *J. Phys. Oceanogr.*, **9**, No. 2, 422–434.

Carmack E.K., Gray C.B.J., Pharo C.H., Daley R.J. 1979. Importance of lake-river interaction on seasonal patterns in the general circulation of Kamploops Lake, British Columbia. *Limnology, Oceanography*, **24**, No. 4, 634–644.

Chekhin L.P. 1987. *Light Regime of Water Bodies.* Petrozavodsk: Academy Science Publ., 131 pp. (in Russian).

Chernykh O.A. 1987. Micro-elements distribution over Lake Ladoga. In: *Lake Ladoga: Criteria of the Ecosystem Status.* St. Petersburg: Nauka Press, pp. 80–83.

Church P.E. 1943. The annual temperature cycle of Lake Michigan. 1. Cooling from late autumn to the terminal point, 1941/42. *Inst. Meteorol. Univ. Chicago Misc. Rep.*, No. 4, 48 pp.

Csanady G.T. 1971. On the equilibrium shape of the thermocline in the shore zone. *J. Phys. Oceanogr.*, **1**, No. 2, 263–270.

Csanady G.T. 1972. The coastal boundary layer in Lake Ontario. Part 1: The spring regime. *J. Phys. Oceanogr.*, **2**, No. 1, 41–53.

Csanady G.T. 1977. Intermittent full upwelling in Lake Ontario. *J. Geophys. Res.*, **82**, 397–419.

Csanady, G.T and G.T. Scott. 1974. Baroclinic coastal jets in Lake Ontario during IFYGL. *J. Phys. Oceanogr.*, **4**, 524–541.

Data Book World Lake Environments. 1988–1993. Otsu: ILEC/UNEP Publ., Vols. 1–4.

Demin, Y.L. and R.A. lbraev. 1989. A numerical method of calculation of currents and sea surface topography in multiply connected domains of the oceans. *Sov. J. Numer. Anal. Math. Modelling*, **4**, 211–225.

Demin, Yu.L. and A.S. Sarkisyan. 1977. Calculation of equatorial currents. *J. Marine Res.*, **35**, 339–356.

Dmitrieva A.A., Ryanzhin S.V. 1976. Hydrometeorologic conditions for occurrence of Langmuir circulation. *Leningrad State University Herald*, **18**, 110–117 (in Russian with abstract in English).

Domanitskii A.P., Dubrovina R.G., Isaeyeva A.I. 1971. *Rivers and Lakes of The Soviet Union*. Leningrad: Gidrometeoizdat., 104 pp. (in Russian).

Ehlin U. 1974. Water dynamics in the coastal zone. *9th Nordic Symposium Water Research*. Nordforsk: Miljovardssekretariatet, No 4, pp. 71–97 (in Swedish).

Elliott G.H, Elliott J.A. 1969. Small-scale model of the thermal bar. *Proceedings 12th Conference Great Lakes Research*. Ann Arbor: International Association of Great Lakes Research, pp. 553–557.

Elliott G.H, Elliott J.A. 1970. Laboratory studies on the thermal bar. *Proceedings 13th Conference Great Lakes Research*. Ann Arbor: International Association of Great Lakes Research, pp. 413–418.

Elliott G.H. 1971. A mathematical study of the thermal bar. *Proceedings 14th Conference Great Lakes Research*. Milwaukee: International Association of Great Lakes Research, pp. 545–554.

Faller A.J., Caponi E.A. 1978. Laboratory studies of wind-driven Langmuir circulation's. *J. Geophys. Res.*, **183**, 3617–3633.

Faller A.J., Woodcock A.H. 1964. The spacing of windrows of Sargassum in the ocean. *J. Marine Res.*, **122**, 22–29.

Farmer D., Li M. 1995. Patterns of bubble clouds organised by Langmuir circulation. *J. Phys. Ocean.*, **25**, 1426–1440.

Fedorov K. 1983. *Physical Nature and Structure of Oceanic Fronts*. Leningrad: Hydrometeoizdat., 295 pp.

Filatov N.N. 1983. *Dynamics of Lakes*. Leningrad: Gidrometeoizdat., 161 pp. (in Russian).

Filatov N.N., D.V. Beletsky and L.V. Zaitsev. 1990. Variability of hydrophysical fields in Lake Onega. "Onega" experiment. Petrozavodsk: Water Problems Department, Karelian Scientific Centre Acad. Sci., 112 pp.

Filatov N.N. 1983. *Dynamics of Lakes*. Leningrad: Gidrometeoizdat., 167 pp. (in Russian with English summary).

Filatov N.N. 1991. *Hydrodynamics of Lakes*. St. Petersburg: Nauka Press, 198 pp. (in Russian with English summary).

Filatov N.N., Beletsky D.V., Zaytsev L.V. 1990. Variability of Hydrophysical Fields in Lake Onega. "Onega" experiment. Petrozavodsk: Water Problems Department, Karelian Scientific Centre Acad. Sci., 112 pp.

Filatov N.N., D.V. Beletsky, Zaitsev L.V. 1990. Variability of hydrophysical fields in Lake Onega. "Onega" experiment. Petrozavodsk: Water Problems Department, Karelian Scientific Centre Acad. Sci., 114 pp. (in Russian).

Filatov N.N., Ryanzhin S.V., Zaitsev L.V. 1981. Investigation of turbulence and Langmuir circulation in Lake Ladoga. *J. Great Lakes Res.*, **7**, No. 1, 1–6.

Filatov N.N., S.V. Ryanzhin, L.V. Zaitsev, 1981. Investigations of turbulence and Langmuir circulations in Lake Ladoga. *J. Great Lakes Res.*, **7**, 1–6.

Forel F.A. 1880. La congélation des lacs suisses et savoyards pendant l'hiver 1879–1880, Lac Leman, Geneva. *L'Echo des Alpes*, **3**, 149–161.

Forel F.A. 1892. *Le Leman: Monographie Limnologique*. Tome 1, *Géographie, Hydrographie, Géologie, Climatologie, Hydrologie*. Lausanne: F. Rouge, Vol. XIII, 543p.

Frey D.G. (ed.) 1966. *Limnology in North America*. Madison, Milwaukee and London: Wisconsin University Press, 734 pp.

Gates D.M. 1980. *Biophysical Ecology*. New York: Springer, 320 pp.

Grenfell T.C., Mayakut G.A. 1977. The optical properties of ice and snow in the Arctic basin. *J. Glaciol.*, **18**, 445–463.

Håkanson L., Peters R.H. 1995. *Predictive Limnology. Methods for Predictive Modelling*. Amsterdam: SPB Academic, 464 pp.

Halbfass W. 1922. *Die seen der Erde*., Erganzungsheft, ed. Petermanns Mitteilungen. Gotha: Justus Perthes, No. 185, 150 pp.

Hatchinson D. 1969. *Limnology*. Moscow: Nauka Press, 591 pp. (in Russian).

Henderson-Sellers B. 1984. *Engineering Limnology*. Boston: Pitman, 356 pp.

Herdendorf C.E. 1982. Large lakes of the world. *J. Great Lakes Res.*, **8**, No. 3, 379–412.

Herdendorf C.E. 1990. Distribution of the World's Large Lakes. In: M.M. Tilzer and C. Serruya (eds.), *Large Lakes. Ecological Structure and Function*. Berlin: Springer-Verlag, 338 pp.

Huang J.C.K. 1971. The thermal current in Lake Michigan. *J. Phys. Oceanogr.*, **1**, No. 2, 105–122.

Huang J.C.K. 1972. The thermal bar. *Geophys. Fluid Dynamics*, **3**, No. 1, 1–25.

Hubbard D.W., Spain I.D. 1973. The structure of early spring thermal bar in Lake Superior. *Proceedings 16th Conference Great Lakes Research*. Ann Arbor: International Association of Great Lakes Research, pp. 735–742.

Hudon C.. 1997. Impact of water level fluctuations on St. Lawrence River aquatic vegetation. *Can. J. Fish. Aquat. Sci.*, **54**, 2853–2865.

Hutchinson G.E. 1957. *A Treatise on Limnology*, Vol. 1. New York: John Wiley and Sons, 1015 pp.

Hutchinson G.E., Loffler H. 1956. The thermal classification of lakes. *Proc. Nat. Acad. Sci.*, **42**, 84–86.

Imberger J. 1985. The diurnal mixed layer. *Limnology, Oceanography*, **30**, No. 4, 737–770.

Iofina I.V. 1992. Water mycoflora composition and functionality. In: *Lake Ladoga: Criteria of the Ecosystem Status*. St. Petersburg: Nauka Press, pp. 179–187.

Jerlov N.G. 1976. *Marine Optics*, 2nd edn. Amsterdam: Elsevier, 250 pp.

Jirka G.H., Ryan P.J., Stolzenbach K.D. 1975. Basic physical processes in heat transport. In: *European Course on Heat Disposal from Power Generation in the Water Environment, Delft Hydraulics Laboratory*. Lecture Notes, Part 1, No. 4, pp. 1–131.

Johannessen O.M., Sandven S., Melentyev V.V. 1996. ICEWATCH– Ice SAR Monitoring of the Northern Sea Route. *Proceedings of the 2nd ERS Applications Workshop, London, UK, 6–8 December 1995*. ESA Sp-383, pp. 291–296.

Kapoustina L.L. 1992. Specific features of bacterioplankton time/space distribution in Lake Ladoga. In: *Lake Ladoga: Criteria of the Ecosystem Status*. St. Petersburg: Nauka Press, pp. 146–149.

Karetnikov S.G., Naumenko M.A. 1993. Using of IR satellite imageries for study of Lake Ladoga thermics. *J. Earth Study Space (Russia)*, No. 4, 69–78 (in Russian).

Kenney B.C. 1977. An experimental investigation of the fluctuating currents responsible for the generation of windrows. Ph.D. thesis, University of Waterloo, Ontario, 200 pp.

Kirillova T.V. 1970. *Radiation Balance of Lakes and Reservoirs*. Leningrad: Gidrometeoizdat., 254 pp. (in Russian).

Kitaev S.P. 1984. *Lakes Production from the Different Natural Zones*. Moscow: Nauka Press, 208 pp. (in Russian).

Kolodochka A. 1997. *On Lowering of Order of Space Region of Lake Ecosystem Modelling*. Petrozavodsk: Karelian Research Centre, pp. 115 (in Russian).

Kondratyev K.Ya. 1965. *Actinometry*. Leningrad: Gidrometeoizdat., 300 pp. (in Russian).

Kondratyev K.Ya., Adamenko V.N., Vlasov V.P., Druzhinin G.V., Kreiman K.D., Pozdnyakov D.V., Rumyantsev V.B., Tikhomirov A.I., Filatov N.N. 1986. *Large Lake Like an Imitation Model of the Ocean*. Leningrad: Nauka Press, 63 pp. (in Russian).

Kondratyev K.Ya., Bobylev L., Pozdnyakov D., Naumenko M. 1996a. Combined application of remote sensing and in situ measurements in monitoring environmental processes. *Hydrobiologia*, **322**, 227–232.

Kondratyev K.Ya., Filatov N.N., Zaitsev L.V. 1989. Estimation water exchange and polluted water by remote sensing. Report of Academy of Sciences of USSR, No. 304, pp. 829–832 (in Russian)

Kondratyev K.Ya., Johannessen O.M., Melentyev V.V. 1996b. *High Altitude Climate and Remote Sensing*. Wiley/Praxis Publishing, 200 pp.

Kondratyev K.Ya., Melentyev V.V., Nazarkin V.A. 1992. *Remote Sensing of Water Area and Watersheds (Micro-wave Methods)*. St. Petersburg: Gidrometeoizdat., 248 pp. (in Russian).

Kotlyakov V.M. 1968. *The Earth Snow Cover and the Glaciers.* Leningrad: Gidrometeoizdat., 479 pp. (in Russian with summary in English).

Kreiman K.D. 1989. The thermal bar simulated by laboratory experiments. *Oceanologia (USSR)*, **29**, No. 6, 935–938 (in Russian).

Kudersky L.K., Jurvelius J., Kaukoranta M. 1996. Fishery of Lake Ladoga—past, present and future. In: Simola H., Viljanen M., Slepoukhina T., Murthy R. (eds.), *Ecological Problems of Lake Ladoga: Causes and Solutions.* In: *Proceedings of the First International Lake Ladoga Symposium, St. Petersburg, Russia, 22–26 November 1993.* Dordrecht: Kluwer Academic, pp. 57–64.

Lake Ladoga: Ecosystem Status Criteria. 1992. Petrova N.A., Torzhevik A.Yu. (eds.). St. Petersburg: Nauka Press, 325 pp. (in Russian).

Lakes in China. Research of Their Environment, Vols. 1, 2. 1995. Beijing: China Ocean Press, 586 pp, 482 pp.

Landsberg H.E. 1961. Solar radiation at the earth's surface. *Solar Energy*, **5**, 95–98.

Langmuir I. 1938. Surface motion of water induced by wind. *Science,* **187**, 119–123.

Leibovich S. 1983. The form and dynamics of Langmuir circulations. *Ann. Rev. Fluid Mech.*, **115**, 391–427.

Leonov A., Filatov, N., Titov. 1995. Lake Ladoga mathematical model of nutrients in Lake Ladoga. In: *Developments in Hydrobiology*, No. 113. Kluwer Academic, pp. 103–108.

Lerman A. (ed.) 1978. *Lakes: Chemistry, Geology, Physics.* Berlin: Springer Verlag, 363 pp.

Lewis W.M. Jr. 1983. Temperature, heat, and mixing in Lake Valencia, Venezuela. *Limnology, Oceanography*, **28**, 273–286.

Lewis W.M. Jr. 1983a. A revised classification of lakes based on mixing. *Canad. J. Freshwater Aquat. Sci.*, **10**, 1779–1787.

List R.J. 1951. *Smithsonian Meteorological Tables*, 6th edn. New York: Smithsonian Miscellaneous Collection, 600 pp.

Malm J, Jonsson L. 1993. A study of the thermal bar in Lake Ladoga using water surface temperature data from satellite images. *Remote Sensing Environ.* **44**, 35–46.

Malm J. 1994. Thermal bar dynamics. Spring thermo- and hydrodynamics in large temperate lakes. Report 1012, Department Water Res. Engin. Lund: Lund University, 350 pp.

Malm J. 1995. Spring circulation associated with the thermal bar in large temperate lakes. *Nordic Hydrology*, **26**, 331–358.

Malm J., Jonsson L. 1993. A study of the thermal bar in Lake Ladoga using water surface temperature data from satellite images. *J. Remote Sensing Environ.*, **44**, 35–46.

Malm J., Jonsson L. 1994. Water surface temperature characteristics and thermal bar evolution during spring in Lake Ladoga. *J. Remote Sensing Environ.*, **48**, 35–46.

Malm J., Mironov D.V., Terzhevik A., Grahn L. 1993. Field investigation of the thermal bar in Lake Ladoga, Spring 1991. *Nordic Hydrology*, **24**, 339–358.

Malm J., Mironov D.V., Terzhevik A., Jonsson L. 1994. Investigation of the spring thermal regime in Lake Ladoga using field and satellite data. *Limnology, Oceanography*, **39**, No. 6, 1333–1348.

Malm J., Terzhevik A., Bengtsson L., Boyarionov P., Glinsky A., Palshin N., Petrov M. 1996. A field study of thermo- and hydrodynamics in three small Karelian lakes during winter 1994/1995. Report No. 3197, Department Water Res. Engin. Lund: Lund University, 108 pp.

McLeish W. 1968. On the mechanisms of wind-slick generation. *Deep-Sea Res.*, **15**, 461–469.

Menon A.S., Dutka B.J., Jurkovich A.A. 1971. Preliminary bacteriological studies of the Lake Ontario thermal bar. *Proceedings 14th Conference Great Lakes Research*. Ann Arbor: International Association of Great Lakes Research, pp. 59-68

Melentyev V.V., Tikhomirov A.I., Kondratyev K.Ya., Johannessen O.M., Sandven S., Pettersson L.H. 1997. Study of freeze-up phase change on large temperate zone inland water bodies and possibilities of their microwave diagnostics. *Earth Obs. and Remote Sensing*, No. 3, 61–72 (in Russian).

Menshutkin V, Astrakhancev G., Yegorova N., Rukhovec L. 1998. Mathematical modeling of the evolution and current conditions of the Ladoga Lake ecosytem. *Ecological Modelling*, **107**, 1–24.

Meybeck M. 1995. Global distribution of lakes. In: Lerman A., Imboden D.M., Gat J.R. (eds.), *Physics and Chemistry of Lakes*. Berlin: Springer-Verlag, pp. 1–36.

Mironov D.V. 1991. Air-water interaction parameters over lakes. In: Zilitinkevich S.S. (ed.), *Modelling Air–Lake Interaction. Physical Background*. Berlin: Springer Verlag, pp. 50–62.

Monin A., Jaglom A. 1968. *Statistical Hydromechanics*. Moscow: Nauka Press, 638 pp. (in Russian).

Monin A., Krasitsky V. 1985. *Features on Surface Layer of the Ocean*. Leningrad: Hydrometeorological Press, 375 pp. (in Russian).

Mortimer, C.H. 1979. Strategies for coupling data collection and analysis with dynamic modelling of lake motions. In: Graf W.H., Mortimer C.H. (eds.), *Hydrodynamics of Lakes*. Amsterdam: Elsevier, pp. 183–222.

Murray J., Sir, Pullar L. 1910. Bathymetrical survey of the Scottish fresh-water lochs. Report of Scientific Results No. 1. Edinburgh: Challenger Office, 785 pp.

Myer G.E. 1969. A field study of Langmuir circulation. *Proceedings 12th Conference Great Lakes Research*. Ann Arbor: International Association of Great Lakes Research, pp. 652–663.

Naumenko M., Karetnikov S. 1994. Airborn remote sensing application for Russian large lakes dynamic structure investigation. *Proceedings of the First International Airborne Remote Sensing Conference and Exhibition*, Vol. 111. Strasbourg, France pp. 86–95.

Naumenko M. 1994. Some aspects of the thermal regime of large lakes: Lake Ladoga and Onega. In special issue, *Water Pollution Res. J. Canada*, **29**, No. 2–3, 423–439.

Naumenko M.A. 1998. Relationships for spatial-temporal variability of thermal processes in large dimictic lakes. DSc thesis, Limnology Institute. St. Petersburg: The Russian Academy of Sciences, 38 pp.

Naumenko M.A., Karetnikov S.G. 1998. Velocity of movement of a spring thermal frontal zone in Lake Ladoga. *Meteorology, Hydrology (Russia)*, No. 4, 107–115 (in Russian).

Nikitin A.M. 1977. Morphometry and morphology of Middle Asia lakes. *Proc. Middle Asia Hydrometeorol. Inst.*, **50**, No. 131, 4–21.

Noble V.E., Anderson R.F. 1968. Temperature and current in the Grand Haven, Michigan vicinity during thermal bar conditions. *Proceedings 11th Conference Great Lakes Research*. Ann Arbor: International Association of Great Lakes Research, pp. 470–479.

North G.R., Cahalan R.F., Coakley J.A. 1981. Energy balance climate models. *Rev. Geophys. Space Phys.*, **19**, 91–121.

Okhlopkova, A.N. 1961. Investigations of circulation in Lake Ladoga with use of the dynamic height method. *Oceanology*, **1**, 1025–1033 (in Russian).

Okhlopkova, A.N. 1972. Currents in Lake Onega. In: Malinina, T.I. (ed.), *Dynamics of Water Masses in Lake Onega*. Leningrad: Nauka Press, pp. 74–114.

Palmer M.D. 1973. Some kinetic energy spectra in a nearshore region of lake Ontario. *J. Geophys. Res.*, **78**, 3585–3597.

Panin G.N. 1985. *Thermo- and Matter Exchange Between Water Body and the Atmosphere under Natural Conditions*. Moscow: Nauka Press, 207 pp. (in Russian).

Patterson J.C. and Hamblin P.F. 1988. Thermal simulation of a lake with winter ice cover. *Limnology, Oceanography*, **33**, No. 3, 323–338.

Petrova N.A. 1990. *Successions of Phytoplankton with the Anthropogenic Eutrophication of Large Lakes*. Leningrad: Nauka Press, 200 pp.

Pienitz R., Smol J.P., Lean D.R.S. 1997a. Physical and chemical limnology of 59 lakes located between the southern Yukon and the Tuktoyaktuk Peninsula, Northwest Territories (Canada). *Canad. J. Fish. Aquatic Sci.*, **54**, No. 2, 330–346.

Pienitz R., Smol J.P., Lean D.R.S. 1997b. Physical and chemical limnology of 29 lakes located between Yellowknife and Contwoyto Lake, Northwest Territories (Canada). *Canad. J. Fish. Aquatic Sci.*, **54**, No. 2, 347–358.

Piotrovich V.V. 1958. *Ice Formation and Melting in Lakes and Reservoirs and Methods of Prediction*. Leningrad: Gidrometeoizdat., 200 pp. (in Russian).

Pivovarov A.A. 1972. *Thermal Conditions in Freezing Lakes and Rivers*. New York: John Wiley, Halsted Press, Israel Program for Scientific Translations, 136 pp.

Pivovarov A.A. 1979. *Sea Thermics*. Moscow: Moscow State University Press, 250 pp. (in Russian).

Podsetchine V., Huttula T., Filatov N. 1995. Water exchange in starigths of Northern Ladoga: Results of Field Studies and numerical simulations. *Phys. Chem. Earth*, **20**, No. 2, 207–213.

Popov A.V., Kondratyev K.Ya. (eds.). 1982. *Problems of Land Hydrology*. Leningrad: Gydrometizdat., 230 pp. (in Russian).

Prescott J.A., Collins J.A. 1951. The lag of temperature behind solar radiation. *Quart. J. Roy. Met. Soc.*, **77**, 121–126.

Rao, D.B. and B.C. Doughty. 1981. Instability of coastal currents in the Great Lakes. *Arch. Meteorol. Geophys. Bioklimatol., Ser. A.*, **30**, 145–160.

Raspletina G.F. 1992. Phosphorus availability in Lake Ladoga. In: Rihter G.D. (ed.), *Lake Ladoga: Criteria of the Ecosystem Status*. St. Petersburg: Nauka Press, pp. 74–87. 1959; *Catalogue of Karelian Lakes*. Moscow-Leningrad: Acad. Sci., 240 pp. (in Russian).

Recommendations on Using Ice Terms for the Description of Ice Parameters and Ice Regime Features of Lakes and Reservoirs. 1989. Leningrad: GGI, 38 pp. (in Russian).

Rodgers G.K. 1965. The thermal bar in the Laurentian Great Lakes. *Proceedings 8th Conference Great Lakes Research*. Ann Arbor: University of Michigan, Publ. 13, pp. 358–363.

Rodgers G.K. 1966. The thermal bar in Lake Ontario, spring 1965 and winter 1965-66. *Proceedings 9th Conference Great Lakes Research*. Ann Arbor: University of Michigan, Publ. 15, pp. 369–374.

Rodgers G.K. 1968. Heat advection within lake Ontario in spring and water transparency associated with the thermal bar. *Proceedings 11th Conference Great Lakes Research*. Ann Arbor: International Association of Great Lakes Research, pp. 942–950.

Rodgers G.K. 1971. Field investigation of the thermal bar in Lake Ontario: Precision temperature measurements. *Proceedings 14th Conference Great Lakes Research*. Milwaukee: International Association of Great Lakes Research, pp. 618–624.

Rodgers G.K., Sato G.K. 1970. Factors affecting the progress of the thermal bar of spring in Lake Ontario. *Proceedings 13th Conference Great Lakes Research*. Ann Arbor: International Association of Great Lakes Research, pp. 942–950

Rossolimo L.L. 1957. Thermal regime of the Lake Baikal. *Proceedings of Baikal Limnological Station*, Vol. XVI, Irkutsk, 551 pp. (in Russian).

Ryanzhin S.V. 1980. Investigation of currents, energy and temporal-spatial scales of Langmuir circulation. Cand. Sci. (PhD) thesis. Leningrad: Leningrad State University, 220 pp. (in Russian)

Ryanzhin S.V. 1982. Transverse dimensions of Langmuir circulating cells in a lake in the absence of a thermocline. *Izv. Atmos. Ocean Phys. (USSR)*, **118**, 814-820 (in Russian; English translation by Plenum Press).

Ryanzhin S.V. 1983. The kinematics of horizontal flows in circulating Langmuir cells. *Izv. Atmos. Ocean Phys. (USSR)*, **119**, 41-46 (in Russian; English translation by Plenum Press).

Ryanzhin S.V. 1989. *Thermal Regime Relationships for Freshwater Lakes of the World*. Leningrad: USSR Geographic Society, 70 pp. (in Russian; with summary and contents in English).

Ryanzhin S.V. 1991. A new version of spline-models for zone description of the Northern hemisphere freshwater lake surface temperatures. *Dokl. (Trans.) Acad.*

Sci. (USSR), **317**, No. 3, 628–634 (in Russian; English translation by *Scripta Technica*).

Ryanzhin S.V. 1991a. Zonal relationships for thermal regime of the Northern hemisphere freshwater lakes. *Water Resource (USSR)*, No 4, 15–29 (in Russian; English translation by Plenum Press).

Ryanzhin S.V. 1992. The global statistical estimates for terrestrial heat flow in freshwater lake bottom sediments. *Dokl. (Trans.) Acad. Sci. (Russia)*, **324**, No. 3, 562–566 (in Russian; English translation by *Scripta Technica*).

Ryanzhin S.V. 1994. Latitudinal-altitudinal interrelationships for the surface temperatures of the Northern hemisphere freshwater lakes. *Ecological Modelling*, **74**, No. 3–4, 231–253.

Ryanzhin S.V. 1994a. How does dispersion and mean value of Langmuir circulation spacing correlate? *Dokl. (Trans.) Acad. Sci.(Russia)*, **336**, No. 5, 692–696 (in Russian; English translation by *Scripta Technica*).

Ryanzhin S.V. 1995. Inter-relationships for spatial-temporal variability of thermo-hydrodynamic processes in freshwater lakes. DSc. thesis. St.Petersburg: St. Petersburg State University, Russia, 232 pp. (in Russian).

Ryanzhin S.V. 1997. Thermophysical properties of lake sediments and water-sediments heat interaction. Report 3214, Department Water Res. Engin. Lund: Lund University, 96 pp.

Ryanzhin S.V., .Mironov D.V. 1985. On the probability and critical conditions for Langmuir circulation appearance. *Izv. Atmos. Ocean Phys. (USSR)*, **121**, 184–190 (in Russian; English translation by Plenum Press).

Ryanzhin S.V., 1990. The Northern hemisphere freshwater lake surface temperatures depending on latitude and above sea level altitude. *Dokl. (Trans.) Acad. Sci. (USSR)*, **312**, No. 1, 209–214 (in Russian; English translation by *Scripta Technica*).

Ryanzhin S.V., Filatov N.N., Michkailov Yu.D., Terzhevik A.Yu. 1981. *Thermo-physical Processes in Deep Lakes*. Leningrad: Nauka Press, 220 pp. (in Russian).

Ryanzhin S.V., Rubtzov D.L. 1986. Characteristics of Lake Sevan temperature and wind regimes and their evolution over recent years. *Water Resource (USSR)*, No. 2, 86–96 (in Russian with summary in English).

Scavia D., Bennett J.R. 1980. Spring transition period in Lake Ontario—a numerical study of the causes of the large biological and chemical gradients. *Canad. J. Fish. Aquat. Sci.*, **37**, 823–833.

Schindler, D.W., Kalf, J., Welch, H.E., Brunskill, G.J., Kling, H., and Kritsch, N. 1974. Eutrophication in the high Arctic Meretta Lake, Corwallis Island (75°N Lat.). *J. Fish. Res. Board Canada*, **31**, No. 5, 647–662.

Schwab D., K. Bedford. 1994. Initial implementation of the Great Lakes Forecasting System: Real-time system for predicting lake circulation and thermal structure. *Water Pollution Res. J. Canada*, **29**, No. 2–3, 203–220.

Scott J.T., Myer G.E., Stewart R.W., Walther E.G. 1969. On the mechanism of Langmuir circulations and their role in epilimnion mixing. *Limnology Oceanography*, **14**, 493–503.

Scott R.W., Huff F.A. 1996. Impacts of the Great Lakes on regional climate conditions. *J. Great Lakes Res.*, **22**, 845–863.

Sellers W.D. 1965. *Physical Climatology*. Chicago: University of Chicago Press, 450 pp.

Serruya C., Pollingher U. 1983. *Lakes of the Warm Belt*. New York: Cambridge University Press, 570 pp.

Simons, T.J. 1973. Development of three-dimensional numerical models of the Great Lakes. *Can. Inland Waters Branch, Sci. Ser.*, **12**, 41.

Simons, T.J. and W.M. Schertzer. 1987. Stratification, currents and upwelling in Lake Ontario, summer 1982. *Can. J. Fish. Aquat. Sci.*, **44**, 2047–2058.

Smirnova N.P. 1968. Radiation balance of Lake Ladoga. In: Kalesnik S.V., Smirnova N.P. (eds.), *Thermal Regime of Lake Ladoga*. Leningrad: State University, pp. 5–72 (in Russian).

Sorokin Yu.I., Fedorov V.K. 1989. Assessments of primary production and organic matter decomposition rates in Lake Onega. *Proc. Inland Waters Biol.*, **19**, 1–8.

Spain J.D., Wernert G.M., Hubbard D.W. 1976. The structure of the spring thermal bar in Lake Superior, II. *J. Great Lakes Res.*, **2**, No. 2, 296–306.

Storm O., Bengtsson R. 1993. The thermal bar in Lake Onega—a study using information from satellite images. M.Sc. thesis, Department Water Res. Engin. Lund: Lund University, 28 pp.

Straškraba M. 1980. The effects of physical variables on freshwater production: analyses based on models. In: Le Gren E.D., McConnell R.H. (eds.), *Functioning Freshwater Ecosystems*. Cambridge: Cambridge University Press, pp. 13–84.

Straškraba M. 1993. Some new data on latitudinal differences in the physical limnology of lakes and reservoirs. In: Boltovskoy A., Lopez H.L. (eds.), *Proceedings International Conference on Limnology*. La Plata: Instituto de Limnología "Dr R.A. Ringuelet", pp. 19–39.

Straškraba M., Gnauck A. 1985. *Aquatic Ecosystems. Modelling and Simulation*. Amsterdam: Elsevier, 309 pp.

Sundaram T.R. 1974. Transient thermal response of large lakes to atmospheric disturbances. *Proceedings 17th Conference Great Lakes Research*. Ann Arbor: International Association of Great Lakes Research, pp. 801–810.

Telesh I.V. Species composition of planktonic Rotifera, Cladocera and Copepoda in the littoral zone of Lake Ladoga. In: Simola H., Viljanen M., Slepoukhina T. and Murthy R. (eds.), *Proceedings 1st International Lake Ladoga Symoposium: Ecological Problems of Lake Ladoga. St. Petersburg, Russia, 22—26 November 1994*. Dordrecht: Kluwer Academic, pp. 181–185.

Thorpe S.A. 1984. The effect of Langmuir circulation on the distribution of submerged bubbles caused by breaking waves. *J. Fluid Mech.*, **142**, 151–170.

Thorpe S.A., Hall A.J. 1982. Observations of the thermal structure of Langmuir circulations. *J. Fluid Mech.*, **114**, 237–250.

Tikhomirov A.I. 1959. On the thermal bar of Yakimvarsky Bay of Lake Ladoga. *Izv. All Union Geograph. Soc.*, **91**, No. 5, 424–438 (in Russian).

Tikhomirov, A.I. 1963. The thermal bar of Lake Ladoga. *Bull. (Izv.) All-Union Geogr. Soc.*, **95**, 134-142 (in Russian).

Tikhomirov, A.I. 1982. *Thermal Regime of Large Lakes.* Leningrad. Nauka Press, 232 pp. (in Russian).

Tregoubova T.M., Koulish N.T. 1992. Organic matter pool formation in Lake Ladoga. In: *Lake Ladoga: Criteria of the Ecosystem Status.* St. Petersburg: Nauka Press, pp. 219–240.

Verner M. 1983. A new Seasat altimeter geoid for the Baltic. *Rep. Finnish Geodetic Inst., Helsinki.*, **83**, No. 4, 30 pp.

Veselova M.F. 1968. The role of river waters in a heat balance of Lake Ladoga. In: Kalesnik S.V., Smirnova N.P. (eds.), *Thermal Regime of Lake Ladoga.* Leningrad: State University, pp. 217–222 (in Russian).

Welander P. 1963. On the generation of wind streaks on the sea surface by action of surface film. *Tellus*, **15**, 67–71.

Weller R.A., Price J.F. 1988. Langmuir circulation within the oceanic mixed layer. *Deep Sea Res.*, **35**, 711–747.

Wetzel R.G. 1975. *Limnology.* Philadelphia: W.B. Saunders, 743 pp.

Yakushko O.F. (ed.) 1983. *Handbook. Lakes of Belarussia*, Part 1. Minsk: State University, 382 pp. (in Russian).

Yakushko O.F. (ed.) 1985. *Handbook. Lakes of Belarussia*, Part 2. Minsk: State University, 259 pp. (in Russian).

Zilitinkevich S.S (ed.). 1991. *Modelling Air-Lake Interaction. Physical Background.* Berlin: Springer Verlag, 132 pp.

Zilitinkevich S.S., Kreiman K.D. 1989. Theoretical and laboratory investigation of the thermal bar. *Oceanologia (USSR)*, **30**, No. 5, 750–755 (in Russian).

Zilitinkevich S.S., Kreiman K.D., Terzhevik A.Yu. 1992. The thermal bar. *J. Fluid Mech.*, **236**, 27-42.

Zilitinkevich S.S., Malm J.G.B. 1993. A theoretical model of thermal bar movement in a circular lake. *Nordic Hydrology*, **24**, 13–30.

Zilitinkevich S.S., Terzhevik A.Yu. 1987. The thermal bar. *Oceanologia (USSR)*, **27**, No. 5, 732-738 (in Russian).

Zilitinkevich S.S., Terzhevik A.Yu. 1989. Correction to the "The thermal bar". *Oceanologia (USSR)*, **29**, No. 5, 755–758 (in Russian).

3

Modern passive and active optical and microwave remote sensing: advanced feasibilities for applications in contemporary limnological studies

3.1 OPTICAL REMOTE SENSING

3.1.1 Passive remote sensing

3.1.1.1 *Optical models of inland waters*

Natural waters are complex physical-chemical-biological media comprising living, non-living, and once-living material that may be present in aqueous solution or in aqueous suspension. Together with air bubbles and inhomogeneities resulting from small-scale water eddies, these components determine the bulk optical properties of natural water bodies.

The principal living organisms present in water columns are *plankton*, a collective term encompassing all vegetable and animal organisms suspended in water (either hovering or floating), unable to resist the current, and not rigidly connected to the confining basin. Plankton include animal organisms (*zooplankton*), algal plant organisms (*phytoplankton*), bacteria (*bacterioplankton*), and lower plant forms such as *algal fungi*. These organisms represent the lowest level of feeding in a system of alimental food chain relationships involving not only plankton (and their essential nutrients), but also higher forms of aquatic life. Phytoplankton consume nutrients (biogenes) from their confining waters and in the presence of subsurface sunlight synthesize these nutrients into organic matter through the process of *primary production*, a topic to which we shall return later. Zooplankton graze on phytoplankton. As consequences of their vital functions and their mortality, zooplankton generate secondary organic matter. Bacterioplankton decompose this organic matter. Due to these low-level food-chain dynamics, a dissolved organic matter (which hereinafter will be referred to as *doc*- dissolved organic carbon) component (termed the *autochtonous doc* component) is indigenous to natural waters. Lakes and marine/oceanic coastal waters are also receptacles of doc inputs from surficial run-off and river discharges which contribute an *allochthonic doc* component. These runoff and river discharge inputs introduce a suspended inorganic matter (*sm*) component to non-case – 1 waters (for ref. see Chapter I).

Zooplankton, as well as insects and higher plants, serve as food for fish that, in turn, extend the food chain to aquatic and terrestrial carnivores and herbivores.

Both *doc* and *sm* (along with the plankton) dictate the manner in which the subsurface light field is distributed in the natural water body and, therefore, should be considered in terms of their specific absorption and scattering of optical cross-section spectra. However, the extreme diversity of organic and inorganic components of inland and marine/oceanic waters, coupled with temporal and spatial variability of the composition of these waters, presents a severe obstacle to such a considera-tion. Thus, it becomes incumbent to select a manageable number of these com-ponents (or surrogates to these components) if an adequate physical model is to be utilized to effectively describe the optical properties of a natural water body under scrutiny.

Based on the main principal of choosing the major optically active components (OAC) in terms of their impact on the spectral distribution of water-leaving radiance $L_w(+0, \lambda)$ in the visible spectrum, some components can be confidently ruled out from the list of OAC. Dissolved gases with the sole exception of oxygen do not absorb in the visible spectrum, but even in waters rich in oxygen, O_2 concentrations prove to be insufficient to have a more or less substantial impact on spectral distribution of upwelling radiance $L_w(+0, \lambda)$ (Kondratyev and Pozdnyakov, 1990).

H_2O. The absorption spectrum of pure water (by pure water we imply a chemically pure substance which is a mixture of several types of water-molecule isotopes, each of different molecular mass) exhibits a dip minimum in the short and middle wavelength regions of the visible spectrum which is followed by an increase in absorption at $\lambda > 600$–650 nm. The absorption in the red part of the spectrum is a peripheral manifestation of much more intense IR absorption bands located at $\lambda > 700$ nm. A pronounced increase of water absorption in the UV spectrum is due to electron transitions occurring within the water molecule.

It was generally believed that the most precise values of water absorption a_w and scattering b_w coefficients were those suggested by Smith and Baker (1981). However, recently Hakvoort (1994) published a revised list of spectral values of water inherent optical properties. According to the revised data, as compared with the data published by Smith and Baker, the water absorption minimum dip is deeper and shifted to shorter wavelengths (Fig. 3.1a).

Water absorption properties vary with temperature: the liquid water absorption coefficient a_w increases with water temperature t in the spectral region 400–780 nm and it decreases in the spectral interval 780–800 nm. With increasing temperature, the absorption maximum shifts to shorter wavelengths. The above temperature-driven variations in a_w could be parameterized via (Hakvoort, 1994):

$$a_w(t) = a_w(t = 20, 1°C) + A(t - 2, 1°C) \tag{3.1}$$

where A is the absorption increment due to temperature increase. The spectral

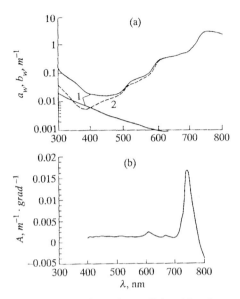

Fig. 3.1. Spectral variations in pure water absorption coefficient (a) and temperature correction coefficient
A (b). 1—from Smith and Baker (1981); 2—from Hakvoort (1994).

variations of A are given in Fig. 3.1b: water molecule absorptivity rises with growing
temperature in the spectral region 400–780 nm, and decreases with growing
temperature at $\lambda \geq 780$ nm, these changes being coupled with a shift of the water
absorption maximum to shorter wavelengths.

Thus, scattering by water molecules becomes insignificant when compared to
absorption by water molecules for $\lambda > \sim 580$ nm: the attenuation of light (c_w) of
wavelengths greater than ~ 580 nm becomes essentially a consequence of molecular
absorption a_w (i.e. $c_w \approx a_w$). However, for values of λ in the range 400–520 nm,
scattering by water molecules (b_w) prevails over absorption by water molecules, so
that in the blue region of the spectrum c_w is primarily dictated by b_w.

Elastic scattering of light on fluctuations of water density displays a spectral
dependency proportional to $\sim \lambda^{-4.3}$, whereas its angular distribution conforms to the
well-known Einstein–Smolushovsky equation (Whitlock et al., 1981). For pristine
oceanic waters, directional scattering at visible wavelengths by water molecules
(described by the volume scattering function) has been modelled by Morel (1980)
and given as:

$$\beta_w(\theta) = \beta_w(90°)(1 + 0.835)\cos^2\theta \tag{3.2}$$

where $\beta_w(\theta)$ = volume scattering function at wavelengths λ and scattering angle θ,
$\beta_w(90°)$ = volume scattering function at wavelengths λ and scattering angle 90°; the
value of $\beta_w(90°)$ has been estimated (at $\lambda = 550$ nm) as 0.93×10^{-4} m^{-1} sr^{-1}.

From eq. (3.2), it is seen that $\beta_w(90°) = \beta_w(0°)$. That is, in pure water medium,
scattering in the backward direction is numerically equal to scattering in the forward

direction. Such is distinctly not the case for natural waters containing absorption and scattering centers other than water molecules.

Dissolved inorganic salts (dis)

Present in natural waters, *dis* undoubtedly have an impact on light absorption and scattering processes in the water column. However, this influence is mainly confined to the UV spectral range ($\lambda < 300$ nm) where are located absorption bands arising from electron transitions within the molecules of these salts. Light scattering in the visible spectrum due to *dis obeys* the same spectral dependence as does light scattering on pure water density fluctuations and accounts (with respect to the latter) for about 20–30% of the total scattering in open marine waters of salinity averaging 35 g/l (by comparison, salinity levels in inland water bodies may vary from ∼0 to ∼42 g/l, and in a fresh water basin like Lake Ladoga the salinity does not exceed several tens of milligram per liter (Viljanen *et al.*, 1996)).

Molecular scattering of light by mid-oceanic pure water plus dissolved salts is generally insignificant when considered in relation to the total light attenuation (absorption + scattering) appropriate to these waters. However, molecular scattering, from the pure water/dissolved salt combination assumes greater significance when specific directional scattering is considered. The ratio $\eta = \beta_{mol}(\theta)/\beta(\theta)$ (i.e. scattering from pure water and dissolved salt molecules/scattering from pure water plus dissolved organic salts plus all other suspended and/or dissolved matter residing in the mid-oceanic water) can reach substantial magnitudes for moderate to large scattering angles. Shifrin (1983) illustrated that, while a value for η of 0.10 is appropriate for mid-oceanic waters for scattering into the forward hemisphere, values of $\eta > 0.85$ are encountered for scattering into the backward hemisphere.

Due to the concentrations of terrestrially derived matter and the concomitant enhanced fertility that contrast case-1 and non-case-1 waters (see Chapter 1), the η values generally appropriate to inland and marine/oceanic coastal waters (except for rare occasions such as, e.g., infertile lakes) are considerably smaller, and the role of dissolved organic salts in impacting a directional nature to $\beta(\theta)$ is dramatically reduced.

Dissolved organic matter (doc)

As a consequence of either photosynthetic activity of algae (autochthonic) or direct inputs of terrestrially derived matter (allochthonic), *doc* is present in marine waters in quantities less than several milligrams of carbon (C) per liter (i.e. by four orders of magnitude less than the concentration of dissolved salts). In inland water bodies *doc* levels are not infrequently as high as 10 mg C/l (Viljanen *et al.*, 1966). The presence of water-soluble humic substances in natural waters is readily apparent in inland and coastal waters (when their presence is not otherwise veiled by suspended particulates) by varying degrees of yellowness that are observable in such waters. This yellowness is a consequence of melanoids composing DOC. However, these melanoids, comprise only a small fraction ∼10–40% of aquatic DOC. As a consequence of this yellow hue, the dissolved aquatic humus is generally referred to as *yellow substance* (*ys*) or *gelbstoff*.

The absorption by *doc* exponentially decreases from short to long wavelengths in the visible spectrum. The observed *doc* absorption in the visible spectrum is a "tail" of electron transition absorption bands located in the UV part of the spectrum. A certain fraction of *doc* can fluoresce with emission maximum at 310–*ca.* 520 nm depending on the nature of *doc* and the excitation wavelength (Coble, 1994).

The light absorption properties of dissolved aquatic humus prove to be a consequence of electronic transitions. *doc* absorption dramatically decreases with increasing λ. The spectral dependence of the absorption coefficient for the colored fraction of *doc* (so-called yellow substance), $a_{ys}(\lambda)$ can be reasonably described by the exponential formula (Zepp, Schlotzhauer, 1981):

$$a_{ys}(\lambda) = a_{ys}(\lambda_0) \exp[-s(\lambda - \lambda_0)] \tag{3.3}$$

where s is a slope parameter that is assumed to be independent of wavelength λ. The relative constancy of the shape of the $a_{ys}(\lambda)$ versus λ curve of eq. (3.3) enables the selection of a reference wavelength λ_0 upon which to compare the yellow substance concentrations of different waters. The selection of $\lambda_0 = 400$ nm is certainly appropriate since it represents the maximum a_{ys} in the visible spectrum. However, λ_0 value of 440 nm, is also frequently encountered in the published literature.

Finally, it should be pointed out that a certain fraction of *doc* (presumably small) can fluoresce with emission maximum at $\sim$310–$\sim$520 nm depending on the nature of doc and the excitation wavelength (Coble, 1994).

Suspended matter

All natural water bodies inescapably contain suspended matter comprised of organic and inorganic matter under the collective term *seston*. Seston is extremely diverse in origin and in composition, and includes mineral particles of terrigenous origin, plankton, detritus (largely residual products of the decomposition of phytoplankton and zooplankton cells as well as macrophytic plants), volcanic ash particles, particulates resulting from *in situ* chemical reactions, and particles of anthropogenic origin.

Suspended inorganic (mineral) matter (sm)

The presence of terrigenous suspended particles is a consequence of river discharge, shore erosion, runoff, long and short-range transport of atmospheric particulates followed by dry deposition. According to Adamenko *et al.* (1991), fine clay particulates rarely exceed 3–4 μm in diameter, silt particles are in the range 5–40 μm, very fine grain sands fall into the range 40–130 μm, and coarse grain sands are in the range 130–250 μm.

In Lake Ladoga, seston concentrations display a range 0.1–12.0 g per m^3, with suspended minerals comprising 50% to 97% of this concentration (*Present state...*, 1987). Suspended mineral concentrations in Lake Ontario have been reported in the range 0.2–8.9 g per m^3 (Bukata *et al.*, 1981).

When water contains large amounts of fine particles of Fe^{++} or/and Mn^{++} hydroxides (which are colored and known to absorb in the spectral range 350–550 nm (Kondratyev and Pozdnyakov, 1990)), a more sophisticated spectral distribution of the SM absorption should be expected.

Suspended organic matter

Suspended organics in natural water bodies incorporate planktonic organisms and detritus particles. As indicated above, the plankton in its turn consist of algal fungi, bacteria (bacterioplankton), micro-algae (phytoplankton) and micro-animal organisms (zooplankton). With the exception of algal fungi (which are transparent and colorless and display the refraction index practically indistinguishable from that of water), all planktonic species are optically active in the visible spectrum as a result of light absorption stemming mostly from pigments either indigenous to aquatic cells (bacterioplankton and phytoplankton) or otherwise acquired through trophic chain interactions (zooplankton) (Kondratyev and Pozdnyakov, 1990; Morel, 1988; Karnaoukhov, 1988).

When deciding on the eligibility of water co-existing constituents for inclusion in the principal OAC group dictating bulk-water optical properties, their absolute and relative concentrations become a decisive criterion. The Lake Ladoga water can be considered as an example. In spring, the bacterioplankton content C_{bp} in the coastal zone may attain values of 1.5–2.0 millions of cells/ml, whereas within the pelagic region, C_{bp} does not exceed 0.1–0.01 millions of cells/ml. In summer, C_{bp} decreases in coastal waters down to $\sim$0.5–0.65 millions of cells/ml and increases up to 10–20 millions of cells/ml in the central part of the lake. In fall, C_{bp} is about 0.6–0.7 millions of cells/ml and is nearly uniformly distributed throughout the lake (Viljanen *et al.*, 1996). The phytoplankton concentration (C_{php}) within the top (0–10 m) layer varies very strongly depending on the vegetation period, and the C_{php} mean value might be expected at the level of several hundreds of thousands cells/l in the pelagic part of the lake and several millions of cells/l in the littoral zone (Viljanen *et al.*, 1996). The concentrations of zooplankton (C_{zp}) in surface waters (0–20 m) of the littoral zone are several hundreds of plankters per cubic meter. Generally, however, C_{zp} is somewhat lower in the littoral zone (Viljanen *et al.*, 1996).

Consequently, in the above example, the concentration ratios of bacterioplankton/phytoplankton/zooplankton for maximal and minimal values of C_{bp}, C_{php}, and C_{zp} are as follows $10^{12}/10^6/10^2$ and $10^{11}/10^5/10^1$. Although, in some vegetation periods (basically of rather short duration), the concentration of phytoplankton and zooplankton could be of the same order of magnitude in inland water bodies (Kondratyev and Pozdnyakov, 1990), it could, nevertheless, be generally assumed that zooplankton might be left outside the group of major OAC.

Detritus

Detritus, as specified above, are suspended organic and partly mineralized particulates which are fragments of dead plankton and zooplankton fecal pellets. The detritus content in water can vary substantially depending on vegetation period and water body trophic level. Some studies indicate (Kondratyev and Pozdnyakov, 1990) that the detritus concentration could be closely related to the phytoplankton content, and is about 10–30% of C_{php}. However, although, these concentration ratios are typical of Lake Ladoga, it could be assumed that they are equally appropriate for other temperate inland fresh water bodies (Petrova, 1990).

Summarizing the above data, it could be suggested that the group of major OAC for non-case-1 waters should be composed of water *per se*, bacterioplankton, phytoplankton, DOC, SM, and, in some cases, detritus.

OAC specific inherent optical properties

The combined processes of absorption and scattering determine the light transfer through a natural water column. Therefore, the bulk natural water can be characterized by so-called inherent optical properties encompassing the coefficient of absorption a, coefficient of scattering b, coefficient of attenuation c, coefficient of backscattering b_b, phase function β. The absorption and scattering acts occurring in the aquatic medium are controlled by the amounts of scattering and absorption that may be attributable to each optically significant component comprising the natural water body. Indeed, being convolutions of aquatic medium optical properties, the inherent optical properties are additive in nature:

$$a = \sum_i^I a_i; \quad b = \sum_i^I b_i; \quad b_b = \sum_i^I b_{b_i}; \quad \beta = \sum_i^I \beta_i, \quad i = 1, 2, 3, \ldots, I \quad (3.4)$$

where I is the number of OAC comprising the natural water body.

The additive equations (3.4) could be further modified via introducing specific absorption a^* and specific scattering b^* (backscattering b_b^*) coefficients (also referred to as absorption and scattering (backscattering) cross-sections):

$$a = \sum_i^I C_i a_i^*; \quad b = \sum_i^I C_i b_i^*; \quad b_b = \sum_i^I C_i b_{b_i}^*; \quad \beta = \sum_i^I C_i \beta_i^*, \quad i = 1, 2, 3, \ldots, I$$

$$(3.5)$$

Therefore, given the specific inherent optical properties, the bulk-water optical properties could be assessed exclusively through *in situ* determinations of the major OAC. It certainly considerably simplifies the problem. However, rigorously speaking, such an approach is only adequate if each of the OAC is uniform in terms of its composition and has a unimodal size distribution. Otherwise, the summation should be performed not only over the major OAC, but also with respect to each specific coefficient over the number of diversity species j appropriate to the i-th component. Moreover, the specific coefficients a_i^*, b_i^*, $b_{b_i}^*$, β_i^*, should meet the prerequisite of being *representative* of at least a definite water body and a definite vegetation period. Ideally, they should also be translatable to a larger variety of both water bodies and application timings. Therefore, it appears, appropriate to discuss the data available to date in the literature on the major OAC specific inherent coefficients of absorption and scattering and their responsiveness to biotic and abiotic factors.

Phytoplankton

Generally, inland and marine/oceanic waters are receptacles of a considerable number of co-existing phytoplankton taxonomic groups which cumulatively are usually referred to as the phytoplankton community or assemblage. Most often, any phytoplankton community is characterized by some *dominant* algal species. For instance, in temperate lakes, depending upon the time of year (i.e. vegetation season) and the trophic status of the water body, the group of dominant species incorporates diatoms, blue–greens, greens, and some others. The composition of phytoplankton communities residing in fresh and marine waters reveals differences both at the level of major floristic groups as well as at lower taxonomic levels. For instance in marine waters such algae as *Ptychdiscus brevis* (which bring about *red tide* blooms) are definitely uncharacteristic of inland water bodies.

Floristic differences, particularly those appropriate to upper levels of the taxonomic pyramid, are at the basis of very substantial variations in the optical properties of algal cells (Fig. 3.2). These variations basically reside in differences in size and geometrical form of cells, their internal (intracellular) structure, chemical composition of the cell envelope, as well as in the type of colonial organization of cells and, particularly important, in the photosynthetic pigment composition (Paramonov, 1995).

Of the three types of photosynthetic pigment (chlorophylls, carotenoids and biliproteins) the first two are characteristic of all green plants, including micro- and macroalgae), whereas the third one is only found in such algae as *Cyanophyta*, *Cryptophyta* and *Rhodophyta*.

Amongst the chlorophylls, only chlorophyll-*a* (*chl-a*) is always present in all green

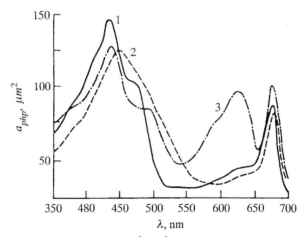

Fig. 3.2. Normalized (to the volume equal to $10^3 \, \mu m^{-3}$) absorption spectra of green algae (1), diatoms (2) and blue–green algae (3) (Paramonov, 1995).

plants. At the same time, chlorophylls-*b* (*chl-b*), -*c* (c_1, c_2: *chl-c*$_1$, *chl-c*$_2$) and -*d* (*chl-d*) are accessory pigments co-existing in various combinations contingent on algal taxonomic type. Chlorophylls absorb light in the short wavelength ($\lambda = 430$–455 nm, the Soret blue band) and in the long wavelength ($\lambda = 625$–690 nm, the α-band) regions of the visible spectrum. For all chlorophylls, absorption in the blue band is more strong that it is in the red band. Unlike chlorophyll-*a*, and chlorophyll-*b*, chlorophylls-*c* (c_1 and c_2) are less strong absorbers in the red portion of the visible spectrum. All chlorophylls are reported to absorb at about 580 nm (the β– band). Interestingly, in the case of *chl-c* this absorption band is almost as strong as the α– absorption band (Weaver, Wrigley, 1994).

Carotenoids are oil-soluble pigments characterized chemically by a long aliphatic polyene chain composed of isoprene units. They fall in two groups—carotenes and xanthophylls. The principal carotenoids are β-carotene, lutein, violaxantene. However, such algae as Syphonales (Chlorophyta) contain α-carotene as the principal carotenoid. Of xanthophylls, lutein is the most common. Both carotenes and xanthophylls absorb light in the spectral region 418–483 nm (Weaver and Wrigley, 1994).

Biliproteins are presented by two groups of pigments—red and blue. They absorb in the spectral region 500–670 nm. Red biliproteins include phycoerythrin (absorbing at 498, 545, 562–565 nm) and phycoerythrocyanin (absorbing at 590, 568 nm). Blue biliproteins include phicocyanin (absorbing at 585, 615, 630, 670 nm) and allophycocyanin (absorbing at 590, 568 nm) (Kondratyev and Pozdnyakov, 1990; Weaver, Wrigley, 1994).

The existence of light absorbing substances (pigments) in the algal cell determines spectral variations in the phytoplankton absorption coefficient a_{php}. Assuming that the algal cell absorption is controlled exclusively by co-existing pigments, it is possible to write:

$$a_{php}(\lambda) = \sum_{j=1}^{n} a_j^*(\lambda) C_j \qquad (3.6)$$

where n represents the total number of pigments considered, $a_j^*(\lambda)$ is the specific absorption coefficient of the *i*th pigment at wavelength λ, C_j is the concentration of this pigment. According to Hoepffner and Sathyendranath (1993), the specific absorption coefficient of each pigment can be estimated by decomposition of a_{php} into several Gaussian bands simulating the absorbing properties of these pigments:

$$a_{php}(\lambda) = \sum_{k=1}^{l} C_k a_k^*(\lambda_{mk}) \exp\left[\frac{(\lambda - \lambda_{mk})^2}{2\sigma_k^2} \right] \qquad (3.7)$$

where λ_{mk} is the position of maximum absorption; σ_k is the width of the *k*th Gaussain band; $a_k^*(\lambda_{mk})$ is the specific absorption coefficient of the *k*th band centered at λ_{mk}; C_k is the concentration of the pigment producing the *k*th absorption band. Since a given pigment may be responsible for more than one absorption band, $l > n$.

Figure 3.3 presents spectral variations in specific absorption coefficients of chlorophylls and carotenoids (Hoepffman and Sathyendranath, 1993) which exhibit

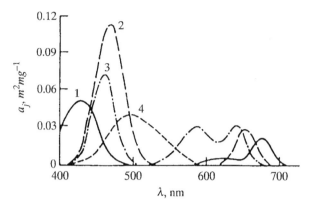

Fig. 3.3. Spectral variations in absorption coefficient of *chl-a* (1) , *chl-b* (2), *chl-c* (3), and carotenoids (4) (Hoepffner, Sathyendranath, 1993).

the following absorption bands: *chl-a* 381.5; 410.8; 433.5; 620.6; 675.6; 699.8 nm, *chl-b* 466.6; 652.9 nm, *chl-c* 459.2; 585.6; 640.7 nm, carotenoids 487.8; 532.0 nm.

The position of absorption bands as well as the spectral values of the specific absorption coefficients of photosynthetic pigments, and consequently, the spectral values of the specific absorption coefficient of phytoplankton reveal a spatial variability driven by a number of reasons. In particular, the optical properties of algal cells are controlled by the water body trophic status since it largely determines, among other things, both taxonomic composition and size distribution of algal cells in the phytoplankton community. According to the extensive studies conducted in various parts of the global ocean, in various vegetation periods (April–October), at different depths (0–180 m) and in water masses of different trophic status $(0.02 < C_{chl} < 25 \, mg/m^3)$ (Bricaud *et al.*, 1995; Allali *et al.*, 1997), the specific absorption coefficient of phytoplankton at all wavelengths decreases with increasing chlorophyll concentration. Based on the analysis of more than 800 spectral values of $a_{php}^*(\lambda)$, the following formulation has been suggested:

$$a_{php}^*(\lambda) = A(\lambda)C_{chl}^{-B(\lambda)} \qquad (3.8)$$

where A and B are wavelength-dependent coefficients (Fig. 3.4). It is reported that $a_{php}^*(\lambda)$ depends on C_{chl} most strongly in the blue and red part of the spectrum. A defendable explanation to the revealed regularity of $a_{php}^*(\lambda)$ decline with increasing C_{chl} should obviously be sought in the interdependence between C_{chl} and the co-existing concentrations of accessory pigments: there is experimental evidence that the content in algal cells of such pigments as nonphotosynthesizing carotenoids, *chl-b* and divinyl *chl-b* is almost always higher in eutrophic than in mesotrophic and oligotrophic waters (Bidigare *et al.*, 1990).

Apart from the above factors impacting spectral variations in $a_{php}^*(\lambda)$, a very strong influence is produced by the presence in phytoplankton cells of chlorophyll decomposition products engendered through the reactions of oxygen hydrolysis. The degradation of *chl-a* into *pheophytin* results in a displacement of the absorption

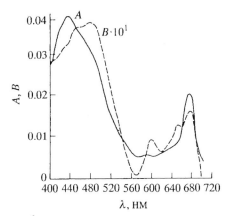

Fig. 3.4. Spectral values of A (m^2/mg) and B in eq. (3–8) (Bricaud *et al.*, 1995).

maximum at 430 nm to *ca*. 415 nm, whereas the long wavelength maximum at 664 nm transforms into a weak band centered at *ca*. 667 nm. A weak acidification of the medium brings about the complete disappearance of the red absorption band. Therefore, any increase in the ratio of the blue to red absorption maxima in the spectral distribution of $a^*_{php}(\lambda)$ may indicate the presence of "dead" or dying phytoplankton in considerable amounts (Sosik and Mitchell, 1995).

The optical properties of algal cells are also greatly influenced by the water column illumination conditions (*viz.* solar light intensity and spectral composition, temporal variations in solar fluxes propagating through the water column) and the optical properties of water *per se* (Mitchell, Kiefer, 1988; Roesler, Perry, 1995). It has been found that phytoplankton cell growth under conditions of low rates of irradiation results in a reduction of absorption bands of $a^*_{php}(\lambda)$ both in the blue and red parts of the visible spectrum. However, the most pronounced changes in the spectral distribution of $a^*_{php}(\lambda)$ occur in the short wavelength region of the spectrum: there is a flattening of the blue maximum which is accompanied by its broadening due to enhancement of absorption by accessory pigments. As a result, the specific absorption coefficient $a^*_{php}(\lambda)$ of algal cells adapted to low levels of irradiation is three to four times less than $a^*_{php}(\lambda)$ of algal cells adapted to high irradiation levels (Mitchell and Kiefer, 1988). This effect becomes particularly pronounced in the case of large cells, because of the package effect conducive to intracellular pigment self-shadowing (incidentally, the package effect also results in a certain decrease in phytoplankton fluorescence).

The above factors actively determining phytoplankton absorption properties are also very much active in terms of controlling phytoplankton scattering characteristics. Algal cell scattering reveals spectral selectivity: within the spectral regions of pigment absorption, there is a decrease in the phytoplankton specific scattering. Indeed, the effect of scattering suppression at wavelengths corresponding to absorption bands has been forecast theoretically in a wealth of studies (see Bricaud *et al.*, 1983).

Interestingly, according to model predictions and direct measurements (Bricaud, Stramski, 1990; Ahn Yu-Hwan *et al.*, 1992), spectral variations in scattering $b^*(\lambda)$ and backscattering $b_b^*(\lambda)$ cross-sections for one and the same algal monoculture can differ very substantially. This effect is particularly strong in the phytoplankton red-absorption spectral region where $b^*(\lambda)$ decreases while $b_b^*(\lambda)$ steadily increases. However, in the blue spectral region, the growth of $b_b^*(\lambda)$ with λ, being characteristic of practically all phytoplankton taxa, is not necessarily accompanied by an appropriate rise (after a slight drop at $\lambda = 430$–$450\,\mathrm{nm}$) of $b_b^*(\lambda)$: on the contrary, the coefficient of scattering $b^*(\lambda)$ continuously diminishes over the remaining portion of the visible spectrum right up to the red boundary. It should also be pointed out that experiments with nine algal monocultures having narrow size distributions (maxima of distributions corresponded to mean cell diameters d falling in the range 1–120 μm) failed to reveal any stable dependence of $b^*(\lambda)$ and $b_b^*(\lambda)$ on d (Ahn Yu-Hwan *et al.*, 1992) which is at variance with the opinion that $b_b^*(\lambda)$ should decrease with increasing cell size (Sathyendranath, 1995).

Bacterioplankton. By the nature of metabolic processes, bacterioplankton could be ascribed to two groups: autotrophic and heterotrophic bacterioplankton. Within this partitioning, bacterioplankton can also be subdivided in chemosynthetic and photosynthetic species. Auto(chemo/photo)trophic bacteria are organisms capable of producing organic matter from carbonic acid, water molecules and mineral salts through photosythetic reactions or otherwise by redox reactions. Heterotrophic bacteria use organic substances present in water as nutrients.

Cells of photosynthetic bacterioplankton contain such pigments as bacteriochlorophylls (which are homologs of phytoplankton *chl-a*) and carotenoids. The presence of photopigments determines the absorption properties of this type of bacterioplankton. Bacterial photosynthesis, unlike phytoplankton photosynthesis, does not result in oxygen release into the ambient water. The hydrogen donators here are organic and innorganic compounds containing hydrogen. Autotrophic photosythetic bacteria reside in anaerobic conditions in meromictic lakes (i.e. the water bodies which remain stratified throughout the year).

In well-aerated upper layers of water bodies (i.e. within the depths attainable for remote-sensing means), bacterioplankton is practically totally composed of heterotrophic chemosynthetic species (photosynthetic bacterioplankton only accounts for a few percent of the bacterioplankton total number counts). This type of bacterioplankton is practically colourless.

The data on optical properties of bacteria cells are scarce. Kopelevich *et al.* (1987) were among the pioneers in quantitatively assessing the contribution of bacterioplankton to the total scattering and absorption in marine waters. They reported that bacteria can account (at concentrations ranging from 0.1 to 0.3 million cells/cm^3) for about 12% of the total absorption ($\lambda = 550\,\mathrm{nm}$) and between 20–30% of the total scattering ($\lambda = 550\,\mathrm{nm}$).

Heterotrophic marine bacterioplankton with a mean equivalent radius $\tilde{d} = 0.55\,\mathrm{\mu m}$ and values of the refraction index and imaginary part of the complex refraction index in the visible spectrum averaging 1042–1068 and 10^{-4}, respectively,

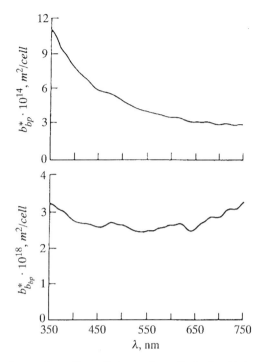

Fig. 3.5. Spectral dependences of specific scattering and specific backscattering coefficients for marine heterotrophic bacterioplankton (Stramski, Kiefer, 1990).

were studied by Stramski and Kiefer (1990). Figure 3.5 illustrates the specific (i.e. normalized to a bacterium cell) coefficients of scattering (b_{bp}^*) and backscattering ($b_{b_{bp}}^*$) reported by Stramski and Kiefer. The specific absorption coefficient at $\lambda = 400$–415 nm was estimated at 3.8×10–15 m^2/cell.

Model calculations indicate that the absorption efficiency factor for bacterio-plankton $Q_{a_{bp}}$ is considerably lower than its counterpart for algae. In particular, the bacterioplankton absorption efficiency factor is 10 times lower than that for blue–green algae in the blue portion of the spectrum and it is 40 times lower than that for large phytoplanktonic cells.

According to Ulloa, Sathyendranath (1992) the coefficients of scattering and backscattering for marine bacterioplankton ($\tilde{d} = 0.50$–$0.65\,\mu$m) at 440 and 550 nm varied in surface waters within the following limits:

	$\lambda = 440$ nm	$\lambda = 550$ nm
b_{bp},m^{-1}	0.020–0.205;	0.01–0.183;
$b_{b_{bp}} \times 10^3, \mathrm{m}^{-1}$	0.086–0.760;	0.08–0.953.

Ulloa, Sathyendranath report that bacterioplankton scattering accounted for 10 to

30% of the bulk-water total scattering coefficient and for 3 to 12% of the bulk-water total backscattering.

However, although both sets of data reported by Stramski and Kiefer (1990) and Ulloa, and Sathyendranath (1992) are very similar, they, are based on very scarce experimental evidence and certainly need further substantiation.

Interestingly, all attempts to reveal any stable correlation between the concentrations of phytoplankton and bacterioplankton have not yet resulted in reliably established relationships. Apparently, it can be considered as an indication that in general bacterioplankton contribution to bulk water total scattering and backscattering in a given water mass does not necessarily relate to its trophic status (Ulloa, Sathyendranath, 1992).

The relative contribution of bacterioplankton to the scattering due to suspended organic particulate matter might be very essential, particularly in the short wavelength region of the visible spectrum. Indeed, the total scattering by hydrosol varies as $\sim \lambda^{-1}$, whereas the spectral variation of bacterioplankton scattering is proportional to $\sim \lambda^{-2}$. Assessments of bacterioplankton contribution to the total scattering at $C_{bp} = 10^6$ cells/cm^3 and $C_{chl+phaeo-a} = 1$ mg/m^3 yielded at 400 and 700 nm 6% and 10%, respectively. As could be expected, in less productive (oligotrophic) waters ($C_{chl+phaeo-a} = 0.1$ mg/m^3), the corresponding contributions proved to be essentially higher: 17% and 28%, respectively (Stramski and Kiefer, 1990).

Dissolved organic carbon (DOC). DOC variability resides in spatial and temporal variations in concentrations of fulvic and humic fractions composing *doc*. The a_{doc} values display a particularly strong variance in a short-wavelength region of the visible spectrum. However, the dispersion of the doc specific absorption coefficient $\sigma^2(a^*_{doc})$ should be essentially lower than the dispersion of the doc absorption coefficient, i.e. $\sigma^2(a_{doc}) \gg \sigma^2(a^*_{doc})$. This enables parameterization of the *doc* specific absorption coefficient a^*_{doc}. For instance, for Lake Ontario the following exponential formula has been suggested (Bukata *et al.*, 1995):

$$a^*_{doc}(\lambda) = 0.173 \exp[-0.0157(\lambda - 400)] \tag{3.9}$$

There is a considerably better agreement in spectral shape (albeit not in the values of coefficients in eq. (3.9)) among the measured cross-sections of *doc* (see Bukata *et al.*, 1995) suggesting that the absorption for *doc* can be considered to be less variable than the absorption and scattering for particulates.

Detritus. Some authors report in the literature that very much like $a_{doc}(\lambda)$, the spectral distribution of absorption by detritus particulates a_{det} is exponential. Therefore, it could be approximated by an equation similar to (3.9) (Roesler, Perry, 1995):

$$a_{det} = a_{det}(\lambda_0) \exp[-S(\lambda - \lambda_0)] \tag{3.10}$$

where λ_0 is usually 390–400 nm, and S is equal to 0.011 ± 0.002 nm^{-1}.

When establishing the detritus *specific* absorption coefficient ($a^*_{det}(\lambda)$), it is important to identify the weight which the absorption coefficient a_{det} should be normalized to. The detritus filtered from water samples is basically "contaminated" by heterotrophic bacterioplankton, and decomposed phytoplankton, which accom-

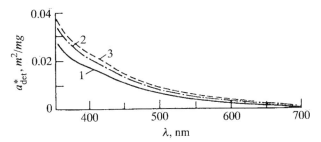

Fig. 3.6. Spectral variations in the detritus-specific absorption coefficient for three locations along the US western coastline: (1)—Pt. Conception , (2)—nearby San Diego, (3)—nearby San Francisco.

pany fragments of dead zooplankton, zooplankton fecal pellets, etc. If the assumption that there is a stable correlation between the amount of detritus and the phytoplankton concentration in the water column is valid, then a_{det} could be normalized to $C_{chl+phaeo}$. Figure 3.6 illustrates a family of averaged spectral distributions of $a^*_{det}(\lambda)$ obtained from a_{det} divided by $C_{chl+phaeo}$ (Sosik, Mitchel, 1995).

We are unaware of data on detritus-specific scattering and specific backscattering coefficients. It is certain that detritus should have a complex, multi-mode size distribution. It is not unreasonable to suppose that the spectral distribution of scattering by coarse detritus particulates is proportional to λ^0, whereas the spectral distribution of scattering by fine detritus particulates is roughly proportional to λ^{-1}. Therefore the net spectral distribution of the detritus specific absorption coefficient will almost be neutral. However, the degree of "neutrality" should depend in each particular case on the coarse-to-fine detritus fraction ratio. By and large, it is possible to expect that both absolute values and spectral distribution of specific $b_{b_{det}}$ should be close to those of phytoplankton.

Suspended minerals (sm)

Terrigenous matter brought to natural water bodies by runoff, shoreline erosion, and river discharge loses its coarse fraction in the vicinity of the coastline. Deep waters of oceans and seas contain only small fractions of suspended particulates with diameters less than 1 μm whereas particles larger than 1 μm are very scarce. The suspended inorganic matter consists of particles which are predominantly amorphous and crystal. The elemental composition of such suspended matter includes silicon, aluminum, and iron, generally in the form of oxides (*Ocean Optics. . .*, 1983). In inland water bodies, suspended sediment concentrations in surface waters prove to be rather high. Not infrequently a large portion of the *sm* coarse fraction is found not only in the coastal but also pelagic zones. The chemical composition of *sm* indicates that these are mainly quartz particulates, as well as particulates of silica clays and calcites. These materials have low imaginary parts of the complex index of refraction (10^{-3}–10^{-4}), and their absorption increases up to the near UV spectral region as well as at $\lambda > 700$ nm (Kondratyev *et al.*, 1983).

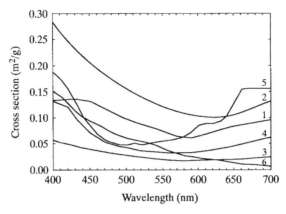

Fig. 3.7. An intercomparison of independent determinations of the absorption cross-section spectra for non-chlorophyll matter: (1) – Bukata *et al.*, 1985; (2) – Bukata *et al.*, 1981; (3) – Gallie, Murtha, 1992; (4) – Mored, Prieur, 1977; (5) – Prieur, Sathyendranath, 1981; (6) – Bowers *et al.*, 1996.

There are few published values for *sm* absorption ($a_{sm}^*(\lambda)$) and backscattering ($b_{b\,sm}^*$) cross-section spectra. Figure 3.7 is an intercomparison of absorption cross-section spectra obtained for non-chlorophyll matter by several workers. Generally, $a_{sm}^*(\lambda)$ decreases steadily with λ (Roesler *et al.*, 1989), but at $\lambda \geq 600$–650 nm it slightly begins to increase. The collected data are suggestive of $a_{sm}^*(\lambda)$ increasing from oceanic to marine and fresh water.

The specific backscattering coefficient of suspended minerals $b_{b\,sm}^*$ depends only slightly on the light wavelength: it decreases with λ but very slowly. Based on these findings, the following formulation has been suggested to describe spectral variations in $b_{b\,sm}^*$:

$$b_{b\,sm}^* = A - B(10^{-5})\lambda \qquad (3.11)$$

where A and B are constants. The A and B values were determined (Whitte *et al.*, 1982) at 0.0316 and 0.844, respectively. However, due to the reasons discussed above, these coefficients are not readily translatable to a wide variety of water bodies, and should rather be considered as area-specific.

Thus, the specific inherent optical properties (a_i^* and $b_{b\,i}^*$) of major OAC are subject to variations driven by physical, chemical and biological factors. The principal question here is to what degree these coefficients are conservative within a given water body and a given vegetation period, i.e. can they be used as input parameters for a hydro-optical model for a given water body or water mass.

One of the answers to this question might be the results reported by Roesler *et al.* (1989) and Roesler and Perry (1995) for a wide variety of natural waters indicating that the specific inherent optical properties of phytoplankton, *doc* and detritus could be characterized by so-called shape factors defining the spectral shape of respective cross-section spectra. The fact itself that such shape factors have been established is implicitly a sign of the existence of *intrinsic* spectral distributions for a_i^* and $b_{b\,i}^*$ whereas their absolute values may vary depending upon site-specific conditions.

Another indication in favor of this premiss came from extensive research reported by Bukata *et al.* (1991): a set of tabulated spectral values of a_i^* and b_{bi}^* for phytoplankton, *sm* and *doc* were established in the course of shipborne measurements run along a transect in Lake Ontario.

Field studies conducted each July for three consecutive years throughout the total area of Lake Ladoga provided further substantiation of the validity of attempts to set up a representative optical model of a natural water body. The measurements in Lake Ladoga provided spectral values of specific coefficients of absorption and backscattering of phytoplankton, *sm* and *doc* (Fig. 3.8, Kondratyev *et al.*, 1990c). Averaged for every year, separately for peripheral and pelagic parts of the lake, these specific coefficients revealed a high degree of identity.

A comparison of the averaged spectral values of a_i^* and $b_{b_i}^*$ for each of the three years showed that the interannual variations in a_i^* and $b_{b_i}^*$ are small. As a result, a hydro-optical model for Lake Ladoga has been suggested for the vegetation summer period. One of the merits of such a model resides in the fact that it enables to numerical assess the spectral values of absorption and backscattering coefficients of bulk water in any part of Lake Ladoga based exclusively on concentrations of major OAC determined *in situ*. On the other hand, such a model allows reproduction of optical properties in the water column with any combination of major OAC. This option proves to be a very valuable tool in developing approaches to inverse problem solution in case of retrieving water-quality parameters from remote sensing data.

3.1.1.2   *Upwelling radiance and its relation to inherent optical properties of natural waters*

When viewing a water surface from an altitude h at a zenith angle θ_v to the surface normal and at an azimuth angle φ_v to the principal plane of the sun, a remote sensing device receives the radiance L_t, which is made of

$$L_t(\theta_v, \varphi_v, z, \lambda) = [L_w(+0, \theta_v, \varphi_v, \lambda) + L_r(\theta_v, \varphi_v, \lambda)]T_a + L^*(\theta_v, \varphi_v, z, \lambda) \qquad (3.12)$$

where L_w, L_r are components arising from light coming out from beneath the water surface and reflected by the water surface in direction (θ_v, φ_v), respectively; T_a is that portion of the surface-leaving radiance which is neither absorbed nor scattered out of the field of view, and known as the diffuse transmittance of the atmosphere; L^* is the path radiance resulting from those sun rays crossing through the field of view of the remote sensor which are scattered towards it by the atmosphere.

Legitimate information about water-quality parameters (*viz.* concentrations of phytoplankton chlorophyll, suspended minerals and dissolved organics, and some others) is solely contained in the component L_w which could be related to water inherent optical properties (the coefficients of absorption a, scattering b, attenuation c, and $\beta(\theta_v, \varphi_v, \theta', \varphi', z, \lambda)$ the volume scattering function of the angle between the unscattered ray traveling in direction (θ', φ') and the scattered ray traveling in direction (θ_v, φ_v) through the coefficient of volume reflectance just below the water surface

$$R(-0, \lambda) = E_u(-0, \lambda)/E_d(-0, \lambda) \qquad (3.13)$$

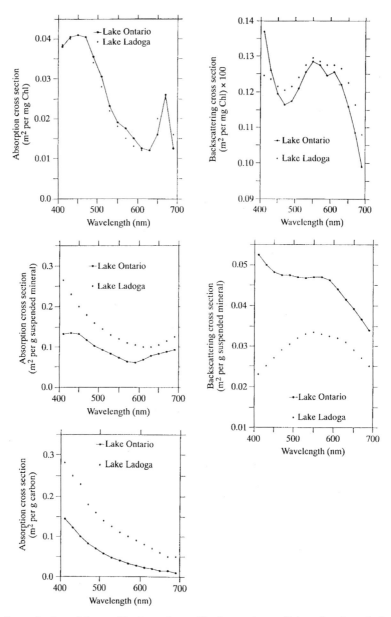

Fig. 3.8. Spectral values of the specific absorption and backscattering coefficients for phytoplankton (*chl*), *sm* and *doc* determined for Lake Ladoga (1) and Lake Ontario (2) waters (Bukata *et al.*, 1995).

where $E_u(-0,\lambda)$ and $E_d(-0,\lambda)$ are the upwelling irradiance and downwelling irradiance just below the water surface.

According to Bukata *et al.* (1995)

$$L_w(+0,\theta_v,\varphi_v,\lambda) = R(-0,\lambda)T_s(f_1 E_{sky} + f_2 E_{un})Q^{-1} \tag{3.14}$$

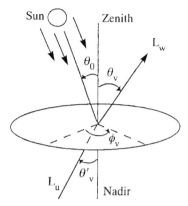

Fig. 3.9. Symbols and a schematic representation of geometry related to solar light propagation through the air–water interface (Myrmehl, Morel, 1996).

where T_s = transmission of nadir radiance through the air–water interface including correction for the n^2 radiance law (n is the relative index of refraction), Q^{-1} = ratio of the upwelling irradiance below the water surface to the upwelling radiance below the water surface, i.e. $Q = E_u(-0, \lambda)/L_u(-0, \lambda, \theta'_v, \varphi_v \to \theta_v, \varphi)$, where $L_u(-0, \lambda, \theta'_v, \varphi_v \to \theta_v, \varphi)$ = subsurface upwelling radiance in direction (θ'_v, φ) which, allowing for refraction at the water–air interface, assures photon trajectory in direction (θ_v, φ), $\theta'_v = \arcsin(\sin \theta_v)/n$ (Fig. 3.9), f_1 = fraction of the downwelling diffuse sky irradiance that is transmitted into the water, f_2 = fraction of the downwelling direct solar irradiance that is transmitted into the water.

Assuming the ocean is a perfect Lambertian surface, and bulk water is devoid of hydrosol (i.e. suspended mineral and organic particulates), Q should be direction and wavelength nearly independent and equal to a value of π. However, in reality the values of Q differ from π. According to Austin (1980), under cloudless skies in case 1 waters (Morel and Prieur, 1977) Q should in fact be maximum and approximately 5.0 for near nadir observations, whereas in near horizontal direction and $\varphi = 0°$, Q is minimal (Myrmehl and Morel, 1966). Factor Q increases with increasing solar zenith (θ_0), reaching its maximum at $\theta_0 = 75°$. It increases with the wavelength λ of incident light (at $\theta_0 \geq 20°$), and grows with the increasing ratio (η) of Raleigh scattering to total scattering in the water column (for $0° \leq \theta_0 \leq 35°$). However, for solar zenith $35° \leq \theta_0 \leq 90°$, Q decreases when η grows. At solar zenith $\leq 20°$, Q displays a very intricate wavelength dependence. As a result, the value of Q can acquire a large spectrum of values ranging between ~ 5.5 to ~ 3.7 (however, according to numerical simulations conducted by Morel et al. (1995), depending on the actual hydrosol loading, the lowest values of Q at $\theta_0 = 0°$ can descend down to 2.4–2.7). Such a considerable variability in Q values as a response to water-surface illumination conditions and the optical status of bulk water predetermines substantial inaccuracies in results of retrieval of water quality parameters from remote measurements.

At the same time, remote sensing generally involves measurements of upwelling radiance rather than upwelling irradiance. Therefore, it is more reasonable to consider the ratio of the upwelling radiance from a given nadir direction to the downwelling irradiance, when measured from a remote platform. This ratio is referred to as the *remote sensing reflectance* R_{rs} (Gordon *et al.*, 1984):

$$R_{rs}(+0, \theta_v, \varphi_v, \lambda) = L_w(+0, \theta_v, \varphi_v, \lambda)/(E_{sky} + E_{sun}) \qquad (3.15)$$

Coefficient R_{rs} can evidently be related to its value beneath the air–water interface, which is referred to as the *subsurface remote sensing reflectance*, R_{rsw} (Jerome *et al.*, 1996):

$$R_{rs}(+0, \theta_v, \varphi_v, \lambda) = R_{rsw}(-0, \theta_v', \varphi_v, \lambda)[(1 - \rho_s(\theta_0, E_{sky}/E_{sky} + E_{sun}))][(1 - \rho(\theta_v'))/n^2] \qquad (3.16)$$

where ρ_s = surface reflectivity for incident downwelling irradiance (a function of the solar zenith angle and the fraction of diffuse irradiance in the incident irradiation), $\rho(\theta_v')$ = internal water–air interface reflectivity for an incident angle of θ_v'.

Due to high sensitivity of both upwelling (E_u) and downwelling (E_d) subsurface vector irradiance to the light field spatial (angular) distribution, measurements of E_u and E_d are usually attended by high demands of horizontality in sensor positioning. However, measurements could be made easier, if instead of the vector irradiance, scalar irradiance is to be measured. A *modified subsurface remote sensing reflectance* R_{rsw}' can be introduced:

$$R_{rsw}'(-0, \theta_v', \varphi_v, \lambda) = L_w(-0, \theta_v', \varphi_v, \lambda)/E_{od}(-0, \lambda) \qquad (3.17)$$

where $E_{0d} = (E_d/\bar{\mu}_d)$ = downwelling scalar irradiance at $z = -0$, $\bar{\mu}_d$ = mean cosine for a downwelling irradiance distribution. In the general case, R_{rsw}' can be given by (Zaneveld, 1995):

$$R_{rsw}'(-0, \theta_v', \varphi_v, \lambda) = f_b(-0, \theta_v', \varphi_v, \lambda)b_b(-0, \lambda)/$$
$$2\pi[-\cos\theta_v' k(-0, \lambda) + c(-0, \lambda) - f_f(-0, \theta_v', \varphi_v, \lambda)b_f(-0, \lambda)] \qquad (3.18)$$

where b_b = backscattering coefficient, b_f = forward scattering coefficient, f_b and f_f = shape factors assuming the scattering function is spatially isotropic and equal to $[b_b/2\pi]$ and $[b_f/2\pi]$, respectively; k = vertical attenuation coefficient for radiance:

$$k(-0, \theta_v', \lambda) \equiv dL_w(-0, \theta_v', \varphi_v, \lambda)/L_w(-0, \theta_v', \varphi_v, \lambda)\, dz.$$

For the vertically upwelling radiance (i.e., $\theta_v' = \pi$), eq. (3.18) becomes:

$$R_{rsw}'(-0, \pi, \lambda) = f_b(-0, \pi, \lambda)(b_b/2\pi)/[k(-0, \pi, \lambda) + c(-0, \lambda) - f_f(-0, \pi, \lambda)(b_f(-0, \lambda)] \qquad (3.19)$$

Under clear-sky conditions, downwelling radiance distribution nearly always shows a well-defined maximum. Near the surface, this normally occurs at the angle of the refracted solar zenith. The cosine of this angle is also very close to the average cosine of the light field, as it dominates the radiance. The angle at which this maximum (θ_m) occurs changes only slowly with depth. Consequently, it is not unreasonable to assume that the light coming from any depth to the remote sensor is a result of light that is backscattered from the radiance maximum or its immediate neighborhood. Thus, hypothesizing that most of the nadir radiance originates from single scattering of light near the maximum radiance, we can write:

$$R'_{rsw}(-0, \pi, \lambda) = \beta(-0, \pi - \theta_m, \lambda)/[k(-0, \pi, \lambda) + c(-0, \lambda) - f_f(-0, \pi, \lambda)b_f(-0, \lambda)]$$

$$(3.20)$$

where $\beta(-0, \pi - \theta_m, \lambda)$ = volume scattering function in direction $(\pi - \theta_m)$.

Further, it is possible to show that, for angles neighboring θ_m, the vertical attenuation coefficient k, denoted as $k_m(-0, \pi, \lambda)$, equals the asymptotic diffuse attenuation coefficient K_∞. In this case, according to (Prieur and Morel, 1971):

$$k_m(-0, \pi, \lambda) = c[1 - 0.52b/c - 0.44(b/c)^2] \qquad (3.21)$$

$$R'_{rsw}(-0, \pi, \lambda) = \beta(-0, \pi - \theta_m, \lambda)/a(-0)[1 + m(-0) + b_b(-0) - 0.05b_f(-0)] \quad (3.22)$$

where $m(-0) \equiv k_m(-0, \pi, \lambda)/a(-0)$.

Unlike case-2 waters characterized by predominance of hydrosol scattering, case-1 waters exhibit strong molecular scattering, and as a result, the contribution of b_b in the denominator of eq. (3.22) might be substantial. However, for the majority of ocean waters

$$a(-0)[1 + m(-0)] \gg [b_b(-0) - 0.05b_f(-0)]$$

and hence:

$$R'_{rsw}(-0, \pi, \lambda) = \beta(-0, \pi - \theta_m, \lambda)/a(-0)[1 + m(-0)] \qquad (3.23)$$

As mentioned above, there is an evident relationship between R'_{rsw} and R_{rsw}:

$$R_{rsw} = L_u(-0)/E_d(-0) = L_u(+0)/E_{0d}(-0)\bar{\mu}_d = R'_{rsw}\overline{\mu_d} \qquad (3.24)$$

and for ocean waters in particular:

$$R_{rsw} = \beta(-0, \pi - \theta_m, \lambda)/a(-0)[1 + m(-0)]\overline{\mu_d} \qquad (3.25)$$

Coefficients R_{rsw} and R'_{rsw}, in their turn, can be related by definition to the diffuse, or otherwise termed, volume reflectance R:

$$R = E_u(-0)/E_d(-0) = QL_u(-0)/E_{0d}(-0)\overline{\mu_d} = QR_{rsw} = QR'_{rsw}/\overline{\mu_d} \qquad (3.26)$$

Diffuse reflectance R can be further expressed via the water inherent optical properties:

$$R \approx \varsigma b_b/a \qquad (3.27)$$

where ς is a parameter depending on the spatial (angular) distribution of the light field and the bulk water volume scattering function. The ς variability analysis, conducted in the single scattering approximation, indicates (Gordon, 1989; Sathyendranath and Platt, 1997) that the diffuse reflectance is bound to increase with increase in both solar zenith and water column turbidity (i.e. with the rising hydrosol abundance in water).

Through Monte Carlo simulations, a number of formulations have been suggested for relating the volume reflectance R to the water inherent properties and atmospheric conditions (*viz.* solar zenith angle, cloudiness, ratio between incident direct and diffuse solar radiation). Gordon *et al.* (1975), being concerned with ocean (case 1) water, used curve-fitting to Monte Carlo simulations and obtained the desired formulation as an expansion into a power series:

$$R = \sum_i r_i X^i \quad (i = 1, 2, 3)$$

where $X = (b_b/a)(1 + b_b/a)$. In this power series the first term is dominating: $r_1 = 0.32$ for clear skies and small solar zenith angles, and $r_1 = 0.37$ for overcast conditions.

Kirk (1984), who was concerned with turbid inland and coastal waters ($(b/a) = 30$), obtained for the case of non-vertical incidence the following relation:

$$R(-0, \theta'_0, \lambda) = (0.975 - 0.629\mu_0)b_b/a \tag{3.28}$$

where $\mu_0 = \cos \theta'_0$, θ'_0 is the in-water solar zenith refracted angle. For overcast conditions.

$$R(-0, \theta'_0, \lambda) = 0.437(b_b/a) \tag{3.29}$$

Through similar Monte Carlo simulations, Jerome *et al.* (1988) examined the dependence of the ζ parameter on water-scattering properties:

$$R(-0, \theta'_0, \lambda) = \begin{cases} (1/\mu_0) \times 0.319 b_b/a & \text{for } 0 \le b_b \le 0.25 \\ (1/\mu_0)[0.0013 + 0.267 b_b/a] & \text{for } 0 \le b_b \le 0.25 \end{cases} \tag{3.30}$$

For overcast conditions $\mu_0 = 0.858$. Jerome *et al.* (1988) point out that the above dependences do not strictly conform to the relevant experimental data at large solar zenith angles, which is perhaps suggestive of a possible second-order relationship between R and b_b/a, and the necessity of adding extra terms to eqs. (3.30).

Kirk (1991) also investigated the variation in volume reflectance with shape of the volume scattering function and derived the following expression:

$$R(-0, \theta'_0, \lambda) = \{(0.9713 - 2.290B)[(1/\mu_0) - 1] + 0.31\}Bb/a \tag{3.31}$$

where B is the backscattering probability, defined as the ratio of scattering into the backward hemisphere (the one trailing the incident flux) to the total scattering occurring in all directions.

As pointed out above, striving to exclude the poorly definable coefficient Q from the R formulations was the prime incentive for introducing the remote sensing reflectance R_{rs} or the subsurface remote sensing reflectance R_{rsw}. However, the analytical expressions for R_{rs} and/or R_{rsw}, being very meaningful in terms of revealing variations of R_{rs} and/or R_{rsw} with hydro-optical properties and spatial (angular) distribution of the underwater light field, prove too complicated to be readily used in practice when retrieving water-quality parameters from remotely sensed data. To overcome this difficulty, attempts have been undertaken to obtain simplified formulations of R_{rs} and/or R_{rsw} through Monte Carlo simulations (Jerome *et al.*, 1996). A wide spectrum of hydro-optical conditions have been explored

$(b_b/a = 1.5; 3.0; 9.0; 19.0$, which are equivalent to the single scattering albedo $w_0 = b/c \, 0.60; 0.75; 0.90; 0.95$, respectively). The volume scattering functions, taken from Petzold (1972), represented typical nearshore, and offshore waters (the B values were 0.0133, 0.0150, 0.0192, 0.0253, 0.0378 and 0.0440). The numerical simulations were run for solar zenith angles 0, 15, 30, 45, 60, 75, 89 degrees as well as for two patterns of incident diffuse radiation, *viz.* an isotropic radiance distribution and a cardioidal radiance distribution (Moon and Spencer, 1942). The following relationship between R_{rsw} and water-inherent properties has been established $(r = 0.99)$:

$$R_{rsw} = -0.000\,42 + 0.112(b_b/a) - 0.0455(b_b/a)^2 \tag{3.32}$$

For a zenith sun $(\theta_0' = 0)$

$$R_{rsw} = -0.000\,53 + 0.112(b_b/a) - 0.0432(b_b/a)^2 \quad (r = 0.99) \tag{3.33}$$

For other solar zenith angles, R_{rsw} could be best expressed as a function of θ_0 and w_0:

$$R_{rsw}(-0, \theta_v', \lambda) = G(\theta_v', w_0) R_{rsw}(0°), \quad 0.65 \leq w_0 \leq 0.95 \tag{3.34}$$

where

$$G(\theta_v', w_0) = G(\theta_v', 0.60) + [G(\theta_v', 0.95) - G(\theta_v', 0.60)](2.3 - 8.0w_0 + 7.0w_0^2) \tag{3.35}$$

$$G(\theta_v', 0.60) = 1.00 - 0.0119\theta_v' + 0.000\,103\theta_v'^2 \tag{3.36}$$

$$G(\theta_v', 0.95) = 1.00 - 0.002\,05\theta_v' + 0.000\,0215\theta_v'^2 \tag{3.37}$$

For the incident isotropic and cardioidal radiance distribution, the Monte Carlo results had an average difference of only 3.4%. Therefore for each water type, the R_{rsw} values for isotropic and cardioidal distributions were averaged to represent the diffuse situation:

$$R_{rsw}(diff) = (1.37 - 215w_0 + 1.89w_0^2) R_{rsw}(0°) \tag{3.38}$$

The value of R_{rsw} averaged over the solar zenith angles 15–89° can be formulated as follows:

$$\overline{R_{rsw}} = -0.000\,36 + 0.110(b_b/a) - 0.0447(b_b/a)^2 \quad (r = 0.99) \tag{3.39}$$

For the diffuse incidence situation

$$R_{rsw}(diff) = 1.045\overline{R_{rsw}} \tag{3.40}$$

From eq. (3.40) it can be seen that the sub-surface remote sensing reflectance for a diffuse incident distribution is only 4.5% higher than the average sub-surface remote sensing reflectance for the six direct incidence conditions.

The above equations are of importance for remote sensing. To a first approximation, R_{rsw} is rather independent of not only solar zenith angle, but also of the proportion of direct and diffuse radiation in the incident radiation distribution. However, R_{rsw} is most sensitive to the inherent optical properties of the bulk water, *viz.* the ratio b_b/a:

$$(b_b/a) = 0.0027 + 9.87 R_{rsw} - 34.5(R_{rsw})^2 + 1.534(R_{rsw})^3 \quad (r = 0.99) \qquad (3.41)$$

Consequently, there is a possibility to use eq. (3.41) in order to develop a retrieval algorithm for inferring water-quality parameters from remotely sensed upwelling radiance $L_w(+0.\lambda)$.

Nevertheless, the R formulations discussed above have been derived assuming that solar light traveling through the water column interacts with the aquatic medium through the mechanisms of absorption and elastic scattering. In reality, the source function in the radiative transfer equation should in general include not only elastic but also inelastic (Raman) scattering on water molecules, as well as the phytoplankton and dissolved organics fluorescence. In addition, the above considerations implicitly assume that the water body under surveillance is infinitely deep whereas, rather often, the upwelling radiance contains a certain contribution due to bottom reflection. It means, that the remotely measured signal ($L_w(+0, \lambda)$) is normally contaminated with "noise", and to achieve better retrieval results, it is necessary to develop approaches providing tools to get rid of this "noise". It could be done if these "extraneous" radiance contributions are parameterized. Consequently, it warrants a close consideration of relevant formulations for Raman scattering, fluorescence and bottom effect.

Raman scattering. Scattering bringing about a change in light frequency due to excitation of vibrational modes in a molecule is termed Raman scattering. In water, the dominant Raman line is produced by the fundamental O–H stretch mode of the water molecule. According to Waters (1995), the Raman band is relatively broad and centered near $3330 \, cm^{-1}$. Since the differential Raman scattering cross-section of water $(d\sigma_{H_2O}/d\Omega)_{90°}$ is not small at all (evidently, the most reliable value of $(d\sigma_{H_2O}/d\Omega)_{90°}$ is *ca.* $8.3 \times 10^{-30} \, cm^2$ molecule^{-1} sr^{-1} at $\lambda_{ex} = 488$ nm in a direction perpendicular to the polarization plane of the incoming light beam, integrated over the Raman emission linewidth, per molecule, and for the Raman frequency shift of $3400 \, cm^{-1}$), it is warranted to assess the possible contribution of the water Raman scattering to total upwelling radiance L_w. Integration of the differential Raman cross-section of water over 4π results in a volume scattering coefficient b_r:

$$d\sigma_{H_2O}/d\Omega = (d\sigma_{H_2O}/d\Omega)_{90°} \frac{[\rho + (1 - \rho)\sin^2\alpha]}{(1 + \rho)} \qquad (3.42)$$

(α = scattering angle, ρ = depolarization ratio of water),

$$b_r(\delta\nu_r) = N \int_{4\pi} \left(\frac{d\sigma_{H_2O}}{d\Omega} \right) d\Omega \qquad (3.43)$$

where $\delta\nu_r$ is the Raman frequency shift, N is the number of water molecules in cm^{-3} (at normal ambient pressure $N = 3.3 \times 10^{22} \, cm^{-3}$ (Fadeev *et al.*, 1982)). A double integration of eq. (3.43) over 4π and the Raman frequency shift $\delta\nu_r$ results in $b_r = 2.6 \times 10^{-4} \, m^{-1}$ at $\lambda_{ex} = 488$ nm. The intensity of the Raman spectrum (i.e. b_r) is proportional to λ_{em}^{-5}, where λ_{em} is the wavelength of the inelastically scattered light

(Sugihara *et al.*, 1984). Consequently, to calculate b_r at other than 488 nm wavelengths, a simple relationship can be used:

$$d\sigma_{H_2O}/d\Omega = d\sigma_{H_2O}/d\Omega)_{\lambda=488\,nm} \left(\frac{\nu_i - \overline{\delta\nu_r}}{\nu_{488} - \overline{\delta\nu_r}}\right)^5 \tag{3.44}$$

where $\nu_i = 1/\lambda_i$, $\overline{\delta\nu_r} =$ mean Raman frequency shift. For calculating b_r at $\lambda < 488$ nm, a relationship inverse to (3.44) can be derived. Sugihara *et al.* (1984) report an expression for determining the wavelength of the Raman band maximum ($\lambda_{em\,max}$) as a function of the impinging light wavelength (λ_{ex}):

$$\lambda_{em\,max}(\lambda_{ex}) = \frac{\lambda_{ex}}{-3357.1 - ^{-7}\lambda_{ex} + 1} \tag{3.45}$$

The Raman scattering phase function could be approximated via (Porto, 1966):

$$\beta_r(\theta) = \frac{3}{16\pi} \frac{1+3\rho}{1+2\rho}\left(1 + \frac{1-\rho}{1+3\rho}\cos^2\theta\right) \tag{3.46}$$

where $\rho =$ depolarization ratio of water ($\rho = 0.17$ at $\overline{\delta\nu_r} = 3400\,cm^{-1}$ (Ge *et al.*, 1995)).

Therefore, the Raman scattering function is symmetric and conforms to the cosine angular distribution which is very similar to the phase function of elastic water scattering (Morel and Gentini, 1991):

$$\beta_w(\theta) = \frac{3}{4\pi} \frac{3}{3+p}(1 + p\cos^2\theta) \tag{3.47}$$

where p = polarization factor (equal to 0.84 for pure water).

Seeking to assess the contribution of Raman scattering on water molecules, Marshall and Smith (1990) have given an approximate solution to the irradiance reflectance problem using the two-stream method for pure water. They found that for small solar zenith angles, a vertically homogeneous water column, and an isotropic Raman scattering function, the irradiation of a horizontal plane z due to Raman scattering can be formulated as follows:

$$E_u^r(\lambda_{em}, z) = \frac{b_r(\lambda_{ex})}{4} E_0(\lambda_{ex}, 0)\left(G_+ \frac{e^{-K(\lambda_{ex})z}}{K^r(\lambda_{em}) + K(\lambda_{ex})} + G_- \frac{e^{-K(\lambda_{ex})z} - e^{-K^r(\lambda_{em})z}}{K^r(\lambda_{em}) - K(\lambda_{ex})}\right) \tag{3.48}$$

where $E_0(\lambda_{ex}, 0) =$ scalar irradiance just beneath the water surface at λ_{ex}; $G_\pm = \frac{K^r}{2a_r}\left(1 \pm \frac{a_r}{K^r}\right)$, the effective Raman diffuse attenuation coefficient $K^r = \sqrt{a_r(a_r + 2b_b)}$, $a_r =$ effective Raman absorption coefficient $a_r = 1.5\,a_w(\lambda_{ex})$ (the use of $a_r = 1.5a_w(\lambda_{ex})$ resulted from an approximation of an exponential integral and the assumption that Raman light is nearly diffuse).

For $z = -0$, eq. (3.48) becomes:

$$E_u^r(\lambda_{em}, 0) = \frac{b_r(\lambda_{ex})}{4} E_0(\lambda_{ex}, 0) \frac{G_+}{K^r(\lambda_{em}) + K(\lambda_{ex})} \tag{3.49}$$

In order to assess the relative contribution of the water Raman irradiance at $z = -0$, a dimentionless ratio $\xi(\lambda_{em})$ can be used:

$$\xi(\lambda_{em}) = E_u^r(\lambda_{em}, 0)/E_u^r(\lambda_{em}, 0)_+ E_u'(\lambda_{em}, 0) \tag{3.50}$$

where $E_u'(\lambda_{em}, 0)$ = upwelling subsurface irradiance arising from elastic scattering of sunlight propagating through the water column.

Analyzing the irradiance spectral distribution of Raman upwelling, Waters (1995) found that the most substantial contribution of Raman scattering to the total water-leaving radiance is in the long wavelength region of the visible spectrum where $\xi \geq 0.2$ ($\lambda \geq 650\,$nm). The presence of a maximum in the spectral distribution of $\xi(\lambda_{em})$ (Fig. 3.10) is due to both relatively low pure water absorption in the spectral range 560–580 nm and high pure water absorption at $\lambda = 720\,$nm.

Addition to pure water of spectrally neutral hydrosol (i.e. with the diffraction parameter $\chi = (2\pi r/\lambda) \geq 30$–40, where r = particle's radius) results in dilapidation of $\xi(\lambda_{em})$ due to enhanced extinction of downwelling solar radiation.

Monte Carlo simulations indicate (Waters, 1995) that for $\theta_0 = 20°$ and phytoplankton concentration $C_{chl} = 0.01\,\mu g/l$ (with the Petzold, 1972 phase function), the fractional (relative) contribution of water Raman scattering to the water-leaving radiance exhibits a slight spatial (angular) structure, and a strong dependence on the wavelength.

The increase of C_{chl} up to 0.1 and further to $1.0\,\mu g/l$ brings about (due to absorption of photosynthetic pigments in the spectral range 410–500 nm) a dramatic

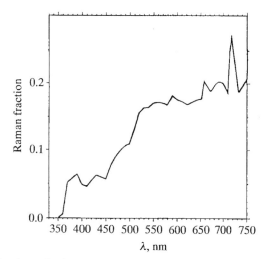

Fig. 3.10. The fractional contribution to water-leaving upwelling radiance from water Raman for pure water (Waters, 1995).

drop in Raman contribution to water-leaving radiance throughout the visible spectrum. This is accompanied by strengthening of the angular dependence of $\xi(\lambda_{em})$, whereas its spectral distribution nearly becomes neutral.

A substantial decrease in $\xi(\lambda_{em})$ should equally be expected from increased content of dissolved organic matter because of its strong absorption in the short wavelength region of the visible spectrum.

Consequently, it could be stated that water Raman contribution to water-leaving radiance is maximum in clear waters. This contribution (which can be as high as 20% in the red portion of the visible spectrum) might be essential for achieving an adequate atmospheric correction if based on the assumption that the upwelling radiance from clear water pixels at $\lambda \geq 670\,nm$ is zero (the well-known method suggested by Gordon Wang (1994). Moreover, in view of the spectral distribution of $\xi(\lambda_{em})$ discussed above, water Raman can bring forth a considerable error when retrieving oligotrophic water-quality parameters through algorithms based on the ratio of radiance at 443 nm (the Raman contribution is minimal) and 550 nm (the Raman contribution is substantial).

Within the two-stream approximation approach, an expression analogous to (3.49) can be derived for the downwelling Raman irradiance just beneath the water surface (i.e. at $z = -0$):

$$
\begin{aligned}
E_d^r(\lambda_{em}, z) = \frac{(b_r(\lambda_{ex}))}{4} E_0(\lambda_{ex}, 0) \Bigg(G_- \frac{e^{-K(\lambda_{ex})z}}{K^r(\lambda_{em}) + K(\lambda_{ex})} \\
+ G_+ \frac{e^{-K(\lambda_{ex})z} - e^{-K^r(\lambda_{em})z}}{K^r(\lambda_{em}) - K(\lambda_{ex})} \Bigg)
\end{aligned}
\tag{3.51}
$$

Using eqs. (3.48, 3.51), and introducing solar downwelling and upwelling irradiances at depth $z(E_d^{solar}(z, \lambda_{em})e^{-K(\lambda_{em})z}$, $E_u^{solar}(z, \lambda_{em})e^{-K(\lambda_{em})z}$, respectively), it is possible to formulate the Raman diffuse reflectance $R(\lambda_{em}, z)$ for small solar zenith angles θ_0:

$$
\begin{aligned}
R(\lambda_{em}, z) = \Bigg\{ E_u^{solar}(z, \lambda_{em})e^{-K(\lambda_{em})z} + \frac{b_r(\lambda_{ex})}{4} E_0(\lambda_{ex}, 0) \\
\times \Bigg(G_+ \frac{e^{-K(\lambda_{ex})z}}{K^r(\lambda_{em}) + K(\lambda_{ex})} + G_- \frac{e^{-K(\lambda_{ex})z} - e^{-K^r(\lambda_{em})z}}{K^r(\lambda_{em}) - K(\lambda_{ex})} \Bigg) \Bigg\} \\
\times \Bigg\{ E_d^{solar}(z, \lambda_{em})e^{-K(\lambda_{em})z} + \frac{b_r(\lambda_{ex})}{4} E_0(\lambda_{ex}, 0) \\
\times \Bigg(G_- \frac{e^{-K(\lambda_{ex})z}}{K^r(\lambda_{em}) + K(\lambda_{ex})} + G_+ \frac{e^{-K(\lambda_{ex})z} - e^{-K^r(\lambda_{em})z}}{K^r(\lambda_{em}) - K(\lambda_{ex})} \Bigg) \Bigg\}^{-1}
\end{aligned}
\tag{3.52}
$$

When $z \to \infty$, eq. (3.52) simplifies to:

$$
R(\lambda_{em}, z \to \infty) \quad \frac{K^r(\lambda_{em}) - K(\lambda_{ex})}{K^r(\lambda_{em}) + K(\lambda_{ex})}
\tag{3.53}
$$

It could be concluded from eq. (3.53) that at large depths in clear ocean waters, a light field in the red portion of the visible spectrum originates from water Raman scattering (as an internal light source). This is very important in the prospective of hydrobiology in general and adequate assessment of deep-water algae photosynthetic activity in particular (Stavn and Weiderman, 1992).

Fluorescence. Fluorescence is a physical process resulting in emission of light by a substance while it is irradiated by light or other electromagnetic radiation. In natural water, two major fluorescent substances generally coexist: dissolved organic matter and phytoplankton chlorophyllous pigments.

Fluorescence of dissolved organic matter (doc). For a pure substance (i.e. a single simple fluorophore), the fluorescence, regardless of the excitation energy, is due to electron transition from the first excited singlet state down to the ground state. As a result, the position of the corresponding fluorescence line does not shift and its width does not change by varying the excitation wavelength, although the strength of the fluorescent line proves to be excitation wavelength dependent. However, when studying variations in the spectral distribution of *doc* fluorescence with the excitation wavelength, three fluorescence maxima have been revealed at $\lambda_{em}^{\max} = 430\,\text{nm}$ ($\lambda_{ex}^{\max} = 340\,\text{nm}$), $310\,\text{nm}$ ($\lambda_{ex}^{\max} = 275\,\text{nm}$), and $450\,\text{nm}$ ($\lambda_{ex}^{\max} = 260\,\text{nm}$) (Coble and Brophy, 1996). The first two maxima arise from the fluorescence of humic substances of complex composition (which display a distinct increase in $\lambda_{em}^{\max}$ when $\lambda_{ex}^{\max}$ shifts to a longer wavelength region) whereas the third maximum falls into the spectral region characteristic of proteins and dissolved free amino acids, and could be tentatively ascribed to the latter. The locations of $\lambda_{em}^{\max}$ and $\lambda_{ex}^{\max}$ along the emission and excitation axes respectively are sensitive to the water mass type: unlike marine *doc* fluorescing bluish light, terrigenous doc emits greenish light. Consequently, variations both in spectral distribution of the excitation light, and the origin of *doc* (prevalence of either allochthonous or autochthonous components) are capable of changing the intensity and spectral composition of the upwelling fluorescent irradiance.

Introducing the volume fluorescence function $b_F^V(\alpha, \lambda_{ex}, \lambda_{em})$ $(\text{m}^{-1}/\text{nm sr})$ (Gordon, 1979)

$$b_F^V(\alpha, \lambda_{ex}, \lambda_{em}) = I(\alpha, \lambda_{em})/\Delta\text{V} \int_{\Delta\lambda_{ex}} E(\lambda_{ex})\, d\lambda_{ex} \tag{3.54}$$

where α = scattering angle with respect to the incident beam, $I(\alpha, \lambda_{em})$ = intensity of radiance emitted from a small volume ΔV at angle α, $E(\lambda_{ex})$ = irradiance in a band of wavelengths $\Delta\lambda_{ex}$ impinging on ΔV, then the coefficient of fluorescence $\phi(\lambda_{ex}, \lambda_{em})$ $(\text{m}^{-1}/\text{nm})$ can be defined as

$$\phi(\lambda_{ex}, \lambda_{em}) \equiv \int_\Omega b_F^V(\alpha, \lambda_{ex}, \lambda_{em})\, d\Omega \tag{3.55}$$

where $d\Omega = \sin\alpha\, d\alpha\, d\phi$, ϕ = azimuth angle.

The quantum efficiency $\phi(\lambda_{ex}, \lambda_{em})$, defined as the rate at which photons are emitted from ΔV with wavelengths between λ_{em} and $\lambda_{em} + \Delta\lambda_{em}$ divided by the rate at which photons with wavelengths between λ_{ex} and $\lambda_{ex} + \Delta\lambda_{ex}$ are absorbed in ΔV is given by

$$\phi(\lambda_{ex}, \lambda_{em}) = (\lambda_{em}/\lambda_{ex})(\phi(\lambda_{ex}, \lambda_{em})\Delta\lambda_{em}/a(\lambda_{ex}) \tag{3.56}$$

where $a(\lambda_{ex})$ is the absorption coefficient (m^{-1}) of the component of substance in ΔV which is resonsible for the fluorescence of intensity $I(\alpha, \lambda_{em})$.

The quantum yield (efficiency) of fluorescence of *doc* varies with λ_{ex}, and according to Green and Blough (1994) displays a pronounced maximum at ~ 380–400 nm, a shoulder at ~ 355 nm and a steep fall down to zero at $\lambda_{ex} \geq 475$ nm. The absolute value of ϕ (Fig. 3.11) depends on the origin of *doc*, i.e. the ratio between fulvic and humic acids. Haves *et al.* (1992) suggested the following formulation for $\phi(\lambda_{ex}, \lambda_{em})$:

$$\phi(\lambda_{ex}, \lambda_{em}) = A_0(\lambda_{ex}) \exp\left\{-\left[\frac{\lambda_{em}^{-1} - A_1\lambda_{ex}^{-1} - B_1}{0.6(A_2\lambda_{ex} + B_2)}\right]^2\right\} \tag{3.57}$$

where A_0, A_1, A_2, B_1, B_2 = regression coefficients given in Ge *et al.* (1995, Table 3.1) for fulvic and humic acids.

When substituting the regression coefficient values from Table 3.1 in Ge *et al.*, (1995) into (2.46), it should be noted the fraction of fulvic acid to total was assumed therein to be 90 %.

The fluorescence phase function

$$\beta_F(\alpha, \lambda_{ex}, \lambda_{em}) \equiv 4\pi b_F^V(\alpha, \lambda_{ex}, \lambda_{em})/\phi(\lambda_{ex}, \lambda_{em}) \tag{3.58}$$

if the fluorescent material in solution (in our case *doc*) is independent of the angle α (i.e. the emission is isotropic), and consequently, $\beta_F(\lambda_{ex}, \lambda_{em}) = 1$.

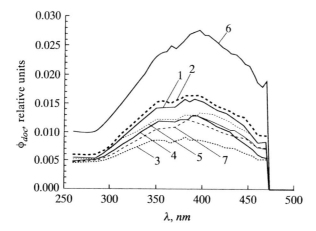

Fig. 3.11. Wavelength dependence of the quantum yield of *doc* extracts from water samples collected near south Florida in spring (1, 2) and fall (3, 4), from the Amazon River mouth (5), the Tamiami River (6), and the Suwannee River (7).

Spitzer and Wernand (1981) report that the irradiance just beneath the water surface ($z = -0$) originating from *doc* fluorescence can be given by

$$E_u^F(0, \lambda_{em}) = 0.5 \int_{\lambda_{ex}} \int_z \delta(\lambda_{em}) D_d(z, \lambda_{ex}) E_d^{solar}(0, \lambda_{ex}) e^{-K_d(\lambda_{ex})z}$$

$$\times \, \phi(\lambda_{ex}, \lambda_{em}) D_u(z, \lambda_{em}) e^{-K_u(z, \lambda_{em})z} \, dz \, d\lambda_{em} \tag{3.59}$$

where D_d, D_u are parameters determining the angular distribution of upwelling and downwelling fluxes (for solar direct radiation $D_d = \mu_0^{-1}$, $\mu_0 = \cos \theta_0'$, θ_0' is as before the refracted solar zenith angle). For isotropic emission, $D_u \approx 2$ (Preisendorfer, 1976). $\delta(\lambda_{em})$ is the spectral distribution of *doc* fluorescence intensity.

It is possible to introduce the volume reflectance coefficient R_F analogous to the irradiance reflectance:

$$R_F = E_u^F(0, \lambda_{em}) / E_d^{solar}(0, \lambda_{ex}) \tag{3.60}$$

The ratio $R_F / R_F + R$ (where R is the diffuse reflectance in the absence of inelastic scattering and *doc* fluorescence) can then be used as a measure of *doc* fluorescence contribution to the total water-leaving irradiance.

According to the numerical simulations conducted by Vodacek *et al.* (1994), the spectral distribution of $E_u^F(0, \lambda_{em})$ displays a maximum in the blue–green to green region between 490 and 530 nm with the width of the peak > 100 nm. The position of the peak emission wavelength varied depending on the combinations of optically active components used in calculations (i.e. water type), whereas the height of the $E_u^F(0, \lambda_{em})$ maximum acquired values ranging from 1.5×10^{12} to 3×10^{12} (photons s^{-1} cm^{-2} 5 nm^{-1}). The vertical profile of $E_u^F(0, \lambda_{em})$ proved to be characterized by a drastic drop over the first tens of centimeters downward from the air–water interface. *doc* fluorescence irradiance decrease with depth is particularly pronounced in the short wavelength region of λ_{em}.

Values of upwelling *doc* fluorescence contribution normalized to downwelling solar irradiance ($R_F = E_u^F(-0)/E_d(-0)$) as a percentage of $R_F + R$ (R is the irradiance reflectance) at wavelengths 415, 445, 510 and 555 (which are close to the SeaWiFS sensor bands) are given in Fig. 3.12 (Vodacek *et al.*, 1994) for four cases: the coefficient of *doc* absorption $a_{dom}(355) = 32.91 \, m^{-1}(1.2)$, $2.06 \, m^{-1}(3)$, $0.69 \, m^{-1}(4)$; the chlorophyll concentration $C_{chl} = 0.2 \, \mu g/l$ (2.4); the suspended minerals concentration $C_{sm} = 0.1 \, mg/l$ (1–3), $0.3 \, mg/l$ (4), $\phi(350) = 1.1 \times 10^{-2}(1-4)$. As can be seen from Fig. 3.12, the maximum contribution of *doc* fluorescence to the water-leaving flux should be expected within the spectral region 450–550 nm.

This contribution which can be as high as $\sim 70\%$ is proportional to *doc* concentration and inversely proportional to bulk water turbidity (i.e. inversely proportional to the *sm* content in the water column).

Fluorescence of phytoplankton chlorophyll

The colour of an algal cell is dictated by the combination of major pigments constituting the cell's pigment system, *viz.* chlorophylls, carotenoids, and phycobilins

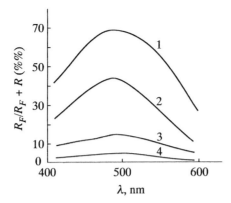

Fig. 3.12. Spectral variation in relative contribution (in %) of *doc* fluorescence irradiance reflectance R_F to total irradiance reflectance $(R_F + R)$ for water samples from Oyster Bay.

(Weaver and Wrigley, 1994). Amongst these pigments, only chlorophylls and phycobilins (allophycocyanin, phycocyanin, phycoerythrin) are fluorophores. Unlike the chlorophylls (first of all chlorophyll-a) which are present in all algae, without exception, phycobilins are characteristic of a small number of algae, among which are blue–greens (or Cyanobacteriae).

The maximum wavelength of chlorophyll fluorescence is 685 nm, of phycoerythrin 570–580 nm, of phycocyanin 650–660 nm, of allophycocyanin 660 nm. The range of excitation wavelengths for chlorophylls extends from $\sim$350 to $\sim$550 nm. Both the position (in general, this is spectral range 425–475 nm) and strength of maxima(um) in the excitation spectral distribution depend on phytoplankton taxonomic group, relative absorbance of the cell photosystem II pigments, the degree of photoadaptation to high/low solar irradiation levels, as well as some other factors (Mitchell and Kiefer, 1986).

The fluorescent flux $F(\lambda_{ex})$ emitted by a spherical planktonic cell with radius r can be given by

$$F(\lambda_{ex}) = E(\lambda_{ex})\phi(\lambda_{ex}, \lambda_{em})Q_a(\lambda_{ex})\pi r^2 Q_a^*(\lambda_{em}) \tag{3.61}$$

where $\phi(\lambda_{ex}, \lambda_{em})$ is the quantum yield of the chlorophyll fluorescence, $Q_a(\lambda_{ex})$ is the absorption efficiency factor: $Q_a(\lambda_{ex}) = 4/3Q_a^*(\lambda_{em})a_{chl}(\lambda_{ex})_r$, $Q_a^*(\lambda_{em})$ is the factor of intracellular reabsorption: $Q_a^*(\lambda_{em}) = (c_i/a_{cm}(\lambda_{em}))a_{chl}^*(\lambda_{em})$, c_i is the intracellular concentration of chlorophyll-a, $a_{cm}(\lambda_{ex})$ is the cellular material absorptivity, $a_{chl}^*(\lambda_{ex})$ is the chlorophyll-a absorption cross-section (i.e. per algal cell). A specific (i.e. per algal cell) coefficient (or else, cross-section) of fluorescence can be expressed as (Bricaud and Morel, 1986):

$$F^*(\lambda_{ex}) = F(\lambda_{ex})/E(\lambda_{ex})4/3\pi r^3 c_i = [\phi(\lambda_{ex}, \lambda_{em})/a_{cm}^*(\lambda_{em})]a_{chl}^*(\lambda_{ex})a_{chl}^*(\lambda_{em}) \tag{3.62}$$

where $a_{cm}^*(\lambda_{em}) = a_{cm}(\lambda_{em})/c_i$ is the cellular material absorption cross-section. The spectral distribution of $\phi(\lambda_{ex}, \lambda_{em})$ is, in general, rather isotropic at $\lambda_{ex} > \sim 520$ nm,

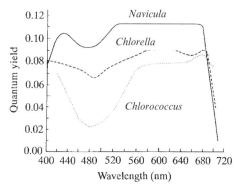

Fig. 3.13. Quantum yield as a function of wavelength in three unicellular algae. (———) *Navicula minima*, diatom; (– – –) *Chlorella pyrenoidosa*, green; (······) *Chlorococcus* spp., blue–green.

and is species specific (Fig. 3.13 (Kirk, 1983)). It also depends on the light availability: the $\phi(\lambda_{ex}, \lambda_{em})$ value decreases by a factor of two when the irradiance of the cell increases by a factor of 15–20 (Mitchel and Kiefer, 1986). It indicates that the major variability in $\phi(\lambda_{ex}, \lambda_{em})$ is predominantly due to variations in specific coefficients $a^*_{chl}(\lambda_{ex})$ and $a^*_{chl}(\lambda_{em})$.

According to Gordon (1979), the absolute values of $\phi(\lambda_{ex}, \lambda_{em})$ in the visible spectrum vary for different species from 1.5% to 14%, averaging $\sim 5\%$ (see also Fig. 3.13).

Since the phytoplankton are generally very irregularly shaped and randomly oriented, and species and size distributions are poorly known, it appears not unreasonable to assume that the phytoplankton fluorescence phase function is isotropic, i.e. $\beta_F(\alpha, \lambda_{ex}, \lambda_{em}) \approx 1$. It would not either be unrealistic to take the fluorescence quantum yield as spectrally invariant over the spectral region 370–690 nm. Under these conditions, the upwelling irradiance at a given depth z generated by the chlorophyll fluorescence could be given as (Gordon, 1979):

$$E_u^F = \delta_0 \xi \frac{1}{2} \int_{\lambda_{ex1}}^{\lambda_{ex2}} a(\lambda_{ex}) \lambda_{ex} E_d^{solar}(z, \lambda_{ex}) \exp[-K(\lambda_{ex})z] f(\lambda_{ex}, \lambda_{em}) \, d\lambda_{ex} \qquad (3.63)$$

where $\delta_0 = (2\pi\sigma^2)^{-1/2} \exp -\dfrac{(\lambda_{em} - \lambda_{0em})^2}{2\sigma^2}$, $\lambda_{0em} = 685$ nm, $\xi = \phi/c(\lambda_{em}) 4\pi \lambda_{0\,em}$.

$$f(\lambda_{ex}, \lambda_{em}) = \frac{K(\lambda_{em})}{K(\lambda_{ex})} \left\{ 1 + \frac{K(\lambda_{em})}{K(\lambda_{ex})} \ln\left[1 + \frac{K(\lambda_{em})}{K(\lambda_{ex})}\right] \right\} \qquad (3.64)$$

Based on eq. (3.63), the phytoplankton fluorescence contribution to water-leaving light can be assessed through the following expression:

$$R(\lambda_{em}) - R^{solar}(\lambda_{em}) = \phi(2\pi\sigma^2)^{-1/2}/2K(\lambda_{em})\mu_0' E_d^{solar}(-0, \lambda_{em})\lambda_{em}$$

$$\times \int_{\lambda_{ex1}}^{\lambda_{ex2}} \lambda_{ex} a(\lambda_{ex}) E_d^{solar}(-0, \lambda_{ex}) f(\lambda_{ex}, \lambda_{em}) \, d\lambda_{ex} \qquad (3.65)$$

Gordon (1979) reports that according to his simulation estimations (for $\phi \approx 0.7\%$) the difference $R(\lambda_{em}) - R^{solar}(\lambda_{em})$ can reach 0.5% at $\lambda_{0em} = 685\,\mathrm{nm}$.

Numerical modelling results (Fisher and Kronfeld, 1990) are suggestive of changes in the planktonic fluorescence angular distribution: initially isotropic, it becomes anisotropic as the emitted light propagates through the water medium and eventually through the water–air interface.

As a result, the fluorescence intensity decreases by more than a factor of two from nadir to horizon directions. As expected, with the increase in solar zenith angle as well as in *doc* concentration, the fluorescent radiance signal rapidly dilapidates. In clear ocean waters, about 90% of upward radiance at $\lambda = 685\,\mathrm{nm}$ originated from the uppermost 2 meter layer. The water-leaving fluorescent signal decreases exponentially with the depth of the phytoplankton abundance maximum.

The sun-stimulated fluorescent signal coming up from high and hyper-eutrophic waters can display at $\lambda > 700\,\mathrm{nm}$ a spectral distribution characteristic of higher plants, *viz.* a steep rise in λ (Chappell *et al.* 1984).

When the phycobilin fluorescence contribution needs to be assessed in the water-leaving radiance (e.g. when blue–greens are predominant in the phytoplankton community), employ the above expressions (3.64, 3.65) which are still applicable with the reservation that the excitation spectral range for phycoerythrin is 420–610 nm, and for phycocyanin and allophycocyanin 520–720 nm (Jadhav, 1987).

Before ending the discussion of upwelling irradiance formation in natural waters, it seems warrantable to examine the plausibility of bottom albedo influence on water-leaving light.

The formulations suggested for diffuse reflectance (eqs. (3.27) to (3.31)) do not provide for the influence of the bottom on water optical status. In fact, when deriving expressions like eqs. (3.27) to (3.31), the water body was considered as a semi-infinite scattering and absorbing layer. However, in the case of sufficiently transparent and shallow waters, such an assumption turns out to be invalid (Spitzer, Dirks, 1987). Indeed, in such waters a certain number of photons traveling downwards find themselves reflected back to the water–air interface instead of absorbed by water molecules at depth (Philpot, 1987; Maritorena *et al.*, 1994; Estep, 1994). It is conducive to the probability that a certain fraction of reflected photons eventually reach the water surface. Thus, the resulting diffuse reflectance just beneath the water surface $R_{tot}(-0, \lambda)$ can be decomposed

$$R_{tot}(-0, \lambda) = R_w(-0, \lambda) + R_{bot}(-0, \lambda) \qquad (3.66)$$

where $R_w(-0, \lambda)$ is the upwelling spectral reflectance due to water, and $R_{bot}(-0, \lambda)$ is the upwelling spectral reflectance due to the bottom optical influence.

An analogous equation can be written for the upwelling irradiance (Maritorena *et al.*, 1994):

$$E_{u\,tot}(-0, \lambda) = E_{uw}(-0, \lambda) + E_{u\,bot}(-0, \lambda) \qquad (3.67)$$

where $E_{u\,bot}(-0, \lambda)$ and $E_{u\,w}(-0, \lambda)$ are the upwelling irradiances produced by the photons which, respectively, both have (at least once) and have not interacted with the bottom.

Considering an infinitely thin layer of water column d at depth z, it is possible to define the upwelling irradiance attenuation as

$$dE_u(z) = b_{b_d} E_d(-0) \exp[-K_d + k)z]\, dz \qquad (3.68)$$

where b_{b_d} is the net backscattering coefficient (or reflectance function) for downwelling light stream (a hybrid property of the water medium, introduced by Preisendorfer, 1961): $b_{b_d} = \sum_i^n b_{b_i}$, where i is the optically active component, n is the total number of optically active components coexisting in the water column, K_d and k are attenuation coefficients for downwelling and upwelling irradiance, respectively.

Integration of eq. (3.68) over the water column 0–z yields:

$$E_u(-0, z) = (K_d + k) b_{b_d} E_d(-0)\{1 - \exp[-(K_d + k)z\} \qquad (3.69)$$

For an infinite water depth ($z \to \infty$), eq. (3.69) reduces to

$$E_u(-0, \infty) = (K_d + k)^{-1} b_{b_d} E_d(-0) = R(-0, \infty) E_d(-0) \qquad (3.70)$$

where $R(-0, \infty)$ represents the reflectance at null depth of the deep ocean, hereafter denoted R_∞.

For a water column limited by the presence of a perfectly absorbing bottom at a depth H, eq. (3.70) reads:

$$E_u(-0, H) = R_\infty E_d(-0)\{1 - \exp[-(K_d + k)H]\} = E_{u_w}(-0) \qquad (3.71)$$

Assuming that the bottom is a Lambertian reflector with an albedo A, the reflected irradiance which reached the water–air interface ($z = -0$) can be defined as

$$E_{u_{bot}}(-0, H) = A E_d(-0) \exp[(-K_d + k)H] \qquad (3.72)$$

Thus, eq. (3.68) becomes

$$E_{u\,tot} = E_d(-0)\{R_\infty[1 - \exp(-(K_d + k)H)] + A \exp[-K_d + k)H]\} \qquad (3.73)$$

If for the sake of simplicity we assume that k equals K_d (which is a fairly rough approximation, of course), and replace them with an operational and unique K coefficient, then eq. (3.73) becomes:

$$\begin{aligned} E_{u\,tot}(-0) &= E_d(-0)\{R_\infty[1 - \exp(-2KH)] + A \exp(-2KH)\} \\ &= E_d(-0)[R_\infty - (R_\infty - A)\exp(-2KH)] \end{aligned} \qquad (3.74)$$

Now the resultant diffuse reflectance $R_{tot}(-0)$ can be defined as

$$R_{tot} = R_\infty + (A - R_\infty)\exp(-2KH) \qquad (3.75)$$

The relationship (3.75) derived by Maritorena *et al.* (1994) proves to be identical to the analytical expression which has been obtained independently by Estep (1994).

The above assumption that $k = K_d = K$ is conducive to inevitable underestimation of $R_{bot}(-0)$. The Monte Carlo simulations undertaken by Maritorena *et al.* (1994) for realistic conditions indicate that the inaccuracy in $R_{bot}(-0)$ arising from this assumption averages $\sim 15\%$. However, it increases in strongly absorbing aquatic media (with the single scattering albedo $\omega = 0.2$) as well as in waters with strong particulate scattering. Field measurements indicate that when the bottom contrast is positive, i.e. when $(A - R_\infty) > 0$ (e.g. it can occur, if the bottom is sandy), eq. (3.75) underestimates $R_{tot}(-0)$ (in spite of the overestimation which should be expected from use of this equation in the case of positive bottom contrast). The underestimate increases with H and for $H = 15\,m$, it equals 10–15% at $\lambda < 600\,nm$ and 40–69% at $\lambda > 600\,nm$.

When using eq. (3.75), a certain $R_{tot}(-0)$ underestimation has also been experimentally revealed for $(A - R_\infty) < 0$ (e.g. it occurs for a bottom covered by brown algae). The underestimates of predicted R values proved to be significant over the full visible spectral range: at $H = 1.3\,m$, the underestimate is $\sim 20\%$ in the spectral range $\lambda < 600\,nm$ and about 30% at $\lambda > 600\,nm$. However, unlike the positive bottom contrast case, it decreases with H.

Hence, utilizing one of the R formulations, the available data on bottom depth, bottom albedo and appropriate bulk water absorption and backscattering coefficients, it is theoretically, possible to assess the optical influence of bottom on the water-leaving radiance.

First of all, it is of interest to specify the conditions under which the bottom optical influence on the water-leaving radiance could be negligible. Obviously, a certain threshold ζ can be set up:

$$(A - R_\infty)\exp(-2KH) \leq \varsigma \tag{3.76}$$

This threshold determines the critical value of ΔR_{tot} which will be detectable by a remote sensor. Since the value of R in natural waters is generally no more than 10%, a cutoff of 1% could not be unreasonably set. Obviously, the lower the water column optical depth and higher bottom albedo, the easier threshold ς could be exceeded (Gordon and Brown, 1974).

Spectral values of the bottom albedo for various algae, coral sand and silica sand are given in Fig. 3.14 (Estep, 1992; 1994). Importantly, coral or silica sand bottom albedo considerably exceeds typical values of $R_\infty(-0)$, this being particularly so in the long wavelength region of the visible spectrum. At the same time, the albedo of a silty bottom is comparable (particularly in the short wavelength region) with $R_\infty(-0)$. However, at longer λ ($\geq 550\,nm$), it overpasses $R_\infty(-0)$, reaching at $\lambda = 700\,nm$ a value of ~ 0.30.

Spectral distributions of albedo of a bottom covered by vegetation display particularly intricate patterns, depending on plant species composition. It also explains liability of albedo of such bottoms to considerable spatial variation. Besides, the actual albedo of vegetation-covered bottoms is generally reinforced in some spectral regions due to sun-induced fluorescence of macroalgae. Higher-aquatic rooted plants, depending on their physiological status, can fluoresce rather strongly in the blue and red spectral regions. However, bottom albedo enhancement in the blue portion of the spectrum is only possible in clear shallow waters where the submerged plants receive sufficiently strong solar radiation with $\lambda < \sim 450\,nm$.

Thus, for natural bottoms, the value and sign of contrast $A - R_\infty$ (even in the case of sandy or silty bottoms which are generally covered by benthic small-sized algae) can exhibit essentially irregular spectral distributions. Later on, we shall explore

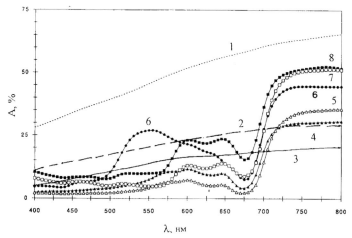

Fig. 3.14. Spectral variations in albedo of different types: 1—coral sand; 2—silica sand, 3—silt, 4–8—algae (4—*Turbinaria*; 5—*Sargassum*; 6—*Boodlea*; 7—*Porolithon*; 8—*Corallinacea*).

possible ways of filtering off the bottom-driven signal from the water-leaving radiance in order to achieve a higher precision of retrieval of water-quality parameters.

3.1.1.3   *IR emissivity of water, foam, and oil spills*

The correct decoding of the data of IR remote sounding of inland water bodies requires knowledge of both the reflecting and emitting properties of lake waters, their variability in the presence of dissolved ions, foam and surface polluting films, as well as changing temperatures and surface roughness. These aspects have been well studied (see Rusk *et al.*, 1971; Kropotkin and Buznikov, 1974). Therefore, in this section we shall confine ourselves to a brief enumeration of the basic results of both theoretical and experimental studies.

The effect of dissolved ions

The presence of salts in lake waters and variations in their concentrations over the water body are characteristic of water masses of different origin. Salinity affects the optical properties of water since the presence of dissociated ions changes the structure of molecules. Bogorodsky *et al.* (1979a) have studied the reflection spectra for large quantities of water solutions of salts and acids as well as natural and artificial sea water in the wavelength interval 2 to 25 μ. The measurements were relative: the emissivity of solutions was compared with that of twice distilled water. Figure 3.15 exemplifies the data of measurements at 10.6 μ.

The measurements have shown that the presence of dissolved ions leads to a shift of the water absorption bands as well as their broadening, the shifting and the character of broadening depending on the kind of dissolved ions. In some cases the dissolved ions can lead to an appearance of special characteristic bands in the

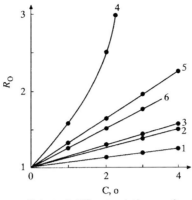

Fig. 3.15. Relative reflection coefficient of different solutions as function of concentration (wavelength $\lambda = 10.6\,\mu m$): 1—sea water (artificial solution), 2—NaCl; 3—KCl; 4—MgSO^4; 5—KJ; 6—MgCl^2.

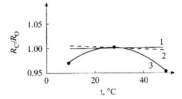

Fig. 3.16. Relative reflection coefficient of fresh water as function of temperature, with wavelength as parameter: 1—0.63 and 1.15 μm; 2—3.39 μm; 3—10.6 μm

reflection spectrum of the solution. At the same time, with the concentrations of dissolved salts even typical of different water masses near the shore and in the center of the lake, the contrast in water emissivity is negligibly small. However, the apparent differences in emissivities of fresh and sea water can serve as the basis for developing the method of IR sounding of the deltas and estuaries of large rivers (Kropotkin and Buznikov, 1974).

The effect of temperature. Temperature, like salt content, affects the inner structure of molecules, therefore a change in the optical properties of water masses with their warming turns out to be substantial, especially in the far-IR spectral region or Fig. 3.16 of (Bogorodsky *et al.* (1979a).

Surface pollution films

The difference in the optical and electrical properties of water and surface pollution films serves as the basis for developing methods for remote sounding of the inland-basin pollution: Fig. 3.17 shows the reflection spectra for liquid water, some oil products, and the water covered with oil film of Bogorodsky *et al.* (1979a). It is seen from Fig. 3.17 that the emissivity of the studied oil products in the 2–22 μ interval varies negligibly, remaining at 4% level.

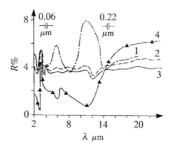

Fig. 3.17. Reflection spectra of fresh water covered by oil slick (1); of oil (2); of diesel fuel (3); of clean fresh water (4)

The revealed decrease in oil product emissivity at 3, 4, 7, and 11–13 μ is determined by the presence of the respective absorption bands typical of hydrocarbons. It also follows from Fig. 3.17 that at around 3 and 6 μ and in the region 10–12 μ (corresponding to the water absorption bands) the contrast of emissivities of water and oil products is at a maximum. This circumstance enables one to recommend these wavelengths for the development of sensors to remotely sound lake pollution. Studies also show that in the case of oil spills the contrast of emissivities of pure and polluted waters due to multiple reflections from the oil film can increase additionally by 20–30%. Moreover, at some wavelengths the emissivity of oil-polluted water can increase due to interference. The calculations show that for this reason the emissivity of oil-polluted water can exceed 10–12 times that of pure water.

Apart from oil spills, biological pollution can substantially affect the lake surface emissivity. Field studies of the spectral emission of duckweed-covered water (2–25 μ) have shown that in the presence of biological objects, additional bands of selective absorption appear in the water reflection spectrum, in particular near 4 and 5 μ, which agree well with the data on chlorophyll absorption bands given in Kondratyev *et al.* (1972).

The presence of foam substantially affects water emissivity at 2–5 μ. From the data of Bogorodsky *et al.* (1979a), the coefficients of the foam-covered lake spectral brightness decrease several times. As in the presence of duckweed, this phenomenon is explained by scattering of incident radiation: the foam-covered water emissivity becomes nearly diffuse, and that of pure water remains specular—Fig. 3.18 of Bogorodsky *et al.* (1979a).

An evaluation of the effect of sea surface roughness is important for correct interpretation of IR survey data. From field test data of Bogorodsky *et al.* (1979a), with near-to-water-surface wind speed 8–10 m/s in conditions of strong sea-surface roughness, the signal becomes 10–20 times weaker than in the case of a calm surface. Similar effects are detected with sounding both in nadir and at small viewing angles (7–10° from the horizon).

Let us briefly discuss the results of studies of the IR contrasts between snow–ice and water surfaces, since these data are the basis for developing methods to retrieve radiation balance of lake eco-systems. They are also important to understand the

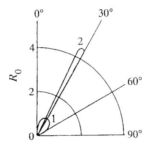

Fig. 3.18. Relative reflection coefficient of fresh water covered by foam as function of angle observation (1); of clean fresh water (2); angle of sounding is 28° (Bogorodsky *et al.*, 1979a).

physical processes in the lake–atmosphere system and to estimate the screening effect of ice and snow cover of lakes on eigen emission of lake water masses. The effect of radiation balance variations on IR radiometer-measured SST has been studied both in field experiments and by aircraft over Ladoga Lake (Bogorodsky *et al.*, 1979b). Figure 3.19 demonstrates a registrogram showing the spatial and temporal variations in radiative temperature of the four characteristic ice-cover areas as a function of their thickness and snowness.

There is a marked increase in temperature of the test ice areas constituting 2°C for the ice with $H = 18.4$ cm and snow with $h = 1.5$ cm, and 1.5°C for the ice with $H = 15.2$ cm and snow with $h = 1.7$ cm, respectively. For thinner fresh-water ice these effects are weakly expressed, which results from the effect of heat supply in the lake water masses.

Thus the studies of many authors testify to the fact that the presence of dissolved salts (up to concentrations inherent to sea waters) slightly changes the emissivity of the inland water bodies—in the whole IR interval it constitutes no more than 5–7%.

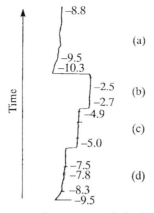

Fig. 3.19. IR temperature of ice cover and snow cover (Lake Ladoga, experimental data), with ice thickness H and snow thickness h as parameter: *a*—stage 0 ($H = 15.2$ cm, $h = 1.7$ cm); *b*—stages 1–111 ($H = 5.0$ cm, $h = 0.0$ cm); *c*—stages 1Y–Y ($H = 8.4$ and 8.9 cm, $h = 0.5$ cm, patches); *d*—stage Y1 ($H = 18.4$ cm, $h = 1.5$ cm).

A change of temperature from 10 to 50°C changes the coefficient of water emission by less than 5%. That is, the effect of both these factors cannot seriously affect the data of the IR survey of lake waters. But an appearance of oil spills and biological products on lake surfaces, as well as the presence of capillary and gravitation waves contribute much to a change in water emissivities, which must be taken into account in analysis of the data of thermal surveys of inland water bodies. At the same time, the revealed features of fresh water emissivity can be used to develop algorithms to retrieve some parameters of the condition of lakes and reservoirs.

3.1.1.4   *Analysis of volume reflectance spectrum responsiveness to variations in in-water parameters and water surface illumination conditions. Development of water-quality retrieval algorithms.*

Using a hydro-optical model composed of the major OAC specific coefficients of absorption and backscattering, as well as the relationships between the remotely measured radiance $L_w(\lambda)$, volume reflectance $R(0, \lambda)$ and the inherent optical properties of water, it is possible to simulate the $R(0, \lambda)$ spectral distributions as affected by the OAC content variations in the water column. Bukata *et al.* (1995) reported on numerical modelling of $R(0, \lambda)$ in a four-component aquatic medium (H_2O *per se*, phytoplankton, suspended minerals (*sm*), *doc*) which has been conducted using the Lake Ontario hydro-optical model and the $R(0, \lambda)$ formulation suggested in Gordon *et al.* (1975) for a cloudless sky and sun zenith angles $\leq 20°$. The results of these simulations indicate that in hypothetical optically simple waters, practically devoid of *doc* and loaded with *sm* in amounts less than $\leq 0.1\,\text{g/m}^3$, an increase in C_{chl} results in a steady (but not linear) decrease in $R(0, \lambda)$ within the spectral region 430–450 nm (so called "chlorophyll dip") whereas the $R(0)$ value at $\lambda = 505$ nm proves to be almost insensitive to variations in C_{chl}. This wavelength can be called "hinge-point". Even a small enhancement in C_{sm} results in a substantial displacement of the "hinge-point" to longer wavelengths. For example, at $C_{sm} = 0.5\,\text{g/m}^3$, the "hinge-point" wavelength moves to ~ 570 nm. This displacement becomes particularly rapid if the hypothetical water also contains a certain amount of DOC. A further increase in C_{sm} and C_{dom} eventually leads to moving the "hinge-point" wavelength beyond 700 nm, and finally to complete disappearance of the "hinge-point", whereas the "depth" of the chlorophyll dip becomes affected not only by the chlorophyll abundance but also by the respective contents of other OAC (C_i) concurrently present in this hypothetical water. It implies that the chlorophyll dip could not be any longer related directly to the chlorophyll content if the other OAC concentrations do not correlate with C_{chl}. Moreover, given a stable correlation between C_{chl} and C_i, the responsiveness of the depth of the chlorophyll dip to C_{chl} dramatically reduces. Waters meeting the conditions of the "hinge-point" existence are referred to as case-1 waters (according to Morel classification). These are waters of open parts of oceans and seas. Conversely, waters characterized by uncorrelated and abundant amounts of AOC, and hence incapable of meeting the conditions of the "hinge-point" existence (these are generally marine coastal and inland waters) are generally referred to as case-2 (or else non-case-1) waters.

These results were further substantiated by model simulations of water-color formation in natural waters (Bukata *et al.*, 1983; Gallie, Murtha, 1993; Jerome *et al.*, 1994; Pozdnyakov *et al.*, 1998a). The upwelling radiance (irradiance) could be subjected to chromaticity analyses which integrate the sensitivity of the human eye with the impinging energy spectrum (Committee on Colorimetry, 1963).

Based on CIE (Commission Internationale de l'Eclairage) colour mixtures for equal energy spectra, such an integration produces tri-stimulus values from which the chromaticity coordinates X (red), Y (green), and Z (blue) may be obtained throughout the visible spectrum (Commission Internationale ..., 1957). Since $X + Y + Z = 1$, two chromaticity coordinates adequately represent colour in a 2-D chromaticity diagram (X–Y, or Y–Z, or Z–X). The loci of these pairs defines an envelope which encompasses all possible chromaticity values. For a "white" spectrum (i.e. $E(\lambda)$ or else $L(\lambda) = $ constant), $X + Y + Z = 0.333$. This defines the achromatic colour or white point. A numerical value of colour is then obtained by drawing a line from this white point through the chromaticity values of the measured spectrum. The intersection of this line with the curve envelope specifies the *dominant wavelength* (λ_{dom}) which is considered as the colorimetric definition of the colour (in our case of the natural water body colour). The distinctiveness of this dominant wavelength is termed "purity" p which is a measure of the magnitude of the contribution of the dominant wavelength to an observed upwelling spectrum (i.e. a spectral purity of 1.0 indicates a monochromatic spectrum at the dominant wavelength, while a spectral purity of 0 indicates a "white" spectrum).

The numerical simulations of the chromaticity coordinate X (analogous results were obtained for coordinates Y, Z) responsiveness to variations in C_{chl}, while C_{sm} and C_{doc} were kept at constant levels, indicate that X depends most essentially on the *sm* and *doc* actual loading: with C_{sm} and C_{doc} increasing, the range of responsiveness of X (Y, Z) to variations in C_{chl} dramatically shrinks, and at $C_{doc} = 8$ mg/l (e.g., this value of C_{doc} is characteristic, e.g., of the littoral zone of Lake Ladoga) it exists no more. This effect becomes particularly pronounced (and, in the first place, this is relevant to coordinate Y) when the *doc* and *sm* concentrations are fairly high. Moreover, these simulations revealed a number of regularities which can be considered as fundamental for understanding natural water colour formation:

1. Low values of λ_{dom} (the blue to turquoise wavelengths 472–500 nm) are associated with low values of water turbidity (i.e. low concentrations of *sm* and *chl*).
2. In the absence of *sm* and *doc*, $472 \leq \lambda_{dom} \leq 500$ nm results from sm concentrations in the range 0–0.9 mg/l.
3. In the absence of both *sm* and *doc*, $472 \leq \lambda_{dom} \leq 500$ nm results from *chl* concentrations in the range 0–5 µg/l.
4. In the absence of both *sm* and *chl*, it is interesting to note that remains unaltered irrespective of the amount of *doc* resident in the water column.
5. λ_{dom} asymptotically approaches *a limit or "end-point"* value as concentration of any of the OAC is increased (the only caveat being that doc requires the

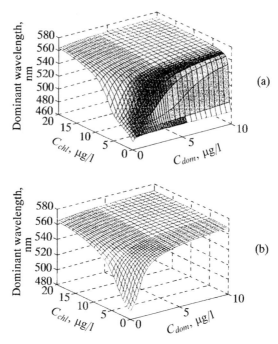

Fig. 3.20. Dependence of λ_{dom} (nm) on variations in C_{chl} (µg/l) and C_{doc} (mgC/l) at $C_{sm} = 0$ (a) and 0.4 mg/l (b).

    minimal presence of either *chl* or *sm*) (Fig. 3.20). This end-point is reddish-brown in color (about 571 nm) (Fig. 3.21).

6. Increase in doc has minimal if any impact on the chl concentration range required to attain the asymptotic approach to λ_{dom}. However, increases in *doc* will dramatically reduce the range of C_{sm} required to attain the asymptotic approach to λ_{dom}.

The above simulations revealed that waters may appear green when they are either heavily loaded with chl or totally devoid of *chl* but contain *doc* in moderate amounts and some minuscule concentration of *sm*. Waters may appear brownish when they are either heavily loaded with *doc* or contain *chl* or *sm* or both in amounts at which the λ_{dom} end-point is attained. These results clearly indicate that in optically complex waters (i.e. waters containing optically competitive organic and inorganic material in considerable amounts), assignment of apparent colour to a definite colorant (*chl, sm* or *doc*) is impossible without some *a priori* information about the water body under scrutiny.

    The spectral purity of pure water rapidly deteriorates as the colorants *sm, chl, doc* are added (the absolute minimum ~0.05 is observed at *sm* = 0, *doc* > 5 mg/l, and *chl* ≈ 1 µg/l, Fig. 3.22). In turbid and/or coloured waters p recovers asymptotically to an end-point which, unlike the λ_{dom} end-point, varies between ~0.33 and ~0.45 depending on the water composition.

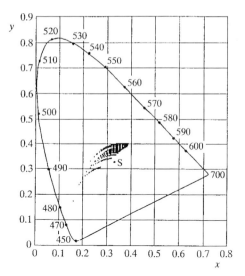

Fig. 3.21. Loci of radiometric colors of simulated aquatic media when concentrations of *chl*, *sm* and *doc* were allowed varying within the limits characteristic of very clean (oceanic) to highly turbid (coastal/inland) waters.

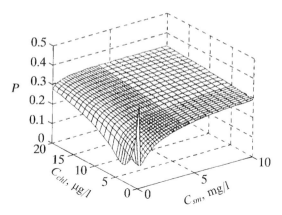

Fig. 3.22. Dependence of p (%) on variations in C_{chl} (µg/l) and C_{sm} (mg/l) at $C_{doc} = 1$ mg C/l.

To recap then, chromaticity can distinguish relatively clear water from more turbid water, and can provide a reasonable approach to inferring that it is one of the OAC concentrations from a water column whose volume reflectance can inarguably be taken as the consequence of a known single colorant. Since study of water chromaticity rests on the analyses of upwelling radiance spectral features, the aforementioned results have a direct relevance to remote sensing.

The optical simplicity of case-1 waters and the chlorophyll dip-related regularities specified above paved the way to constructing parametric algorithms based on statistically established relationships between C_{chl} and the ratio of the upwelling radiance at the wavelength of the chlorophyll dip (λ_i) to the upwelling radiance at the hinge-point wavelength (λ_j). The use of such a radiance ratio provided normalization of the water surface-leaving radiance L_w in the spectral region of the chlorophyll dip by the value of L_w at the hinge-point wavelength at which, in the first approximation, L_w could be assumed independent of the content of OAC in the water column (with the only exception of the water *per se*) and believed to be a function of the intensity and spectral composition of the incident radiation:

$$C_{chl} = A[L_w(\lambda_i)/L_w(\lambda_j)]^{-B} \tag{3.77}$$

where A and B are regression coefficients.

As long as at small but not infinitesimal concentrations of *doc* and *sm*, the increase in chlorophyll content results in a displacement of λ_i and λ_j to longer wavelengths, various λ_i/λ_j pairs (443/520 nm; 443/550 nm; 520/550 nm; 520/670 nm (Gordon, Morel, 1983; Sathyendranath, Morel, 1983; Sturm, 1983) were tried for a given combination of C_{chl}, C_{doc}, and C_{sm} to fit best the eq. (3.77).

When case-1 waters contain not only *chl* but also somewhat increased amounts of suspended minerals and *doc*, all components contribute to changes in the spectral distribution of the upwelling radiance at all wavelengths in the visible spectrum. Utilizing similar pairs of wavelengths, a wealth of retrieval expressions analogous to (3.77) were generated to infer the concentration of total suspended matter (*seston*). Naturally, the values of regression coefficients A and B proved to be area-specific.

When phytoplankton is abundant, the absorption in the blue part of the spectrum becomes so strong that the legitimate signal (i.e. L_w) becomes reduced down to noise level. It turned out to be an incentive to use some longer wavelength pairs. For instance, based on the results of field measurements off the US eastern and western coastlines, as well as in the Sargasso Sea, the following algorithm has been suggested for processing Coastal Zone Color Scanner (CZCS) data relating to case-1 waters (McClain, Yen Eueng-nan, 1994):

$$C_{chl} = 1.1298 \left[\frac{L_w(443)}{L_w(550)}\right]^{-1.71} \tag{3.78}$$

At the same time, NASA workers believe that the pair $L_w(520)/L_w(550)$ remains valid not only for case-1 waters but also for case-2 waters:

$$C_{chl} = 3.3266 \left[\frac{L_w(520)}{L_w(550)}\right]^{-2.40} \tag{3.79}$$

However, a constraint is laid upon the use of eq. (3.78): it should be used with the proviso that C_{chl} as calculated through eq. (3.78) ($C_{chl(3.78)}$) falls into the range $\leq 1.5\,\text{mg/m}^3$. The retrieval algorithm (3.78) could still be used when $C_{chl(3.78)} > 1.5\,\text{mg/m}^3$, but only if C_{chl} as calculated through (3.79) proves to be less than $1.5\,\text{mg/m}^3$; otherwise, eq. (3.79) is to be used. This complication of the retrieval procedure could be, according to McClain, Yen Eueng-nan (1994), circumvented by utilizing a "three-channel" algorithm:

$$C_{chl} = 5.56 \left[\frac{L_w(443) + L_w(520)}{L_w(550)}\right]^{-2.252} \tag{3.80}$$

However, Gordon Wang (1994) believe that the $L_w(443)/L_w(550)$ ratio-based algorithm is more efficient when used in the form of a third-order polynomial in r_L:

$$\lg 3.33 C_{chl} = -1.2 \lg r_L + 0.5(\lg r_L)^2 - 2.8(\lg r_L)^3 \qquad (3.81)$$

where $r_L = (1[L_w(443)]_N/2[L_w(550)]_N)$, and $[L_w(\lambda)]_N$ is the normalized water-leaving radiance at λ; i.e.:

$$tL_w(\lambda) = [L_w(\lambda)]_N \cos\theta_0 \exp\left[-\left(\frac{\tau_r}{2} + \tau_{oz}\right)\left(\frac{1}{\cos\theta_0} + \frac{1}{\cos\theta}\right)\right] \qquad (3.82)$$

where t is the diffuse transmittance of the atmosphere, τ_r is the Rayleigh optical thickness (molecular scattering), τ_{oz} is the ozone layer optical thickness, θ_0 and θ are respectively the sun zenith angle and angle of incidence of sun rays onto a flat water surface.

A very similar algorithm has been suggested by Sturm (1993) which was called the "European method". Based on the case-1 waters model developed by Morel (1988), a regression has been established between $C_{chl}(mg/m^3)$ and the radiance ratio at the CZCS first and third spectral channels as well as at the CZCS second and third spectral channels:

$$C_{chl} = \exp(0.768 - 2.61 \ln X + 0.791(\ln X)^2 - 0.388(\ln X)^3) \qquad (3.83)$$

$$C_{chl} = \exp((1.395 - 8.739 \ln Y + 5.7286(\ln Y)^2 - 27.9504(\ln Y)^3) \qquad (3.84)$$

where $X = R_s(443)/R_s(550)$; $Y = R_s(520/R_s(550)$; $R_s(\lambda)$ is the subsurface reflection coefficient. Algorithms (3.83) and (3.84) were used for retrieval of C_{chl} pertaining to case-1 waters with, respectively, moderate and high phytoplankton chlorophyll concentrations. The value of $R_s(443)$ was used as a criterion for the transition from algorithm (3.83) to algorithm (3.84): if it becomes very small approaching a predefined threshold ($= 0.4$), then a transition from eq. (3.83) to eq. (3.84) should be done.

However, in practice the value of $C_{chl} = 2 - 3\,mg/m^3$ can be applied as a *concentration* threshold. A limit value (R_{lim}) of $R_s(550)$ is suggested as another criterion for automated differentiation between case-1 and case-2 waters:

$$R_{lim} = \exp(1.05 - 0.02 \ln X - 0.429(\ln X)^2 + 0.094(\ln X)^3). \qquad (3.85)$$

Identification of case-2 waters could also be attained provided the value of C_{chl} determined from (3.84) is less than $10\,mg/m^3$. According to Sturm (1993), the assessment of the chlorophyll content in case-2 waters can also be accomplished by using polynomial algorithms based on the ratios $X = R_s(443)/R_s(550)$, $Y = R_N(520)/R_s(550)$. However, it seems very much unlikely in light of the considerations presented above when discussing spectral variations of $R(0, \lambda)$ and displacements of the hinge-point driven by increasing concentrations of OAC.

The complexity of case-2 water composition and optical properties prompted a further development of water quality retrieval techniques.

Still staying in the "rut" of the regression algorithm approach, but already applying a hydro-optical water model based on specific inherent hydro-optical coefficients of phytoplankton and suspended minerals, Tassan (1994) suggested a system of algorithms for OAC retrieval. The author claims that he established steady regression relationships between major OAC for case-2 waters (which seems surprising in view of the present paradigm of case-2 waters). The following relationships allegedly hold for the Gulf of Naples and the northern margins of the Adriatic Sea:

$$\lg C_{sm} = -A_1 + B_1 \lg C_{chl} \tag{3.86}$$

$$\lg[a_{doc}(440)] = -A_2 + B_2 \lg C_{chl} \tag{3.87}$$

where A, A_2, B_{12}, B_2 are regression coefficients equal to 0.25; 0.026; 1.20; 1.28; and 0.57; 0. 59; 0.47; 0.38 for the waters of the Gulf of Naples and the northern margins of the Adriatic Sea, respectively; $a_{doc}(440)$ is the *doc* absorption coefficient at $\lambda = 440$ nm (spectral variations in a_{doc} were assumed exponential (eq. (3.3)) with $s = -0.014$ nm^{-1}:

$$a_{doc}(\lambda) = a_{doc}(440) \exp[-0.014(\lambda - 440)] \tag{3.88}$$

The retrieval algorithms were developed in a polynomial form:

$$\lg C_{chl} = 0.0664 + 0.04621 \lg(X_{chl}) - 4.144 \lg(X_{chl})^2$$
$$(0.025 \leq C_{chl}(\text{mg/m}^3) \leq 1.0) \tag{3.89}$$

$$\lg C_{sm} = 1.83 + 1.26 \lg(X_{sm}) \qquad (0.07 \leq C_{sm}(\text{mg/m}^3) \leq 0.56) \tag{3.90}$$

$$\lg a_{doc}(440) = -3.00 - 1.93 \lg(X_{doc}) \qquad (0.01 \leq a_{doc}(440)(\text{m}^{-1}) \leq 0.065) \tag{3.91}$$

where $X_{chl} = [R(\lambda_2)/R(\lambda_5)][R(\lambda_1)/R(\lambda_3)]^{-1.2}$; $X_{sm} = [R(\lambda_5)+R(\lambda_6)][R(\lambda_3)/R(\lambda_5)]^{-0.5}$; $X_{doc} = [R(\lambda_1)/R(\lambda_3)][R(\lambda_2)]^{0.5}$, $\lambda_1-\lambda_5$ are the wavelengths at which the SeaWiFS spectral channels are centered: 412, 443, 490, 510, 555, 670 nm respectively.

For waters rich in chlorophyll (1–40 mg/m^3), suspended minerals (0.56–4.6 g/m^3), and $doc(a_{doc}(440) \sim 0.2$ m$^{-1})$, algorithms (3.89)–(3.91) have the following regression coefficients:

$$\lg C_{chl} = 0.36 - 4.38 \lg(X_{chl}) \tag{3.92}$$

$$\lg C_{sm} = 1.82 + 1.23 \lg(X_{sm}) \tag{3.93}$$

$$\lg a_{doc}(440) = -4.36 - 6.08 \lg(X_{doc}) \tag{3.94}$$

where $X_{chl} = [R(\lambda_2)/R(\lambda_5)][R(\lambda_1)/R(\lambda_3)]^{-0.5}$; $X_{sm} = [R(\lambda_5)+R(\lambda_6)][R(\lambda_3)/R(\lambda_5)]^{-1.2}$; $X_{doc} = [R(\lambda_1)/R(\lambda_3)][R(\lambda_2)]^{0.25}$.

When choosing R pairs relevant to various spectral bands of SeaWiFS channels, Tassan had in view that each selected pair should have high sensitivity to the parameter to be retrieved and low sensitivity to fluctuations in the dependencies of C_{sm} and C_{dom} on C_{chl} (eqs. (3.86; 3.87)). For example, for X_{chl} the wavelengths λ_2 and λ_5 are close, respectively, to the maximum and minimum of chlorophyll absorption (i.e. the spectral influence of chlorophyll on R is the most pronounced), λ_1 and λ_3 are located at both sides of the chlorophyll absorption maximum ($R(\lambda_1)$, $R(\lambda_3)$ are compensating terms). For X_{sm}, λ_5 and λ_6 correspond to a spectral region of weak spectral influence (absorption) of phytoplankton chlorophyll and *doc* and to a high level of light scattering by the suspended minerals, λ_3 and λ_5 are located within the spectral region of *chl* and *doc* influence ($R(\lambda_3)$, $R(\lambda_5)$ are the compensating terms). For X_{doc}, λ_1 and λ_3 fall into the spectral region of respectively strong and weak absorption by *doc* ($R(\lambda_2)$ is the compensating term).

Numerical experiments run by Tassan (1994) to assess the sensitivity of algo-rithms (eqs. (3.89), (3.91)) to fluctuations in concentrations of *doc* and suspended minerals relative to the respective values specified by regressions (eqs. (3.87)–(3.88)) indicate that the retrieval errors do not exceed 20–30%. However, regardless of such optimistic assertions, it appears clear that the application of retrieval algorithms (eqs. (3.89)–(3.91)) to arbitrarily chosen water areas belonging to case-2 waters will unavoidably require a fine tuning of regression coefficients to local conditions as well as a thorough verification of retrieval results against relevant ground truth data.

Grew (1981) found that ratio $G_{mn}(\lambda_i) = L(\lambda_i)^2/(L(\lambda_{i-m}) \times L(\lambda_{i+n}))$, where $i = 1, 2, 3, \ldots, k$ represent the center channel or band number and m, n, the adjacent band number, varied significantly less with solar elevation, sea state, cloud cover than the sensor measured radiances *per se*. Based on this premiss and adhering to the usual two-band (blue–green) ratio R methodology, Campbell and Esaias (1983) suggested the following chlorophyll retrieval algorithm:

$$\lg C_{chl} = a - b \lg \frac{L(490)^2}{L(460)L(520)} \tag{3.95}$$

where $a = 1.43$, $-b = 10.02$. Campbell and Esaias found that their form of the algorithm is a measure of the spectrum curvature with respect to λ. Allegedly, this algorithm, while effectively eliminating variations due to changes in incident irradiance, enhances spectral features of the water medium. Since the pure water upwelling irradiance displays a strong negative curvature at 490 nm, the increasing content of chlorophyll-bearing algae tend to decrease monotonically this negative curvature. Hoge and Swift (1986) analyzed the suitability of a number of satellite ocean color sensor bands (CZCS, (ERS-I), (NOAA-K), (SPOT-3) and suggested the corresponding spectral bands in the blue–green portion of the spectrum. However, it is evident from the above discussion that the spectral curvature algorithm is only appropriate to case-1 waters.

It is necessary to point out that in the case of eutrophic and hyper-eutrophic waters, the development of chlorophyll retrieval algorithms is sometimes based on the use of the red spectral region. In the spectral distribution of L_w a minimum at $\sim 670\,nm$ (the red absorption of chlorophyll) is followed by a maximum at 690–720 nm (its actual position moves to the longer wavelength region as the chlorophyll concentration increases). This maximum is due to fluorescence of chlorophyll and phytoplankton reflectance enhancement in the red and near IR portions of the spectrum (Gower, Borstad, 1993). The following regression algo-rithm based on the above combined effect has been suggested (Yakobi *et al.*, 1995):

$$C_{chl} = -A + B[L_w(\lambda_{max})/L_w(\lambda = 670\,nm)] \tag{3.96}$$

where A and B are regression coefficients, λ_{max} is the wavelength of the red maximum in the spectral distribution of L_w.

When hyperspectral remotely sensed data are available for a given water surface, the assessment of C_{chl} could be done from the integral value of the fluoresced radiance, if the integral specific fluorescence of the phytoplankton assemblage $\Phi^*\,(\mathrm{m}^2\,\mathrm{mg}^{-1})$, is known (i.e. the value of the phytoplankton specific fluorescence integrated over the wavelengths encapsulated by the phytoplankton fluorescence band). Given the spectral distribution of integral specific fluorescence for individual species of phytoplankton (Φ^*_i) (see Ahn Yu-Hwan *et al.*, 1992), it is possible to infer the content of each of the components (C_i) of the phytoplankton assemblage by applying the multivariate optimization technique (see Kondratyev *et al.*, 1990) to the effect of obtaining (through varying C_i) the best fit between measured and modelled fluoresced radiance.

Analyzing the adequacy of retrieval of C_{chl} from the red spectral region data, it is worthwhile mentioning that the maximum in question is located within a region of strong water absorption. Therefore, the remotely recorded intensity of this signal is mainly due to planktonic cells residing in the near-surface layers. It naturally explains the fact that red spectrum algorithms are basically developed for high productive waters with characteristic bloom events.

Amongst the limitations inherent in these approaches is the diurnal variability in phytoplankton fluorescence, the neglect of both this and also of the effect of photoinhibition at high levels of incident radiation can seriously effect the results of chlorophyll content retrieval (see Kondratyev, Pozdnyakov, 1990). Finally, since space sensors like CZCS and SeaWiFS record radiance signals not on a continuous basis, but only in several wavebands, and the spectral band 680–720 nm is not generally included, the red band-based approaches can not find an immediate application for the presently operative space sensors.

An alternative approach actively developed in recent years consists in applying nonparametric methods of water-quality parameter retrieval.

One of them is based on the utilization of a set of optical cross-sections of inherent water properties as a basic model of natural waters (Kondratyev, Pozdnyakov, 1990; Roesler, Perry, 1995). Given such a hydro-optical model and an adequate para-meterization of the volume reflectance coefficient $R(0, \lambda)$, the retrieval of the desired water constituent concentration could be accomplished by applying well-known mathematical procedures, such as the Levenberg–Marquardt multivariate optimiza-tion technique (Marquardt, 1963), the method of statistical regularization (Tikhonov, Arsonin, 1979), the maximum likelihood method (see Doubovick *et al.*, 1994), the linear matrix inversion technique (Hoge, Lyon, 1996) and some others.

Within the framework of the multivariate optimization technique, the volume reflectance coefficient is presented in a vector form (Bukata *et al.*, 1995):

$$R[a^*(\lambda), b_b^*(\lambda)C] \sim \sum b_b^*(\lambda)C / \sum a^*(\lambda)C \qquad (3.97)$$

where

$$a^*(\lambda) = [a_w(\lambda), a^*_{chl}(\lambda), a^*_{sm}(\lambda), a^*_{doc}(\lambda), \ldots]$$

$$b^*_b(\lambda) = [b_{b_w}(\lambda), b^*_{b\,chl}(\lambda), b^*_{b_{sm}}(\lambda), 0, \ldots]$$

$$C = (1, C_{chl}, C_{sm}, C_{doc}, \ldots)$$

Given the measured spectrum $\{S_j\}$, consisting of a set of irradiance reflectance values at discrete wavelengths λ_j, the weighted residuals between the measured and modeled irradiance reflectance $R[a^*(\lambda), b^*_b(\lambda)C]$ can be written as

$$g_i(C) = \{S_j - R[a^*_j, b^*_{bj}, C]\}/S_j \tag{3.98}$$

The multidimensional least-squares solution (at wavelengths corresponding to $\{S_j\}$) could then be found by minimizing the function of residuals over the concentration vector C:

$$f(C) = \sum_j g_j^2(C) \tag{3.99}$$

A search for the absolute minimum of function $f(C)$ absolute minimum is further carried out utilizing the Levenberg–Marquardt finite-difference algorithm appropriate for such multivariate non-linear least-squares optimization applications (Marquardt, 1963). The result thus obtained can then be interpreted as the desired concentration vector C for the location where the spectrum $\{S_j\}$ has been recorded (a more detailed description of this method and technology of searching for the absolute minimum of $f(C)$ can be found elsewhere (Kondratyev et al., 1990c)).

A very important advantage of such an approach is that it provides for *contemporaneous* retrieval of the concentrations of *all* OAC included in the hydro-optical model used to calculate the irradiance reflectance $R(0, \lambda)$. Moreover, this approach is applicable to a wide range of OAC concentrations, including hydro-optical situations when C_{doc} is in excess of 5–10 gC/m^3 (i.e. the situations which, as discussed above, are really formidable for remote retrieving of phytoplankton chlorophyll). It is also an important advantage of this technique that it does not need the existence of correlation between the concentrations of major OAC as a prerequisite. This approach has been used to successfully retrieve OAC concentrations in Lake Ontario and Lakes Ladoga and Onega (Bukata et al., 1985, Kondratyev et al., 1990).

Incidentally, the method of multivariate optimization can equally be used for inferring, from remotely sensed data, OAC specific coefficients of adsorption and backscattering (Bukata et al., 1985, Kondratyev et al., 1990c).

According to Roesler and Perry, (1995), the vectors of specific coefficients of absorption and backscattering could be presented as a sum of products:

$$a(\lambda) = \sum_{i=1}^{I} a_i(\lambda) = \sum_{i=1}^{I} M_i \tilde{a}_i(\lambda) \tag{3.100}$$

$$b_b(\lambda) = \sum_{j=1}^{J} b_{b_j}(\lambda) = \sum_{j=1}^{J} M_j \tilde{b}_{b_j}(\lambda) \tag{3.101}$$

where a_i and b_{b_j} are coefficients of absorption and backscattering of the ith and jth components, I and J are the total number of respectively absorbing and scattering components in the water column, M_i and M_j are the weights of absorption and bacscattering coefficients of each of the components, $\tilde{a}_i(\lambda)$ and $\tilde{b}_{b_j}(\lambda)$ are dimensionless shape factors of ith and jth components, respectively.

Shape factors, like OAC specific absorption and backscattering coefficients, should be determined *a priori*: based on statistically extensive data sets, coefficients $a_i(\lambda)$ and $b_{b_i}(\lambda)$ can be obtained, following which they are divided by their spectrally integrated values $\tilde{a}_i(\lambda)$ and $\tilde{b}_{b_j}(\lambda)$ and then statistically averaged. The retrieval procedure of vectors a, b_b involving the multivariate optimization technique is, in principal, the same as the one described above for the retrieval of C.

Another nonparametric approach offering a solution to the inverse problem of C retrieval from remotely sensed data is based on minimizing the χ^2 difference between modelled and Rayleigh corrected remotely sensed radiances (Doerffer and Fisher, 1994). A two-flow model is calibrated with a set of radiance data which, in turn, are simulated with a matrix-operator radiative transfer model developed by Fisher and Doerffer (1987). For modeling light transfer through the water column, specific coefficients of absorption and attenuation a_i^*, c_i^* are used. A simplex optimization method (Nelder and Mead, 1965) is applied to minimize the value of χ^2 by varying the concentration vector C and the aerosol optical depth τ_a. When χ^2 becomes ≤ 0.3 (it requires about 100 iterations), the corresponding values of C and τ_a are taken as the result of retrieval. This method has been tested against CZCS data on the North Sea. One of the most important assets of this approach is that it automatically allows for atmospheric correction of radiances coming from case-2 waters (when the atmospheric correction algorithm suggested by Gordon and Wang (1994) proves to be inefficient because of absence of reference pixels within the water-body imagery, for which $L_w \approx 0$ at the long-wave boundary of the visible spectrum).

Recently, neural network simulating techniques (NNST) have been explored in terms of their capacity to retrieve concentrations of OAC from a wide variety of natural waters. Unlike the multivariate optimization technique, the statistical regularization or simplex optimization method which are very time-consuming procedures, NNST are much quicker and hence are good candidates for operational processing of space images of large water areas.

Since the range of concentrations of water constituents in the targeted water bodies is generally limited, the possible reflectance spectra can be computed in advance by one of the appropriate radiative models (see § 3.1.1.2). Establishing an inverse relationship between the concentration vector C and corresponding computed reflectances, the results of such simulations can be stored in the form of a table or a neural network and used for retrieval of remotely sensed data. For this specific application, a neural network is used only as a multiple nonlinear regression technique to parameterize an inverse relationship between $C = \Sigma C_i$ and the computed reflectances. Actually, the construction of such a look-up table for the required combinations of C_i and respective reflectances is hardly practicable since such a table would be too large to be effective. However, if an approximation function derived from such a table could be developed, then it would be possible to affect the desired interpolation and ultimately reduce the computing time to emulate the inverse model.

There are many different types of neural network but one of the most commonly used in remote sensing is the multi-layer perceptron (MLP) (Atkinson and Tatnall, 1997): the MPL generally consists of three types of layer. The first is the input layer where neurons are elements of a feature vector which might consist of, for example, radiances at certain wavelengths or radiances subjected to a principal component transformation (Fisher *et al.*, 1997). The second type of layer is the internal or "hidden layer". The third type of layer is the output layer, the number of neurons in the output layer equals the number of parameters to be determined. Each neuron in the network is reconnected to all neurons in both preceding and subsequent layers by connections with associated weights.

Input signals are transferred to the neurons in the next layer in a feed-forward manner. As the signal propagates from neuron to neuron, it is modified by the appropriate connection weight. The receiving neuron sums the weighted signals from all neurons to which it is connected in the previous layer. The total input that a single neuron receives is weighted in the following way:

$$net_j = \sum \omega_{ji} o_i \qquad (3.102)$$

where ω_{ji} are the weights between neuron i and neuron j, and o_i is the output from neuron i. The output from a given neuron j is then obtained from:

$$o_j = f(net_j) \qquad (3.103)$$

The function f is usually a non-linear sigmoid function. It is applied to the weighted sum of inputs before the signal reaches the next layer. When the signal reaches the output layer, the network output is produced. The created network should be first trained so that it can generalize and predict outputs from inputs that it has not processed before. A training pattern is fed into the neural network and the signals are fed forward. After that, the network output is compared with the truth output, the error is then computed and back-propagated through the network. As a result, the connection weights are modified following the generalized delta rule:

$$\Delta\omega_{ji}(n+1) = \eta(\delta_j o_i) + \alpha\Delta\omega_{ji}(n) \qquad (3.104)$$

where η is the learning rate parameter, δ_j is an index of the error change rate, α is the momentum parameter. Training is continued until the output error reaches a desired level. The trained neural network is then tested against some verification data to assess the network performance efficiency.

Schiller and Doerffer (1999) have reported their experience in creating a neural network specialized for solving the inverse problem in remote sensing of natural waters. They showed that such a specialized network is capable of operationally retrieving the concentration vector C in hypothetical (modelled) waters even when very broad limits of variations in individual C_i are imposed.

Pozdnyakov and Lyaskovsky (1999) have carried out a comparative study of a variety of retrieval algorithms suggested in the literature in terms of their feasibility to meet water-quality retrieval challenges posed by the SeaWiFS mission. Based on

the Lake Ladoga hydro-optical model, a wide set of irradiance reflectance spectra have been generated to cover all realistic combinations of C_{chl}, C_{sm}, and C_{doc}. These data
sets were then used to assess the workability of some parametric algorithms (eqs. (3.78)–(3.79); (3.81); (3.83)–(3.84); (3.89–3.94)) and nonparametric algorithms (multivariate optimization and neuron network simulating techniques).

Following Schiller and Doerffer, to approximate the inverse model a feed forward (error backpropagation) network was chosen (the Stuttgart University Neural Network Simulator (SNNS, 1995)). The neural network was composed of five layers of "neurons": there is an input layer, three hidden layers, and an output layer. The number of units in the input layer equalled the number of input values (i.e. six reflectances at SeaWiFS wavelengths in the visible spectrum). The output layer unit number corresponded to the number of output values (i.e. concentrations of major optically active components—*chl, sm, doc*). The hidden layer units overall number is problem-dependent. An optimum was found for our network: 6 input units, 12 units in the first hidden layer, 15 units in the second hidden layer, 9 units in the third hidden layer, and 3 units in the output layer.

The training pattern (generated useing the OAC absorption and backscattering cross-sections for Lake Ladoga and the *R* formulation developed by Jerome *et al.* (1988)) was presented to the designed neural network in a random order. Error minimization was continued until the error function was down to 3.0 corressponding to the average error of 1.2%.

As a result of presenting a test pattern to the neural network, an average error of 1% has been attained. At this step, the configuration of the neural network was considered appropriate for conducting retrieval experiments with the generated data.

The Levenberg–Marquardt multivariate optimization procedure code has been developed according to principal ideas outlined in Kondratyev *et al.* (1990c).

The results of comparison of retrieval algorithm robustness showed, first of all, that all examined parametric algorithms proved to be absolutely incapable of predicting OAC concentrations under conditions of random and uncorrelated combinations of major OAC which is illustrated in Fig. 3.23 for the algorithms suggested by Tassan (1994) (eqs. (3.89)–(3.94)).

The most gratifying results have been obtained by the Levenberg–Marquardt method (Fig. 3.24). However, this is a very time-consuming procedure, and could not really be a successful candidate for operational processing of SeaWiFS images of large water areas.

The neural network simulating approach displayed less accurate results of retrieval in comparison with the Levenberg–Marquardt method (Fig. 3.25). Nevertheless, first, it provides quite acceptable accuracy, and second, it is incomparably faster than the Levenberg–Marquardt method.

Accuracy of retrieval, irrespective of the algorithm applied, can be increased provided necessary corrections are made to compensate for influences produced by Raman scattering, chlorophyll and dissolved organics fluorescence, and bottom albedo (when appropriate) on the spectral distribution water-leaving radiance. As discussed in the previous section, the Raman scattering contribution might be worth

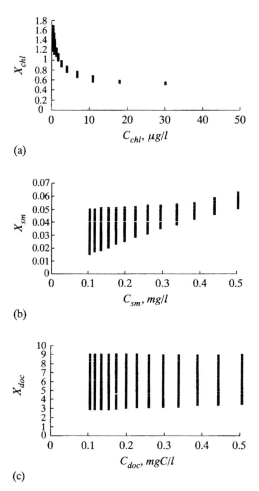

(a)

(b)

(c)

Fig. 3.23. Results of regressing the concentrations of *chl* (a), *sm* (b), and *doc* (c) on parameters X_{chl}, X_{sm}, X_{doc}, respectively, in algorithms suggested by Tassan (1994).

considering in the case of clear waters. Conversely, the fluorescence effects are important in colored and productive waters. The relevant expressions were analyzed in § 3.1.1.2. Obviously, it can be done through iterative procedures: at the first stage, tentative OAC concentrations are to be determined from the "raw spectra" upwelling radiance, then, based on these tentative OAC concentrations, alterations to spectra arising from fluorescence can be identified and subtracted from the "raw" spectra and finally the corrected spectra should be again subjected to the retrieval procedure.

An analogous approach could be pursued when assessing possible alterations to water-leaving radiance spectra stemming from the optical impact of the bottom albedo. Estep (1994) analyzed bottom-driven spectral changes to case-1 water-leaving radiance and the resulting alterations to chlorophyll-content retrieval

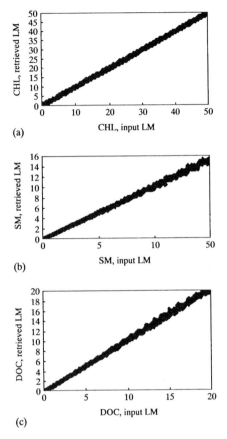

Fig. 3.24. Results of utilizing the Levenberg–Marquardt procedure for retrieval of concentrations of *chl* (*a*), *sm* (*b*), and *doc* (*c*).

utilizing expression (3.77). For shallow waters with depth H averaging 5 m and chlorophyll concentrations ranging between 0–(2–3) μg/l, retrieval error may be as high as half or one order of magnitude if the bottom is covered with either sand or silt, or grass. In moderately productive waters (i.e. $C_{chl} \leq 3$ μg/l) the highest error arises from a grassy bottom. However, in waters loaded with *chl* in excess of 3–4 μg/l, the most substantial retrieval error should be expected for sandy bottoms. Neglect of bottom optical influence on water-leaving radiance can result in enormous (for case-1 waters) chlorophyll retrieval errors: depending on the bottom coverage type, the retrieved values of C_{chl} can be as high as 35–80 μg/l (such amounts could be exclusively characteristic of eutrophic and hypereutrophic waters!) whereas the actual *chl* concentrations were between 7 to 10 μg/l. Analogous results have been obtained for $10 < C_{chl} < 20$ μg/l, except for a sandy bottom whose optical influence in such waters is somewhat reduced. With increasing bottom depth, differences in optical influence produced by sandy, silty or grassy bottoms become less pro-

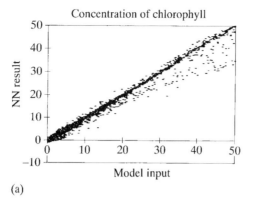

(a)

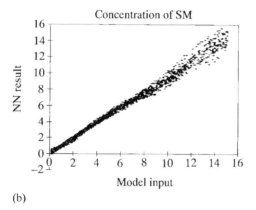

(b)

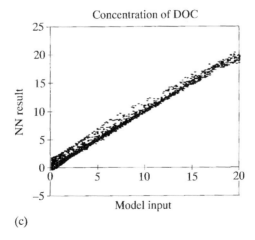

(c)

Fig. 3.25. Results of utilizing the neural network simulator for retrieval of concentrations of *chl* (a), *sm* (b), and *doc* (c).

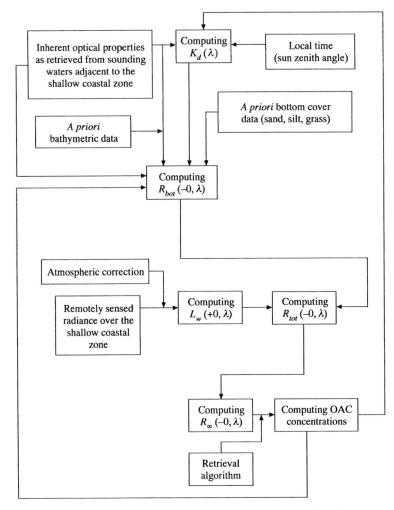

Fig. 3.26. A flow-chart of bottom influence assessment for remote sensing of shallow waters.

nounced, and the absolute retrieval error diminishes, so that at $H = 20$ m the bottom optical impact on retrieval results becomes insignificant. In case-2 waters, which are rich in absorbing and scattering matter, optical influence of bottom should be considerably reduced.

However, irrespective of water type, when sounding shallow basins, bottom influence corrections should be made. A relevant procedure (Fig. 3.26) has been suggested by Pozdnyakov and Kondratyev (1997a).

As shown above, to correct water-leaving radiance for bottom reflectance, information on bottom spectral albedo and depth, as well as sun zenith angle is required. Moreover, hydro-optical properties of the water body under surveillance should be known. The latter could be assessed from the results of retrieval of the concentration vector C relating to the area adjacent to the shallow coastal zone. When choosing the value of bottom albedo, the prevailing bottom type should be taken into consideration. Based on these data, the diffuse attenuation coefficient $K_d(\lambda)$ can be computed for local time (solar zenith) allowing for eventual assessment of the $R_{bot}(-0, \lambda)$ value. Subtraction of $R_{bot}(-0, \lambda)$ from $R_{tot}(-0, \lambda)$ gives $R_\infty(-0, \lambda)$, from which the concentration vector C can be deduced by applying an appropriate retrieval algorithm (see above). Utilization of an iteration procedure to adequately assess the value of $R_{bot}(-0, \lambda)$ will provide for higher accuracy of OAC retrieval. Of course, it could be attained if the iteration procedure proves to be converging.

   Concluding this section, it should also be pointed out that the accuracy of OAC retrieval from airborne spectral measurements of upwelling radiance can be further improved if water surface surveillance is conducted under some optimal conditions. Monte Carlo simulations indicate (Kondratyev *et al.*, 1998) that favorable conditions for inferring water-quality parameters in case-2 waters are attainable with sun zenith angles about $35°$ and calm water surface. In view of pronounced anisotropy in upwelling radiance azimuth distribution and sun glint, the preferable azimuth viewing angles are in the plane normal to the sun plane or nearby. Finally, the legitimate signal is the highest in the longer wavelength region of the visible spectrum. Taking all of this into account, these recommendations are liable to increase signal-to-noise ratio, and thus enhance the discriminative capability of remote-sensing tools.

   To recap then, the problem of retrieving parameters to measure marine coastal and inland water quality, albeit very complicate, does not seem hopeless and is in principle solvable, but requires the availability of adequate hydro-optical models, application of sophisticated retrieval procedures and an accurate assessment of possible interfering factors.

3.1.1.5   *Remote-sensing assessment of phytoplankton primary productivity: methodological aspects*

Mainly due to phytoplankton photosynthetic activity, the global ocean stores about 50 times more carbon than the air and each year oceans and atmosphere exchange around 15 times as much carbon dioxide as human activities produce. Phytoplankton produce more biomass than all other terrestrial ecosystems combined, and take up more than half of atmospheric carbon dioxide (Noever *et al.*, 1994). At the same time, carbon fluxes are believed to be amongst the principal players controlling the planet's climate status (Kondratyev, 1987). That is why one of the major challenges of contemporary global programs, such as JGOFS (1988), IGBP (1988), LOICZ (*Report . . .*, 1996), is assessing global carbon fluxes between atmosphere and ocean. At the same time, the primary production rate dynamics is a crucially important indicator of a water body's trophic status and ecological state. Brought together, these considerations naturally explain the ever-growing interest within the scientific community in exploring possible ways of assessing algae primary production on an operational basis.

The overall progress achieved in the areas of remote determination of chlorophyll (SeaWiFS, 1987), hydro-optical modelling (Pozdnyakov and Kondratyev, 1997a, b; Sathyendranath and Platt, 1997; Bukata *et al.*, 1995), algae primary production formulations (see, e.g., Pozdnyakov *et al.*, 1998b) and increasing concern prompted by reports about anthropogenic rupture of the natural carbon cycle (Kondratyev, 1987) behoves us to analyze closely the feasibility of remote sensing to assess algae primary production in natural non-case-1 waters (i.e. coastal marine/oceanic and inland waters) which generally are, on one side, very productive, and on the other side, optically very complex. This optical complexity of non-case-1 waters presents a lot of specificity in methods and techniques of remotely inferring water quality parameters (Pozdnyakov and Kondratyev, 1997b).

3.1.1.5.1   *Model formulations of algae primary production*

Two basic approaches are widely used to numerically simulate algae primary production. The first approach relies on the use of production-light intensity curves describing photosynthetic rate as a function of photosynthetically active irradiance propagating through the water surface down to the lower boundary of the photic zone over the time interval of interest, usually a day (see, e.g., Vollenweider (1966)). The other approach depends on physiological (quantum yield of photosynthesis ϕ) and optical properties of phytoplankton (pigment-specific absorption coefficient a_{chl}^*) as well as the chlorophyll biomass in the photic zone (Kirk, 1986).

Production–light intensity functions. As noted above, production-light intensity functions describe the photosynthetic response of phytoplankton to irradiance. The simplest formulation is a straight line:

$$P' = \alpha' E \tag{3.105}$$

where P' is the biomass-normalized photosynthetic rate, E is the irradiance, α' is the slope of the line.

As detailed in Talling (1957); Jassby and Platt (1976); Platt (1969), the relationship between the rate of photosynthesis P and solar photosynthetically available radiation (PAR) may be calculated from the extant chlorophyll concentration C_{chl} and two physiologically meaningful photosynthesis parameters (transfer coefficients) P_{max}^* and α^* (the asterisk indicates that both coefficients are normalized to the aquatic chlorophyll concentration, a surrogate for biomass). P_{max}^* is the maximum rate of photosynthesis, and α^* is the slope of the production versus irradiance curve.

The initial slope α^* is a direct consequence of the photochemical aspect of algal photosynthesis (a function of the type and quantity of indigenous photosynthetic pigments). Unlike P_{max}^*, α^* is not sensitive to water temperature. However, this parameter proves to be highly space and time variable: according to thorough investigations performed by different workers (see, e.g., Lee *et al.*, 1996a), α^* can vary by a factor of 4 in the same season for different regions, by a factor of 4 for the same region in different seasons, and by a factor of 5 for different regions and seasons.

The near-horizontal asymptotic portion of the curve at high irradiance levels represents the maximum rate of the enzymatic quantity of cellular enzymes and the specific temperature at which primary production is determined.

Because the quantum yield of photosynthesis decreases with increasing irradiance, the photosynthetic rate is a saturation function of irradiance. An empirical and simple saturation function was suggested in Webb *et al.* (1974):

$$P' = P'_{max}[1 - \exp(-\alpha'E/P'_{max})] \tag{3.106}$$

However, this formulation, however, does not account for primary production dependence on some factors, such as water temperature, nutrition depletion and the closely related effect of wind-induced mixing in the euphotic zone (Sakshaug and Slagstad, 1992). However, it ignores the possible variability in P along the vertical in optically highly inhomogeneous waters.

Platt *et al.* (1991) express the depth dependence of primary production on available subsurface irradiance by the relation:

$$P(z) = \frac{C_{chl}(z)\alpha^*E(z)}{[1 + (\alpha^*E(z)/P^*_{max})^2]^{1/2}} \tag{3.107}$$

where $P(z)$, P^*_{max}, α^*, and $E(z)$ are as previously defined. Although not obligatory, it is not unreasonable to assume, at least for case-1 waters, that coefficients α^*, P^*_{max} are depth invariant. However, Denman and Cargett (1988) observed that the physical barrier imposed by a shallow thermocline resulted in disparate phytoplanktonic photoadaptation profiles within each of the layers (i.e. distinct slopes of the P versus PAR curves). They also observed that the diurnal cycle of vertical motion of algae cells within the upper layer generally proceeds slowly enough for cells to adapt to the high irradiances near the water–air interface, yet rapidly enough to redistribute the photosynthesis per unit chlorophyll roughly evenly throughout the epilimnion.

Equation (3.107) accommodates the assumption that slope α^* is linear and that $E(z)$ is PAR. For the general case of a known spectral depth distribution of PAR, $E(z, \lambda)$, and a known non-linear slope function $\alpha^*(z, \lambda)$, such wavelength dependency should also be accommodated. Further, the subsurface irradiance field, $E(z, \lambda)$, is a function of in-water refracted solar zenith angle θ', and can be partitioned into a direct sunlight component, $E_{d\,sun}(z, \lambda, \theta')$, and a diffuse skylight component, $E_{d\,sky}(z, \lambda)$. It is the daylight progressions of functions $E_{d\,sun}$ and $E_{d\,sky}$ that must be considered.

To incorporate a depth-dependent and non-linear slope function, a spectral and angular dependence of the subsurface PAR light field, and a solar and sky partitioning of $E(z, \lambda)$ into the photosynthesis-light consideration, eq. (3.107) may be rewritten as (Platt *et al.*, 1991):

$$P(z) = \frac{C_{chl}(z)\Lambda(z)}{[1 + (\Lambda(z)/P^*_{max})^2]^{1/2}} \tag{3.108}$$

where

$$\Lambda(z) = \sec\theta' \int \alpha^*(z,\lambda) E_{d\,\mathrm{sun}}(z,\lambda,\theta')\,d\lambda + 1.2 \int \alpha^*(z,\lambda) E_{d\,\mathrm{sky}}(z,\lambda)\,d\lambda \qquad (3.109)$$

Integration of eq. (3.108) over the photic zone for daylight hours will then yield the phytoplankton photosynthesis (approximate primary production) for the aquatic region being remotely monitored.

Vollenweider (1966) has shown that the quotient P/P_{max}, where P_{max} is the maximum rate of photosynthesis per unit volume $(P_{\mathrm{max}} = P^*_{\mathrm{max}} C_{chl})$, could be parameterized as follows:

$$\frac{P}{P_{\mathrm{max}}} = \frac{(E/E_k)}{\{[1+(E/E_k)^2][1+(aE/E_k)^2]^n\}^{1/2}} \qquad (3.110)$$

where a, n parameters: an increase in a brings about a general lowering of the curve (Fig. 3.27) at all irradiances, while an increase in n causes the curve to slope more sharply about the point of maximum curvature.

Except in the special case where either a or n equals zero, the curves exhibit three main features: a nearly linear relationship between photosynthetic rate and irradiance at low light intensities, a peak rate for optimum irradiance, and inhibition at higher irradiances. E is the downwelling irradiance integrated over the PAR spectral region (400–700 nm). The point at which the initial and asymptotic horizontal slopes intersect (Fig. 3.27) is designated as E_k.

As easily seen from Fig. 3.27, parameters P_{max} and E_k are not directly measurable if there is any kind of photosynthetic production inhibition by intense light. The measurable parameters are P^*_{opt} (normalized to the aquatic chlorophyll concentration) and E'_k which are related to the basic parameters via the following

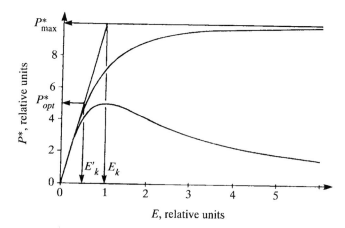

Fig. 3.27. A production–light intensity diagram illustrating the notations of E_k, E'_k, P^*_{max}, and P^*_{opt}. E_k is the light intensity at which extrapolation of the nearly linear portion of the production–light intensity curve would intersect the asymptote $P^* = P^*_{\mathrm{max}}$. E'_k is the light intensity at which the extrapolated line would intersect the line $P^* = P^*_{opt}$.

expressions:

$$P_{\max} = \delta P_{opt}^* \tag{3.111}$$

and

$$E_k = \delta E_k' \tag{3.112}$$

where δ is a constant whose magnitude depends on a and n:

$$\delta = [(1 + E_{opt}^2)(1 + a^2 E_{opt}^2)^n]^{1/2}/E_{opt} \tag{3.113}$$

$$E_{opt} = \left\{ \left[(1 - n)a + \sqrt{a^2(n-1)^2 + 4n} \right] \Big/ 2na \right\}^{1/2} \tag{3.114}$$

The Fee integration of the Vollenweider primary production model over time and depth yields (Fee, 1969):

$$P_\Sigma = P_{opt}\delta \int_{-\zeta}^{+\zeta} \int_{0.01 E_d(z=-0,t)/E_k}^{E_d(z=-0,t)/E_k} \frac{dy\, dt}{K_d(z,t)\{(1+y^2)(1+a^2 y^2)^n\}^{1/2}} \tag{3.115}$$

where $E_d(z = -0, t)$ is PAR just below the water surface at time t, ζ is half the day length, $K_d(z, t)$ is the downwelling irradiance attenuation coefficient at time t and depth z, $y = E(0, t)/E_k$, E_k is the light saturation parameter ($E_k = P_{\max}^*/\alpha^*$) when a or $n = 0$. The limits of integration in eq. (3.115) are for t the times of local sunrise and local sunset, and for y the values corresponding to the surface and depth of the 1% irradiance level, below which depth no significant contribution to primary production is considered to occur. It is noteworthy that the irradiance at depth z for a subsurface refracted angle θ' and the incident radiation comprising both a direct and diffuse component may be written:

$$E_d(z, \theta' \equiv t) = E_d(z = -0, \theta' \equiv t) \exp(-K_d(z, \theta' \equiv t)z) \tag{3.116}$$

where $\theta' \equiv t$ emphasizes the identity between the local time and incoming radiation incidence angle, and points to the possibility that integration of eq. (3.115) over time be replaced with integration over θ'.

Consequently, the application of this approach for assessing primary productivity of a given algae community requires information on parameters determining the production–light intensity curve (a, n, E_k, and $P_{\max}$) appropriate for this specific assemblage of phytoplankton cells growing under concrete environmental conditions (water temperature, nutrients availability, etc.).

Explicitly physiological approach

As mentioned above, the second approach defines primary production P at depth z as a function of chlorophyll specific absorption $a_{chl}^*(\lambda)$, its biomass at depth z (concentration of chlorophyll C_{chl} in milligrams per cubic meter), quantum yield of photosynthesis $\phi(z)$ (in mol C per Einsteins (Ein) absorbed)* and quantum scalar

---

* To convert irradiance expressed in W/m^2 into quanta/m^{2}s, it should be multiplied by λ/hc, where c is the speed of light in m/s, h is Planck's constant equal to 6.63×10^{34} J-s/quanta, λ is the wavelength in meters (1 ein $= 6.02 \times 10^{-23}$ quanta).

irradiance $E_0(\lambda, z)$ (in Ein $\times$ m^{-2}nm^{-1}) at depth z:

$$P(z) = \phi(z) \int_\lambda a_{chl}^*(\lambda) C_{chl}(z) E_0(\lambda, z) d\lambda \qquad (3.117)$$

(the wavelength range of integration is 400–700 nm).

The value of ϕ is assumed to be independent of wavelength, but affected by the scalar irradiance level: at low, vanishing irradiance, it reaches a maximal value ϕ_{max}, and then decreases when irradiance decrease. Several formulations have been suggested. According to Kiefer, Mitchel (1983):

$$\phi(E_0) = \phi_{max} \frac{E_\phi}{E_\phi + E_0} \qquad (3.118)$$

where ϕ_{max}, a constant $= 0.06$ mol C (Ein absorbed)$^{-1}$, is maximum quantum efficiency, E_ϕ, a constant $=$ Ein/day, is irradiance at which the quantum efficiency is equal to $\phi_{max}/2$, E_0 is the spectrally integrated (400–700 nm) scalar irradiance.

Platt *et al.* (1980) introduced a dimensionless function $f(x)$ describing the evolution of ϕ with irradiance $[\phi = \phi_{max} f(x)]$:

$$f(x) = x^{-1}(1 - e^{-x})e^{-\beta x} \qquad (3.119)$$

where β is a parameter controlling photosynthesis inhibition ($\beta = 0.01$), x is a dimensionless irradiance $=$ (photosynthetically usable radiation/photosynthetically usable radiation for which $\phi = \phi_{max}/2$). Function $f(x) = 1$ when $x = 0$, and tends toward 0 when x tends toward ∞. Importantly, this function could be tuned for actual water temperature (Berthon and Morel, 1992).

Zaneveld and Kitchen (1993) suggested an alternative representation:

$$\phi(E_0) = \phi_1 - \phi_2 \ln[E_0(\text{PAR})] \qquad (3.120)$$

where ϕ_1 and ϕ_2 are constants: $\phi_1 = 0.055$ and $\phi_2 = 0.0109$.

Returning to eq. (3.117), it can be further rewritten in the form:

$$P(z) = \int_\lambda \alpha(\lambda, z) C_{chl}(z) E_0(\lambda, z) \, d\lambda \qquad (3.121)$$

where $\alpha(\lambda, z) = \phi(z) a_{chl}^*(\lambda)$ is the rate of photosynthesis (in mol C mg chl (Ein m^{-2})$^{-1}$) which is commonly referred to as the *action spectrum* (Kirk, 1983).

The effect of photoinhibition on primary production could be formulated by introducing in (3.118) an exponential term:

$$\phi(E_0) = \phi_{max} \frac{E_\phi}{E_\phi + E_0} \exp(-\nu E_0) \qquad (3.122)$$

where ν is a constant $= 0.01$ (Ein/m^2/day)$^{-1}$—a value based on Platt *et al.* (1980).

Consequently, eq. (3.121) can finally be presented in the following form:

$$P(z) = \frac{E_\phi \exp(-\nu E_0)}{E_\phi + E_0} \int_\lambda \alpha(\lambda) C_{chl} E_0(\lambda, z) \, d\lambda \qquad (3.123)$$

or else:

$$P(z) = \phi_{max} \frac{E_\phi \exp(-\nu E_0)}{E_\phi + E_0} \int_\lambda a_{chl}^*(\lambda) C_{chl} E_0(\lambda, z) \, d\lambda \qquad (3.124)$$

Downwelling irradiance attenuation and the depth of the photic zone

The general expression for the attenuation of subsurface irradiance (E_d) down-welling along the path length $(Z_{\theta'})$ of the principal direction (θ'') of subsurface propagation to a particular irradiance level is Beer's Law, which may be expressed as

$$E_d(Z_{\theta'}, \theta', \lambda) = E_d(Z_{\theta'} = -0, \theta', \lambda) \exp(-K_d(\lambda) Z_{\theta'}(\theta'')) \qquad (3.125)$$

where K_d is the irradiance attenuation coefficient.

Replacing $Z_{\theta'}$ by the vertical depth z in eq. (3.125) yields:

$$E_d(z, \theta', \lambda) = E(z = -0, \theta', \lambda) \exp(-K_d(\theta', \lambda) z) \qquad (3.126)$$

where $K_d(\theta')$ is the vertical irradiance attenuation coefficient.

Since $Z_{\theta'} = z$ only for those conditions when the sun is directly overhead (i.e. $\theta' = 0$), the value of experimentally determined vertical irradiance attenuation coefficient proves to be variable with the time at which the subsurface irradiance profile is performed. Therefore, extending the use of invariant $K_d(\theta')$ throughout the daylight hours (assuming the water column is vertically homogeneous) defines an appropriate diurnal variation in subsurface irradiance levels E_d. Obviously, this is a factor potently capable of affecting model evaluation of primary production when resorting to the first approach specified above (eq. (3.115)).

Using a Monte Carlo simulation of photon propagation through natural waters, Kirk (1984) has presented empirical relationships relating the vertical irradiance attenuation coefficient to time of day and water type:

$$K_{d\,sun}(\theta') = (\cos \theta')^{-1} [a^2 + (0.425 \cos \theta' - 0.190) ab]^{1/2} \qquad (3.127)$$

$$K_{d\,sky} = 1.168 [a^2 + 0.162 ab]^{1/2} \qquad (3.128)$$

where $K_{d\,sun}(\theta')$ and $K_{d\,sky}$ are the vertical irradiance attenuation coefficient for, respectively, direct and diffuse components of the incident solar radiation; a and b are, respectively, absorption and scattering coefficients of the water (for simplicity, wavelength stipulation, while understood, is not explicitly expressed).

For incident irradiation comprised of a diffuse sky fraction η, the resultant vertical irradiance attenuation coefficient $K_d(\theta')$ could be formulated as

$$K_d(\theta') = \eta_w K_{d\,sky} + (1 - \eta_w) K_{d\,sun}(\theta') \qquad (3.129)$$

where η_w is the fraction of the subsurface downwelling irradiance that is diffuse:

$$\eta_w = \eta(1 - \rho_{sky}) / [\eta(1 - \rho_{sky}) + (1 - \eta)(1 - \rho_{sun}(\theta))] \qquad (3.130)$$

θ = solar zenith angle prior to refraction at the atmosphere–water interface, ρ_{sky} and $\rho_{sun}(\theta)$ are Fresnel reflectivities of sky and solar irradiance (directly propagating from the zenith angle θ), respectively.

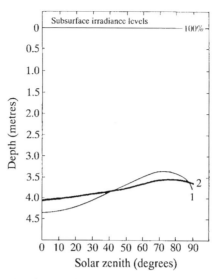

Fig. 3.28. The depth of the 1% subsurface irradiance level as a function of solar zenith angle for a water mass defined by $\omega = 0.60, K_d(0°) = 1\,\mathrm{m}^{-1}$ (1) and $\omega = 0.90, K_d(0°) = 1\,\mathrm{m}^{-1}$ (2) (Bukata *et al.*, 1989).

Bukata *et al.* (1989) have shown that for a water mass defined by a single scattering albedo $\omega[\omega = b/c; (c = a + b)]$, the vertical irradiance attenuation coefficient K_d varies with θ and exhibits two features: (i) as ω increases, the solar zenith angle dependence of K_d decreases, and (ii) the relative value of K_d, for all values of ω, increases with increasing solar zenith angles up to *ca.* 70° at which point the relative value of K_d (normalized to $K_d(0°)$) starts rapidly decreasing due to the increasing percentage of diffuse radiation in the total incident radiation observed for large solar zenith angles. As a result, the depth of the 1% subsurface irradiance level (defining the lower boundary of the photic zone) reveals variations with θ similar to those of $K_d(\theta')$, and these variations degrade as ω increases (Fig. 3.28). As seen from Fig. 3.28, in relatively non-turbid waters ($\omega = 0.60$) for which the photic depth is *ca.* 4 m, vertical migrations of photic depth during daylight hours can attain an amplitude of about 1 meter (i.e. 25%), although in case of more turbid waters ($\omega = 0.90$) they prove to be less pronounced. These results strongly point to the necessity of applying eqs. (3.125)–(3.129) when evaluating primary production not only through modelling (e.g. eq. (3.115)) but also via *in situ* measurements (for correct lowering of incubation bottles down to the photic depth level during daylight hours).

Relationships between vector and scalar irradiance throughout the photic zone

Global radiation, comprising a direct solar and diffuse sky irradiance downwelling on the air–water interface, possesses inherent geometrical properties related to arrival directions of the impinging radiation.

    The subsurface downwelling irradiance $E_d(z)$ determined from such global radiation values will retain corresponding directional biases, and is often referred to as *vector* irradiance.

A photosynthesizing phytoplankton cell suspended in water receives light energy not only from the downwelling irradiance, but also from the upwelling irradiance to which it is subjected. The *scalar* irradiance is the total energy per unit area arriving at a point from all directions when all directions are equally weighted. Thus, $E_0(z)$ is the integration of the radiance distribution at a point over 4π space. When divided by the speed of light in water, E_0 becomes a measure of the energy density at a given point in the aquatic medium. Consequently, it is *scalar* irradiance $E_0(z)$ rather than *vector* irradiance $E_d(z)$ which is more representative of the cell's light regime.

From Gershun (1936), the scalar irradiance is related to the vector upwelling and downwelling irradiances $E_u(z)$ and $E_d(z)$ and the total absorption coefficient $a(z)$ by the relationship:

$$E_0(z) = -\frac{1}{a(z)}\frac{d}{dz}[E_d(z) - E_u(z)] \tag{3.131}$$

Prieur and Sathyendranath (1981) express the E_0/E_d ratio as

$$\frac{E_0}{E_d}(R) = \frac{1}{\bar{\mu}_d} + \frac{R}{\bar{\mu}_u} \tag{3.132}$$

where $\bar{\mu}_d$ and $\bar{\mu}_u$ are mean cosines for the downwelling and upwelling irradiance fields, respectively.

From curve-fitting to the Monte Carlo computer simulations for a homogeneous water column, Jerome *et al.* (1988, 1990) determined the ratio E_0/E_d as a function of volume reflectance $R = E_u/E_d$ and solar zenith angle θ for various depths z within the water column.

For the depth of the 100% downwelling irradiance $(z_{100} \equiv z = -0)$:

$$\frac{E_0}{E_d}(R, \theta) = \left(\frac{1.068}{\mu'} - 0.068\right)(1 + 3.13R(0°)) \tag{3.133}$$

where $R(0°)$ is the volume reflectance for vertical incident radiation, μ' is the inverse of the cosine of the in-water angle of refraction θ'. According to Bukata *et al.* (1995),

$$R(0°) = R(\theta', \eta_w)/(1.165\eta_w + (1 - \eta_w)/\cos\theta') \tag{3.134}$$

and η_w is attainable from eq. (3.130).

In the case of totally diffuse illumination of the water surface, for the depth of the 100% downwelling irradiance (z_{100}), the sought-after ratio takes the form:

$$\frac{E_0}{E_d}(R) = 1.177(1 + 3.13R) \tag{3.135}$$

Appropriate expressions have been obtained for the depth of the 10% down-welling irradiance (z_{10}), and for the depth of the 1% downwelling irradiance (z_1) both for sunny and overcast conditions (Bukata *et al.*, 1995). However, calculations of the scalar irradiance vertical profile can be substantially simplified if, after having determined the E_0/E_d ratio at z_{100}, the coefficient of scalar irradiation attenuation

(Bannister, 1990) is applied:

$$K_0 = (a/\mu')[1 + (0.247\mu')(b_b/a]^{1/2} \qquad (3.136)$$

where a and b_b are absorption and back-scattering coefficients, and μ' is as above the cosine of the incident photon flux as measured just beneath the air/water interface. Hence:

$$E_0(z) = E(z_{100} \equiv z = -0)\exp(-K_0 z) \qquad (3.137)$$

The coefficient of scalar irradiation attenuation K_0 can be assumed either vertically invariant or defined by specific vertical variations in a and b_b. The inherent optical properties of water (which a and b_b belong to) are additive in nature and so are basically controlled by all major optically active components. However, in many cases the vertical inhomogeneity in a and b_b arises from a non-uniform distribution of chlorophyll in the vertical. The latter could be represented in a generalized form:

$$C_{chl}(z) = C_{0chl} + \frac{h}{\sigma(2\pi)^{1/2}}\exp\left[-\frac{(z - z_m)^2}{2\sigma^2}\right] \qquad (3.138)$$

where C_{0chl} is the baseline pigment concentration, z_m is the depth of the chlorophyll minimum, σ defines the breadth of the peak, and h determines the total biomass above the baseline. Consequently, in this case the vertical profile of the attenuation coefficient K_0 will exhibit a dependence analogous to (3.138) but certainly differently scaled. However, when the optical inhomogeneity of the water column in the vertical is not exclusively controlled by chlorophyll but also by, say, suspended minerals and/ or dissolved organics, then the depth dependence of K_0 will be accordingly modified.

3.1.1.5.2   *Estimating primary production from remote sensing*

Remote sensing of natural waters is based on capturing by remote sensor the radiance signal coming up from the water surface. After subtraction of the Fresnel reflection component, the resultant radiance signal is a convolution of processes of solar light interactions with the water column. Consequently, the legitimate optical signal reaching the remote sensor carries information not solely about a surface layer, it also contains contributions from layers deeper in the water column, although strongly damped by aquatic medium attenuation.

Gordon and McCluney (1975) determined that for a sensor viewing the ocean at any wavelength λ, 90% of the recorded signal originates from within a so-called *penetration depth* given by the inverse of the downwelling irradiance attenuation coefficient $K_d(\lambda)$. The penetration depth corresponds to one attenuation length, i.e. the depth at which the downwelling irradiance falls to *ca.* 37% of its value just beneath the air–water interface. Thus, satellite observations over natural waters are restricted to the uppermost attenuation length. As assumed above, the photic zone is considered to extend to the depth of 1% irradiance level. Therefore the photic zone, extends to 4.605 attenuation depth.

In relatively clear case-2 waters (with the North American Great Lakes as an example) the values of K_{PAR} are generally in the range $0.1\,\mathrm{m}^{-1}$ to $0.5\,\mathrm{m}^{-1}$ resulting in penetration depths of 10 to 2 m being commonly observed in open waters. The corresponding photic zones in these waters would extend to depths in the range *ca.* 46 m–9 m. For relatively turbid case-2 waters, the values of K_{PAR} generally are in the range $0.5\,\mathrm{m}^{-1}$ to $1.2\,\mathrm{m}^{-1}$ resulting in penetration depths in the range $2\,\mathrm{m}^{-1}$ to $0.8\,\mathrm{m}^{-1}$ and photic depths in the range 9 m to 4 m (Bukata *et al.*, 1995).

As discussed in the previous section, algorithms are available that convert the optical signal to give a chlorophyll concentration that is representative of the upper attenuation depth (Pozdnyakov and Kondratyev, 1997b). At the same time, satellite techniques have been developed to evaluate the downward solar radiation at the surface (Gautier *et al.*, 1980; Dedieu *et al.*, 1987).

However, for an adequate evaluation of primary production in the water column, information is required about *total* chlorophyll content ($\langle C_{chl} \rangle_{tot}$) within the euphotic zone (i.e. the zone at the depth of which z_{eu} the downwelling irradiance equals 1% of the irradiance just beneath the water surface).

Since remote sensing means, in principle are incapable of providing the required information on $\langle C_{chl} \rangle_{tot}$, the latter can be found either by establishing a regression relationship between remotely determined chlorophyll concentration $C_{chl\,rs}$ and $\langle C_{chl} \rangle_{tot}$, or by applying to $C_{chl\,rs}$ statistical profiles of C_{chl} characteristic of the water body under surveillance (Morel, Berthon, 1989; Brooks, Torke, 1977).

Berthon and Morel (1992) used a model to estimate phytoplankton primary production which rests on the photosynthetically stored radiant energy $\tilde{E}$ content in the euphotic zone:

$$P = (1/39)\tilde{E} = (1/39)\bar{E}_{par}(z = +0)\xi\langle C_{chl} \rangle_{tot}\psi^* \qquad (3.139)$$

where $\bar{E}_{par}(z = +0)$ is the photosynthetically available irradiance just above the water surface and integrated over daylength $(\mathrm{W/m}^2)$, ψ^* is the cross-section for photosynthesis per unit of area C_{chl} $(\mathrm{m}^2/\mathrm{g}\ chl)^*$, ξ is the ratio of photosynthetically active pigments to total pigments (including the pheopigments)

$$\xi = C_{chl}/C_{chl} + C_{pheo}, \qquad (3.140)$$

the conversion factor $(39\,\mathrm{kJ\,g}^{-1}$ of C fixed) is taken from Platt (1969), $\langle C_{chl} \rangle_{tot}$ is as above:

$$\langle C_{chl} \rangle_{tot} = \int_0^{z_{eu}} C_{chl}(z)dz. \qquad (3.141)$$

To adjust this model to remote-sensing needs, the vertically integrated biomass $\langle C_{chl} \rangle_{tot}$ is replaced by an empirical nonlinear expression which involves the chlorophyll concentration retrieved from remote measurements $(C_{chl\,rs})$, and a factor F and an exponent γ derived from regression analysis. In vertically homogeneous or well-mixed waters, γ is nearly 0.5, but the existence of nonuniform profiles (particularly with deep pigment maxima) results in somewhat lower values of γ.

---

*There is experimental evidence (Berthon, Morel, 1992) that the mean value of ψ^* for the euphotic zone is about $0.088\,\mathrm{m}^2\mathrm{g}$ chl-$a)^{-1}$ as opposed to the value of 0.07 (relevant units) previously reported by Platt (1996).

Many workers have suggested area- and season-specific algorithms for retrieval of phytoplankton primary production (see Pozdnyakov *et al.*, 1998). These algorithms are based on very general regression relationships established between surficial and euphotic columnal concentrations of chlorophyll, as well as some other parameters dictating primary production rate (incident irradiation averaged over the vegetation season, water temperature, photosynthesis efficiency, etc.).

Lee *et al.* (1996 a, b) report that they developed an algorithm for estimating primary production at depth from remote sensing which uses an analytically derived phytoplankton (chlorophyll) absorption cross-section at 440 nm ($a_{chl}^*(440)$) and the bulk water absorption coefficient $a(\lambda)$ based on remote measurements of above water upwelling and downwelling radiances. The general scheme of the algorithm encompasses eqs. (3.123)–(3.124). The photoinhibition parameter v in equations (3.123)–(3.124) was set at 0.01 Ein/m^2/day^{-1}, a value based on Platt *et al.* (1980).

The Lee *et al.* algorithm uses the coefficient of downwelling vector irradiance attenuation $K_d \approx a/\mu_d$, where a is the bulk water absorption coefficient and is a sum of three components: $a_w + a_{doc} + a_{ph(chl)}$, a_w is pure water absorption, a_{doc} is the dissolved organic and detritus absorption, $a_{ph(xhl)}$ is absorption by phytoplankton (chlorophyllous pigments).

To retrieve the bulk absorption coefficient, a relationship between the ratio $E_u(z = +0, \lambda)/E_d(z = +0, \lambda)$ and inherent water properties was used (Lee *et al.*, 1996 b):

$$R_{rs}(\lambda) \approx \frac{0.17}{a_w(\lambda) + a_{dom}(\lambda) + a_{ph(chl)}} \left[\frac{b_{b_w}(\lambda)}{3.4} + X \left(\frac{400}{\lambda} \right)^Y \right] \qquad (3.142)$$

where X, Y are coefficients, b_{b_w} is the water backscattering coefficient.

Averaging of K_d over the euphotic depth is attained via multiplying K_d by an empirically found coefficient 1.08. Remotely assessed concentration of chlorophyll was decreased by a factor of 0.8 to account for the contribution of pheopigments. μ_d was given a value of 0.83 for high-latitude cloudy days. E_ϕ and a_{chl}^* (averaged over the PAR spectral region) were given values of 10 Ein/day and 0.016 m^2/mg *chl-a*, respectively. Integration of $E_0(\lambda)$ over depth was performed down to the lower boundary of the euphotic zone (i.e. where $E_0(z, \lambda) = 0.01 E_0(z = -0, \lambda)$).

Pozdnyakov *et al.* (1998) (see also Bukata *et al.*, 1995)) suggested a protocol for estimating primary production in non-case-I (i.e. marine coastal and inland) waters from remote measurements based on the $E_0 - R$ formulism discussed above and a hydro-optical model developed for Lake Ladoga (Kondratyev *et al.*, 1990c). A generalized flow chart outlining parameters and sequential operations involved in remotely estimating primary production in turbid waters is presented in Fig. 3.29.

The recorded upwelling radiance spectrum $L_u(z = +0, \lambda, \theta)$ is converted into the subsurface volume reflectance spectrum $R(z = -0, \lambda, \theta)$ just below the air–water interface:

$$R(z = -0, \theta) = \frac{Q[L_u(+0, \theta) - 0.021\alpha E_{sky} - f_2 E_{sun}]}{0.544(0.944 E_{sky} + f_4 E_{sun})} \qquad (3.143)$$

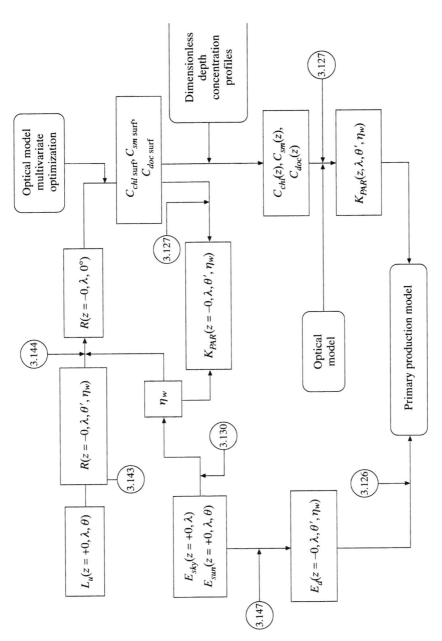

Fig. 3.29. Generalized flow chart outlining parameters and sequential operations involved in remotely estimating primary production in non-case-I waters. The mathematical formulism is described in the text.

where Q = ratio of the upwelling irradiance below the water surface to the upwelling nadir radiance below the water surface, α = spectrally dependent ratio of the downwelling zenith sky radiance to the downwelling sky irradiance,

   f_2 = reflection of solar irradiance into the field of view of the sensor due to surface waves,
   f_4 = reflection of solar irradiance due to surface waves,
   E_{sun} = spectrally dependent downwelling irradiance from the sun,
   E_{sky} = spectrally dependent downwelling irradiance from the sky.

The subsurface volume reflectance just beneath the air-water surface interface $R(z = -0, \lambda, \theta)$ will, of course, be remotely estimated under variable conditions of solar zenith angle θ, and relative fractions of diffuse and direct downwelling global radiation. Thus, it is necessary to normalize the subsurface volume reflectance to a standard solar zenith angle and a standard downwelling radiation. Consequently, the estimated subsurface volume reflectance just beneath the air–water interface can be partitioned into components responsive to direct and diffuse components of global radiation:

$$R(z = -0, \theta, \eta_w) = (1 - \eta_w)R_{sun}(z = -0, \theta) + \eta_w R_{sky}(z = -0) \qquad (3.144)$$

where η_w is described by eq. (3.130), and is the fraction of the total subsurface downwelling irradiance that is diffuse; the fraction that is direct would then be given by $(1 - \eta_w)$; $R_{sun}(z = -0, \theta)$ is the volume reflectance just beneath the air–water interface resulting from the direct fraction of global radiation (it is dependent on θ or rather on θ' = the in-water refraction angle associated with a solar zenith angle of θ); $R_{sky}(z = -0)$ is the volume reflectance just beneath the air–water interface resulting from the diffuse fraction of global radiation (it is independent of θ).

According to Bukata *et al.* (1995), the subsurface volume reflectance $R_{sun}(z = -0, 0°)$ observable when the sun is vertically overhead (i.e. $\theta = 0°$) may be related to the remotely estimated subsurface volume reflectance $R(z = -0, \theta', \eta_w)$ through eq. (3.134). From such subsurface volume reflectance values that are zero-angle normalized and appropriate cross-section spectra for major optically active components in turbid waters, the near-surface chlorophyllous pigment concentrations (along with co-existing concentrations of suspended minerals (*sm*) and dissolved organics (*doc*)) can be estimated making use of one of the available retrieval algorithms (see, e.g., Pozdnyakov, Kondratyev, 1997). Multiplying the retrieved concentration of chlorophyll by the coefficient ξ (eq. (3.140)) will result in evaluation of the abundance of *chl-a* in the top layer of the water column.

For a calm water surface, the downwelling subsurface irradiance $E_d(z = -0, \theta', R)$ can be obtained either directly from on-site measurements or indirectly estimated from above surface values of downwelling irradiance $E_d(z = +0, \theta, R)$ through the relationship:

$$E_d(z = -0, \theta', R) = \left[1 + \frac{R(\theta')\rho_u(\theta')}{1 - R(\theta')\rho_u(\theta')}\right](1 - \rho_d(\theta))E_d(z = +0, \theta, R) \qquad (3.145)$$

where $\rho_d(\theta)$ = coefficient of reflection of above-surface downwelling irradiance for solar zenith angle θ, $\rho_u(\theta') = $ coefficient of reflection of subsurface upwelling irradiance for solar zenith angle θ'.

It has been shown (Bukata et al., 1995) that

$$\rho_u(\theta') = 0.271 + 0.249\cos(\theta') \tag{3.146}$$

for direct incident solar radiation, and $\rho_u = 0.561$ for a cardioidal diffuse incident radiation distribution. Values of $\rho_d(\theta)$ can be calculated from Fresnel's equation, while ρ_d for a diffuse cardioidal distribution can be taken as 0.066 (Jerlov, 1976).

Baker and Smith (1990) suggested a simplified expression for $E_d(z = -0)$:

$$E_d(z = -0) = \frac{1}{1 - \rho_u R(\theta')} t(\theta) E_d(z = +0, \theta) \tag{3.147}$$

where $t(\theta)$ stands for global transmittance through the air–water interface:

$$t(\theta) = 1 - [\rho_{\text{sun}}(1 - \eta) + \rho_{\text{sky}}\eta] \tag{3.148}$$

The value of ρ_{sky} is estimated at 0.066 for a uniform sky radiance distribution, and 0.052 for completely overcast skies (assuming a cardioidal distribution); ρ_w is given the value of 0.47. Fresnel's formula for unpolarized light gives the reflectance of a direct solar beam.

Evaluation of sunlight propagation downwards to the euphotic zone lower limit (z_{eu}) requires information on the vertical profiles of the major optically active components, such as *chl*, *sm*, and *doc*.

Surficial concentrations of *chl*, *sm*, and *doc* can be retrieved in non-case-I waters from the upwelling irradiance employing sophisticated techniques such as the multivariate optimization technique, simplex method, neural networks, and some others described elsewhere (Pozdnyakov, Kondratyev, 1977; Atkinson, Tatnall, 1997), and briefly described in § 3.1.1.3. However, the problem remains because for adequate assessment of column primary production, first, a vertical profile of *chl* concentration is required, and second, for determining z_{eu} the information on column content variations in other optically active components is mandatory.

In an attempt to overcome the variabily in depth profiles of chlorophyll concentration in inland and coastal waters, it might be worth considering an approach suggested for case-I waters by Morel and Berthon (1989). This approach considered that two extreme aquatic configurations could accommodate the chlorophyll concentration profiles of oceanic waters. The first extreme is taken to be stratified oligotrophic oceanic waters wherein nutrient depletion is observed in the near-surface layer. In this case the maximum chlorophyll concentration would be downward-driven. The second extreme is taken to be well-mixed waters wherein sufficient irradiance levels have established eutrophic conditions. In this case the vertical distribution of chlorophyll $C_{chl}(z)$ throughout the photic zone would be quite uniform. Understandably, a wide range of intermediate chlorophyll profiles within these extremes are found encounterable in natural waters. However, it is perhaps not unreasonable, to reduce this wide range of profiles to a limited number of generic chlorophyll depth distributions pertinent to specific types of water columns (and on through specific water bodies) depending on the trophic status of

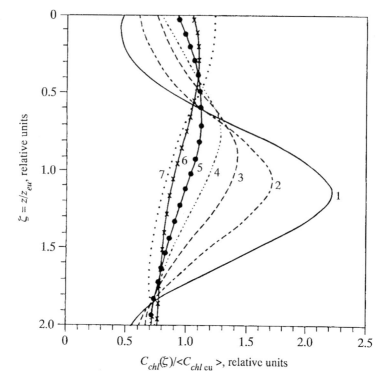

Fig. 3.30. Seven generic dimensionless chlorophyll depth profiles for case-I waters (Morel, Berthon, 1989).

the water, the aquatic vegetation growth cycle, and the intensity and duration of wind-induced vertical mixing.

Statistically analyzing over 4000 oceanic station data sets, Morel and Berthon (1989) illustrated the presence of seven generic classifications of oceanic waters. These profiles are schematically illustrated in Fig. 3.30.

The seven profiles are plotted in terms of a dimensionless chlorophyll concentration $C_{chl}(\zeta)/\langle C_{chl\,eu}\rangle$, and a dimensionless depth $\zeta = z/z_{eu}$, i.e. depth reckoned as a fraction of photic depth, and chlorophyll concentration reckoned as a fraction of total photic zone chlorophyll:

$$\langle C_{chl}\rangle = z_{eu}^{-1}\int_0^{z_{eu}} C_{chl}(z)\,dz \tag{3.149}$$

A numerical analysis of the $C_{chl}(\zeta)/\langle C_{chl\,eu}\rangle$ profiles in Fig. 3.30 as a function of remotely determined (i.e. surficial) concentration of chl (denoted C in equations below) suggests the following parameterization (Morel, Berthon, 1989):

$$C_{chl}(\zeta)/\langle C_{chl\,eu}\rangle = C_b + C_{max}\exp\{-[(\zeta - \zeta_{max})/\Delta\zeta]^2\} \tag{3.150}$$

where C_b is a background over which is superimposed a Gaussian curve with a maximum value given by C_{max} occurring at ζ_{max}, and having a broadness controlled by $\Delta\zeta$; C_b, C_{max}, ζ_{max}, $\Delta\zeta$ being polynomial functions of c.

Brooks and Torke (1977) seemingly found the same kind of regularities in the *chl* vertical profiles in Lake Michigan. They report that for a multi-year period of observations in Lake Michigan chlorophyll concentration increased uniformly at all depths during spring reaching 3–4 mg/m^3 by late May. Thermal stratification was followed by development of a subthermocline chlorophyll peak between 10 and 30 m that reached 8.5 mg/m^3 by late July. The major subthermocline peak collapsed in mid-August but was followed by two lesser peaks at depths of 10 and 30 m. Autumn mixing dispersed these peaks in the mixed layer, increasing the chlorophyll content of the epilimnion at a time when integral chlorophyll levels were declining. At the fall overturn, chlorophyll concentrations were uniformly distributed at approximately 1 mg/m^3.

However, in non-case-1 waters, the levels of high energy available for photosynthesis and the depth of the photic zone are not totally controlled by the chlorophyll content $\langle C_{chl\,eu} \rangle$ but also by the presence of other optically responsive aquatic components that do not necessarily co-vary with $C_{chl}(z)$. At the same time, an analysis of field data in large temperate lakes in the northern hemisphere points to the fact that vertical homogeneity of suspended minerals and dissolved organics may very often be a not unreasonable assumption (Petrova, 1990).

Based on these data, Kondratyev and Pozdnyakov (1995) suggested using three types of *chl* vertical profiles for large temperate lakes: uniform in spring, submerged peak in mid-summer, and near surface peak in early-to-mid-autumn (profiles 6, 1, and 7 in Fig. 3.30). Distributions of *sm* and *doc* are believed to be vertically homogeneous.

Thus, vertical profiles of bulk water absorption and backscattering coefficients can be determined down to z_{eu}. In its turn, it allows calculation of the vertical profile of K_d or K_0 (eqs. (3.127)–(3.129), (3.136)), then, the scalar or vector irradiance vertical profiles (eqs. (3.116), (3.133), (3.137)), and finally, the irradiance profiles applied to any chosen primary production model (i.e. eqs. (3.108), (3.109), (3.115), (3.111)–(3.124)).

Closing up this section, some remarks ought to be made concerning persisting problems in remote assessment of primary production in natural turbid waters. Leaving aside the issues of remote quantitative evaluation of chlorophyll and other surficial concentrations of optically responsive substances (which constitute a separate topic), emphasis should be placed here on the parameters of two basic models describing the photosynthetic activity of algae.

When employing models based either on algal physiological parameters (the chlorophyll quantum yield of photosynthesis ϕ, and PAR specific absorption coefficient a_{chl}^*) or production-light intensity curves (the initial slope a', and P_{max}^* normalized to chlorophyll content), special attention should be drawn to their space and time variability, as well as translatability of the locally determined values of these parameters to other water areas and/or water bodies.

Pozdnyakov and Kondratyev (1997b) analyzed the available data on a_{chl}^* variability as affected by phytoplankton volumetric concentration, and taxonomic composition. These variations can be very substantial, and some typical values of a^*_{chl} which would be readily applicable to water masses/water bodies much differing in geographical location and vegetation cycle are hardly possible. Moreover, even vertical invariability in a^*_{chl} should be considered in many cases (particularly in case-1 waters) as a rather rough approximation.

There is direct evidence (Lee *et al.*, 1996a) that physiological status of algae, as well as their taxonomic composition render extensive departures of ϕ from some "typical" value. The same can be stated about the production-light intensity curve parameters (Platt and Sathyendranath, 1988).

This indicates that, first, more research is needed to establish probabilistic characteristics of the above parameters for a variety of water basins and vegetation seasons, and second, it is considerably more reliable, at least to date, to perform remote determinations of aquatic primary production on a local scale, and much caution should be used in applying primary production assessment results to water areas/ water bodies other than those for which photosynthesis parameters have been actually obtained.

3.1.2 Active (lidar) remote sensing

The above discussion about the feasibility of *passive* optical remote sensing to diagnose aquatic environments revealed both advantages and limitations inherent in this approach. At the root of limitations in the volnerability of passive optical remote sensing to water surface illumination conditions (very strong dependence of water-leaving radiance on sun zenith and viewing angles, its sensitivity to partial cloudiness), as well as to optical complexity of natural waters (above all, non-case-1 waters), and water surface roughness.

The deficiencies of passive remote sensing have prompted developing some alternative or maybe supplementary approaches capable of either overcoming or at least mitigating the problems which appear insurmountable for passive optical remote sensing of natural waters. As a result, potential active (lidar) optical remote sensing was thoroughly examined and then successfully exploited by several scientific teams in the US and Europe for operational surveillance of marine, coastal/brackish and inland waters.

3.1.2.1 *Spectrometric analyses of laser-induced fluorescence of natural/ anthropogenic fluorophors*

In natural waters there are two major fluorophors: phytoplankton and dissolved organic matter.

When discussing phytoplankton pigment systems, in §§ 3.1.1.1 and 3.1.1.2 it was pointed out that mainly two kinds of pigments, *viz.* chlorophills and phycobilins, are natural flourophors capable of emitting light under the influence of incident solar radiation.

In vivo fluorescence is not a conservative property of phytoplankton, since its intensity per unit pigment concentration is a physiological variable that is strongly affected by variations in the photosynthetic characteristics of phytoplankton in response to environmental conditions such as light or nutrient availability.

The rate of *in vivo chl-a* fluorescence emitted by water suspension of algae can be described analogously to eq. (3.117):

$$F = \int_{\Delta\lambda} Ea^*_{chl} C_{chl} \phi_F \, d\lambda \qquad (3.151)$$

where E is the radiation available for photosynthesis, and ϕ_F is the fluorescence quantum yield of *chl-a*. Integration in eq. (3.151) should be conducted over the fluorescence band wavelengths ($\Delta\lambda$).

E may be sunlight or excitation flux produced by an artificial light source. The natural and artificial excitation light sources span a broad range of E values: solar photosynthetically available radiation (PAR) at the water surface varies between 0 at night and $\sim 450 \, W/m^2$ during the day at temperate latitudes, whereas the radiation E emitted by artificial sources such as pulsed lasers ranges from 0.1 to $10^9 \, W/m^2$.

Variability in a^*_{chl} has already been discussed in § 3.1.1.2. The concentration of *chl-a* varies from $0.02 \, mg/m^2$ in strictly oligotrophic waters to more than $50 \, mg/m^3$ in eutrophic coastal/estuarine and inland waters (leaving aside hypereutrophic waters where C_{chl} might acquire values of $100 \, mg/m^3$ and even more).

Obviously, the quantum yield of phytoplankton fluorescence is algal species dependent. Moreover, variations in parameter ϕ_F are caused by variations in photosynthetic quenching and other processes that compete for the available excitation energy. According to Gordon (1979), the absolute values of $\phi(\lambda_{ex}, \lambda_{em})$ in the visible spectrum vary for different species from 1.5% to 14%, averaging $\sim 5\%$ (see also Fig. 3.13, § 3.1.1.1.2). In the first and rather rough approximation, ϕ_F can be taken as spectrally invariable in the visible.

An expression analogous to eq. (3.151) can be written for *doc* fluorescence:

$$F = \int_{\Delta\lambda} Ea^*_{doc} C_{doc} \phi_{doc} \, d\lambda \qquad (3.152)$$

where ϕ_{doc} is the fluorescence quantum yield of *doc*.

In § 3.1.1.2, variability in ϕ_{doc} has been discussed, and it was shown that the quantum yield (efficiency) of fluorescence of *doc* varies with λ_{ex}, and according to Green and Blough (1994) displays a pronounced maximum at ~ 380–$400 \, nm$, an inflection point at $\sim 355 \, nm$ and a steep fall down to zero at $\lambda_{ex} \geq 475 \, nm$. The absolute value of ϕ_{doc} (Fig. 3.11, § 3.1.1.1.2) depends on the origin of *doc*; i.e. the ratio between fulvic and humic acids (see § 3.1.1.1.2.). Results reported by Kondratyev *et al.* (1988) indicate that the specific fluorescence intensity humic acids is four times higher than that of fulvic acids. Concentration of *doc* (measured in mg C/l) in natural waters varies from ~ 0.1 in oligotrophic waters to several tens in eutrophic and hypereutrophic waters.

Fluorescence of aquatic fluorophors can be equally expressed by utilizing the relevant cross-section for fluorescence σ_F:

$$F = \int_{\Delta\lambda} E\sigma_F C \, d\lambda \qquad (3.153)$$

where C is the fluorophor concentration.

There are some factors which can alter the value of σ_F. For chlorophyll-a, these include long- and short-term light history, nutrient and age effects and water temperature.

Long-term light history impact is twofold: algal fluorescence in near surface layers is inversely related to solar irradiation, and the ratio of fluorescence to *chl-a* concentration is 80% larger at night than during the day (Browell, 1977).

The transient change in fluorescence yield of *chl-a in vivo* includes several phases: the value of σ_F, first, rises instantaneously to an initial level I, then rises to a higher level II in 20–50 msec, remains constant or decreases slightly for a brief period before rising to a peak value after 0.25 to 1 sec; within 1 to 2 sec, it decreases to the steady-state level. Blasco's (1975) analysis indicates that when an algal culture reaches its stationary growth phase, the ratio of fluorescence to *chl-a* concentration increases: the exhaustion of nutrients affects photosynthetic activity which allows more energy to be dissipated through fluorescence. The change in the ratio of fluorescence to *chl-a* concentration can be over 100 percent for both nitrate and phosphate-exhausted algal cultures.

Water temperature impact on the efficiency levels of algal fluorescence is believed to be a relatively unimportant parameter, at least within the realistic boundaries of water temperature variations within one and the same water body during phytoplankton vegetation seasons.

Vodacek and Philpot (1987) analyzed the *doc* σ_F dependence on such factors as variability in the molecular weight composition of *doc*, water temperature, pH, and quenching due to metal ions present in water.

Generally, *doc* fluorescence is inversely related to temperature due to increased collisional quenching. If no thermal degradation occurs, this effect is reversible, and is accompanied by a certain shift of *doc* fluorescence band central wavelength λ_c to longer wavelengths.

Interference with *doc* fluorescence spectra due to variations in pH is thought to be a result of ionization or protoionization of hydroxyl and carboxyl groups driven by changes in water pH. This alters the electronic state of *doc* molecules, and hence, their fluorescence efficiency. The results obtained by Vodacek and Philpot (1987) are suggestive of a pronounced enhancement of *doc* fluorescence strength with increasing water pH, whereas *doc* fluorescence spectral distribution seems to be practically insensitive to growing pH.

Metal ions can interfere with the *doc* fluorescence process by enhancing nonradiative loss of absorbed light energy so that the fluorescence is lessened. The changes in *doc* fluorescence depend more on the concentration of metal ions relative to C_{doc} rather than on the absolute metal content in the water column. The metal ions impact on *doc* fluorescence is not exclusively confined to subduing its strength: it also produces a shift of the *doc* fluorescence band central wavelength λ_c to shorter wavelengths. The effect of *doc* fluorescence quenching due to metal ions proves to be highly dependent on water pH: it is lower at low and high values of pH, peaking at pH in the range 5.0–5.5.

The σ_F dependence on molecular composition of *doc* is essential: low molecular weight, weakly fluorescent labile *doc* is readily available to the biota and hence generally less abundant. Higher molecular weight *doc* is more stable, relatively less available to the biota, and often highly fluorescent. Thus, varying percentages of labile low molecular weight and stable high molecullar weight fractions of *doc* will likely determine the resultant fluorescence spectra (see eq. (3.57), § 3.1.1.2).

3.1.2.1.1   Equation for lidar sounding of a water column

The active method discussed here relies on the use of lidars (light detecting and ranging system) incorporating a pulsed laser inducing water constituent fluorescence, and a detection unit with a telescope for recording fluorescence signals from the water surface (Fig. 3.31).

It was shown (Bristow *et al.*, 1981), that remotely sensed fluorescence power at any wavelength λ can be given by a general expression of the form:

$$F_{\lambda_{em}} = \left(\frac{P_{\lambda_{ex}}}{h^2} \right) \frac{C_F \sigma_F d_\lambda}{(k_{\lambda_{em}} + k_{\lambda_{em}})} \qquad (3.154)$$

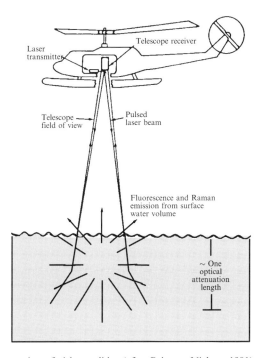

Fig. 3.31. Principle of operation of airborne lidar (after Bristow, Nielsen, 1981).

where P is the power of the laser excitation source at wavelength λ_{ex}, h is the fluorescent target-to-collector lens distance, C_F is the concentration of the fluorophor under investigation, σ_F is the fluorescence excitation cross-section of the fluorophor, d_λ is a constant accounting for known or measurable environmental and lidar system factors, and $k_{\lambda_{ex}}$, $k_{\lambda_{em}}$ are effective optical attenuation coefficients at the excitation (laser) and fluorophor emission wavelengths, respectively.

A number of assumptions are made in deriving this expression, and these are discussed by Bristow *et al.* (1981). The most noteworthy of them are: (a) the total water depth must be well in excess of any of the optical depths given by reciprocal values of $k_{\lambda_{ex}}$, $k_{\lambda_{em}}$, (b) attenuation coefficients $k_{\lambda_{ex}}$, $k_{\lambda_{em}}$ are assumed vertically invariable (i.e. fluorescent and nonfluorescent OAC concentrations remain constant throughout the water column), (c) the fluorescence cross-section σ_F remains invariant over the targeted water sample volume; σ_F must also remain unvarying for all sample locations.

Since eq. (3.154) contains such terms as the power of the laser excitation source, the lidar height over the water surface and the factor accounting for environmental (first of all, aquatic medium optical properties) and lidar system factors which are liable to vary during sounding runs, it brings about a good deal of uncertainty in retrieval results. By taking the ratio of the remotely sensed fluorescence emission intensity or power level to the corresponding water Raman scattering (see § 3.1.1.2), a parameter can be obtained that, to a first approximation, varies only as the concentration of the homogeneously distributed fluorophor and is independent of variation in optical attenuation coefficients $k_{\lambda_{ex}}$, $k_{\lambda_{em}}$.

Indeed, a similar expression can be written for remotely sensed water Raman signal (R) strength measured at a wavelength λ_R at which inelastically scattered light peaks (to avoid confusion, wavelengths of maximum laser emission and fluorofor emanation previously denoted as λ_{ex}, λ_{em} will be referred to as λ_L and λ_F respectively from now on):

$$R = \left(\frac{P_{\lambda_L}}{h^2}\right) \frac{C_W \sigma_R d_{\lambda_R}}{(k_{\lambda_R} + k_{\lambda_L})} \qquad (3.155)$$

where C_W is the concentration of water molecules which for all natural aquatic media is constant ($C_W = 3.33 \times 10^{22} \text{cm}^{-3}$, Fadeev *et al.*, 1982), σ_R is the Raman excitation cross-section for the OH-vibrational stretching mode of water (see § 3.1.1.1.2 for the absolute value of σ_R), d_R is a constant similar to d_λ, and k_{λ_R} is the effective optical attenuation coefficient at the Raman emission wavelength.

Using eqs. (3.154) and (3.155), the above power ratio can be obtained:

$$\frac{F}{R} = \frac{C_F \sigma_F d_F}{C_W \sigma_R d_R} \left(\frac{k_{\lambda_L} + k_{\lambda_R}}{k_{\lambda_L} + k_{\lambda_L}}\right) \qquad (3.156)$$

Since σ_F, d_{λ_F}, C_W, σ_R, d_{λ_R} are all considered to be constant, eq. (3.156) can be simplified to read:

$$\frac{F}{R} = C_F \zeta \left(\frac{k_{\lambda_L} + k_{\lambda_R}}{k_{\lambda_L} + k_{\lambda_F}}\right) \qquad (3.157)$$

where ζ is a constant encompassing the above-mentioned assumed-to-be-constant factors.

Clearly, expression (3.57) can only be useful for developing algorithms for retrieval of a sought after fluorophor, if the ratio $(k_{\lambda_L} + k_{\lambda_R})/(k_{\lambda_L} + k_{\lambda_F})$ remains invariable or at most changes only slowly when significant variations occur in the individual k values. Given this assumption being valid, eq. (3.157) can be reduced to:

$$\frac{F}{R} = \xi C_F \qquad (3.158)$$

where ξ is an unknown factor which could be determined exclusively through field airborne measurements over a definite water body or water mass.

Importantly, water Raman normalization largely determines the linearity of relationship (3.158). Laboratory measurements revealed (Bristow *et al.*, 1981) that when optical attenuation in the water column is dictated predominantly by *chl*, the *chl* fluorescence signal F measured by the laser fluorosensor tends towards *saturation*. Indeed, with increasing C_{chl}, coefficients k_{λ_L} and k_{λ_F} become higher with the result of producing a corresponding reduction in F. In the limit, the optical attenuation-driven reduction in F becomes compensated by a corresponding increase in C_{chl}. In the absence of water Raman correction, F remains constant regardless of further enhancement of columnal *chl* content. However, since the water Raman signal steadily falls with increasing C_{chl}, and does not exhibit any saturation at high C_{chl}, the ratio F/R acquires a linear form.

3.1.2.2 Retrieval algorithm development: basic considerations, and challenges inherent in optical complexity and spatial/temporal variability of natural waters.

Stemming from the fundamental idea of normalizing the fluorescence-driven response signal by water Raman, eq. (3.158) is in fact a basis for establishing operational retrieval algorithms for inferring water-quality parameters from lidar surveillance data. Since in deriving eq. (3.158), it was assumed that the value of factor ξ is not contingent upon C_F, eq. (3.158) implies that the lidar retrieval algorithm should be a linear function of C_F with a slope dictated by ξ.

However, ξ is a complex function of numerous variables, including the fluorescent properties (σ_F) of the substance to be detected, and hydro-optical properties of the water column ($k_{\lambda_F}, k_{\lambda_L}, k_{\lambda_R}$).

Brown *et al.* (1977) reported spectral values of σ_F for three phytoplankton species representing blue–greens, greens, and golden brown algae (Fig. 3.32) were later confirmed by Farmer *et al.* (1979).

Browell (1977) believes that the optimized cross-sections ($\times 10^{-21}$ m^2/molecule) for four excitation wavelengths are those given in Table 3.1.

The issues relating to the *doc* fluorescence cross-section (which is a product of *doc* absorption per *doc* molecule and fluorescence quantum yield) have already been thoroughly discussed above in § 3.1.1.1.2. Fadeev *et al.* (1979) report that cross-section σ_F values laser-induced *doc* fluorescence vary between $(10^{-16}-10^{-18})$cm^2/

Table 3.1. Optimized cross-sections for green, golden-brown, red and blue–green algae at four excitation wavelengths.

Algal group	482 nm	520 nm	562 nm	640 nm
Green	1.13	0.28	0.16	0.39
Golden-brown	0.42	0.28	0.14	0.10
Red	0.09	0.22	0.64	0.13
Blue–green	0.03	0.03	0.06	0.60

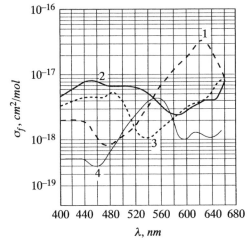

Fig. 3.32. Variation in fluorescence cross-section σ_F with excitation wavelength for four species of phytoplankton (after Brown *et al.*, 1977).

mol sr, and so by one to two orders of magnitude higher than phytoplankton σ_F values.

Consider the ratio $(k_{\lambda_L} + k_{\lambda_R})/(k_{\lambda_L} + k_{\lambda_L})$ dependence on hydro-optical properties. Obviously, the constancy of this ratio is best attained if the lased wavelength is chosen in such a manner that $\lambda_F \approx \lambda_R$: in this case $k_{\lambda_F} = k_{\lambda_R}$, and the ratio $(k_{\lambda_L} + k_{\lambda_R})/(k_{\lambda_L} + k_{\lambda_L})$ is strictly equal to 1. However, such a choice entails the problem of separating bands arising due to water Raman and fluorescence. That is why it seems appropriate to choose such a λ_L which would result in a reciprocally close but separable position of both bands (Fig. 3.33).

The effective coefficient of laser beam attenuation k was assessed by Kim *et al.* (1977) to be between c and K where, as before, c is the beam attenuation coefficient and K is the diffuse light attenuation coefficient. Numerical simulations suggest that, in clearest waters, effective attenuation coefficients at the laser and fluorescence (or Raman) wavelengths approach the respective beam attenuation coefficients at low *chl* concentrations and approach the respective diffuse attenuation coefficients at high *chl* concentrations. For intermediate concentrations, the best approximation

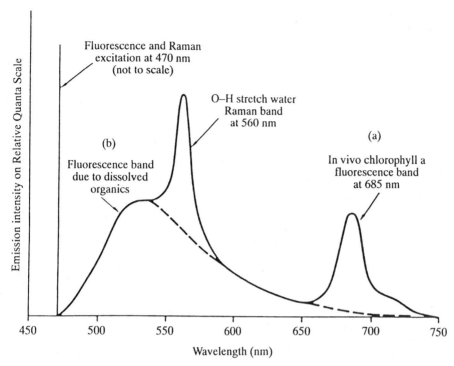

Fig. 3.33. Schematic showing water fluorescence (a—due to phytoplankton, b—due to *doc*) and Raman emission parameters.

was found to be K at laser wavelength and c at the fluorescence (or Raman) wavelengths (Poole, Esaias, 1982).

It reveals one of the clues when choosing λ_L. Due to *doc* and *chl* absorption in meso- and eutrophic inland waters, the beam lased in the short wavelength region of the visible spectrum (see the previous section) is bound to be already absorbed in upper layers of the water column thus reducing lidar capability to sound deeper layers (according to Poole, Esaias (1982)),* the depth above which 90% of the total signal observed by the lidar receiving telescope (z_{90}) is inversely proportional to $k_{\lambda_L} + k_{\lambda_i}$:

$$z_{90} \sim \frac{2.3}{k_{\lambda_L} + k_{\lambda_i}} \tag{3.159}$$

where i stands either for F or R.* Consequently, when sounding phytoplankton, it is preferable to place λ_L in the "transparency window" of meso- and eutrophic waters which is in the blue-green–green spectral range. At the same time, the laser excitation

---

*Incidentally, since the signal-integration depths for *chl* fluorescence and water Raman scattering are not identical, a potential error can arise in using the Raman normalization procedure in clear and vertically stratified waters exhibiting a *chl* maximum at depth below $z_{90,F}$ but above $z_{90,R}$ (Poole, Esaias, 1982).

wavelength should not be far away from the *chl* excitation absorption band (430–440 nm) to obtain a sufficient intensity of the *chl* fluorescence signal. Moreover, placing λ_L in the blue spectral region will induce higher levels of *doc* fluorescence which in the case of *chl* detection will be a hampering component.

To recap then, the laser wavelength at 532 nm (the second harmonic of YAG:Nd^{3+} laser radiation) seems to be appropriate. At this excitation wavelength, water Raman signal peaks at ~ 653 nm, whereas *chl* fluorescence is maximum at 685 nm.

Conforming to the above criteria, this choice can present, however, some problems when sounding fresh water environments abundant in blue–greens—the only inland algae species containing phycobilin pigments in significant amounts. Many of the ubiquitous surface bloom-forming species such as *Microcystis aeruginosa*, *Anabaena flos-aquae* and *Aphanizomenon flos-aquae* contain only phycocyanin which is known (see the previous sections) to strongly absorb at 620 nm and emit at 650 nm. Although being fairly far off 620 nm, $\lambda_L = 532$ nm is sufficient to produce a strong signal at 650 nm (i.e. at water Raman emission maximum). Indeed, Bristow, Zimmermann (1989) report that although the fluorescence excitation cross-section at 487 nm for phycocyanin emission at 653 nm in *Microcystis aeruginosa* is generally about 1% of the peak value at 620 nm, excitation at 487 nm results in phycocyanin and *chl-a* emission bands of about equal amplitude. Consequently, the presence in water of such species is capable of dramatically distorting the response signal in the water Raman spectral band.

At the same time, radiation lased at 532 nm can be a tool to identify the presence of specific algal species. Species in other blue–green families such as *Oscillatoria* may contain varying amounts of phycocyanin and phycoerythrin or exclusively phyco-erythrin whose absorption peaks at 550 nm. Phycoerythrin fluorescence maximum occurs at 570 nm. Thus, remote detection in a freshwater environment of phyco-erythrin emission, exclusively or in addition to that of phycocyanin, would therefore constitute a positive indication of the presence of an *Oscillatoria* bloom. Several workers explored lidar remote sensing feasibility to detect and quantify spatial distribution of phycobilins over marine/coastal/estuarine waters and reported very promising results (Hoge and Swift, 1983; Hoge et al., 1986; Houghton et al., 1983; Bristow and Zimmerman, 1989; Babichenko et al., 1993).

The coincidence of water Raman and phycocyanin fluorescence band position could, of course, be avoided through shifting λ_L off 532 nm to shorter wavelengths. Bristow and Zimmermann (1989) used $\lambda_L = 487$ nm, Babichenko et al. (1993) used $\lambda_L = 445, 480$ and 515 nm. Although it certainly improves lidar analytic capabilities, the problems mentioned above arise. Moreover, shifting λ_L to considerably shorter wavelengths will inevitably result in a much stronger interference of *doc* fluorescence with legitimate signals.

Finally, Fig. 3.32 offers another incentive to choose λ_L within the range 500–530 nm: the σ_F values for major taxonomic groups of algae differ less significantly (by comparison with other spectral regions) averaging $\sim 2 \times 10^{-18}$ cm^2/molecule. It certainly simplifies retrieval procedures (making retrieval algorithms more robust) in

the case of water bodies with significant spatial phytoplankton taxonomic inhomogeneity.

Nonetheless, even an optimal choice of λ_L cannot completely eliminate changes to
the retrieval algorithm slope depending on the combination of dominant phytoplankton species indigenous to specific water bodies (Brown *et al.*, 1977). The size
distribution of dominant species is also an important player since it determines to a
certain extent the algal cross-section due to the effect of packaging (see § 3.1.1.1).
This factor is also capable of influencing the retrieval algorithm slope ξ.

As mentioned above, algae fluoresce more strongly at night than during the day.
Moreover, being weaker, day-time fluorescence of algae is pretty variable inversely
following diurnal variations in solar incident radiation. Dedicated studies conducted
by Nieke *et al.* (1997) strongly suggest that morning hours (up to 90 min after sunrise
during summer) are preferable for lidar soundings and quantification of *chl*
concentrations.

Based on general considerations, it is clear that when changes in water column
optical attenuation are totally caused by variations in phytoplankton population
density, increase in F should be accompanied by a concurrent fall in the water
Raman signal R (see § 3.1.1.2). Interestingly, the other case in which changes in
optical attenuation are due to changes in concentrations of non-chlorophyllous
material such as suspended sediments and dissolved organics is characterized by
situations in which changes in F are directly related to changes in R, i.e. a decrease in
R is accompanied by a corresponding decrease in F. Analyzing the retrieval
algorithm dependence on the presence of suspended terrigenous matter along with
chl in the sounded water column, Poole and Esaias (1983) showed through modeling
that inorganic matter (Calvert soil) in concentrations less or equal to 10 mg/l hardly
alters the slope ξ. They also showed that the water Raman signal can be a measure of
water turbidity due to suspended matter (both organic and inorganic) but there are
severe constraints to utilization of this criterion: the relation $R \sim k^{-1}$ can be assumed
linear for $c < 6\text{–}8\,\text{m}^{-1}$.

However, when considering water bodies with high levels of trophy, which are
generally heavily dissolved and laden with suspended mater, it appears not
unreasonable to expect pronounced variations in the retrieval algorithm slope ξ.
Indeed, such waters display a characteristic spectral distribution of the diffuse
attenuation coefficient K: it exhibits a minimum at about 560 nm (Fig. 3.34).

The symmetry of this transparency "window" envelope increases with *chl*, *doc* and
sm loading. Obviously, if λ_R falls into the transparency window, water Raman signal
will experience less attenuation than the chl fluorescence signal, this inequality being
a function of water trophy and turbidity status (TTS). As a result, the F/R ratio also
proves to be dependent on TTS. Even if λ_R occurs on the left side of the transparency
window (i.e. in a shorter wavelength region), this effect could still be felt given that
the respective values of K at λ_R and λ_F are significantly at variance.

Unlike phytoplankton remote sensing, detection of *doc* by lidar sounding requires
considerably shorter excitation wavelengths (see § 3.1.1.2.), and generally lasers
operating in the ultraviolet spectral region are employed. An exhaustive study
undertaken by Bristow and Nielsen (1981) suggests that the F/R ratio is less useful

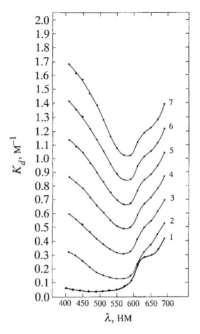

Fig. 3.34. Spectral irradiance attenuation coefficients for Lake Erie (Bukata *et al.*, 1995). Numbers 1 to 7 designate the increase in turbidity from pure water to waters rich in OAC.

for building the *doc* retrieval algorithm in comparison with FR/R, where F_R is the *doc* fluorescence intensity at λ_R (see Fig. 3.33). Earlier in this section, the major factors determining the *doc* retrieval algorithm slope ξ were discussed.

Recently, Mittenzwey *et al.* (1997), in an attempt to obviate problems inherent in employing the water Raman normalization technique, suggested a new approach consisting in simultaneously measuring fluorescence signals coming from the water column (F_{bulk}) and uppermost layers (F_{surf}). Based on the lidar remote-sensing equation, and arranging that the difference between excitation and fluorescence wavelength $\lambda_F - \lambda_L$ is large (e.g. greater than 100 nm) the authors showed that for waters relatively rich in *doc* (i.e. when the optical properties of waters are strongly controlled by *doc*, and hence exhibit an essentially increasing absorption towards shorter wavelengths), the quotient $(F_{bulk}/F_{surf})^{-1}$ is directly proportional to k_{λ_L} given the validity of the approximation $k_{\lambda_L} \gg k_{\lambda_F}$:

$$(F_{bulk}/F_{surf})^{-1} = k_{\lambda_L} d \qquad (3.160)$$

where d denotes the thickness of the first layers. F_{surf} can be measured by means of a time-gated sensor, setting a short time-gate (e.g. using a gate of 1 ns, the fluorescence F_{surf} comes from the first 10 cm of the water column). Consequently, the total attenuation coefficient at the laser wavelength k_{λ_L} can be measured by lidar fluorescence directly.

Conducted for binary and tertiary aquatic media (water $+ chl$; water $+ chl +$ humic acid; water $+ chl +$ ligninesulfonic acid), laboratory experiments indicated that the quotient $(F_{bulk}/F_{surf})^{-1}$ is free from the disturbing influences caused by light attenuation in the water column at Raman and fluorescence wavelengths.

Good correlations between conventional and relevant lidar-derived attenuation coefficients were achieved as described by squared correlation coefficients in excess of 0.95. In conformity to eq. (3.160), a varying fluorescence quantum yield does not disturb the $(F_{bulk}/F_{surf})^{-1}$ ratio. Moreover, since this approach does not involve a normalization of the fluorescence to the water Raman signal, and considers both fluorescent and nonfluorescent substances loading the water column, it presents an advantage for lidar monitoring of eutrophic waters showing only weak water Raman intensity. However, this approach cannot be considered as universal and is only applicable to specific aquatic environments with the constraints identified above.

3.1.2.3 Some other applications of lidar sounding in limnological studies

3.1.2.3.1 Aquatic higher-plant stand mapping

Being an indigenous component of aquatic environments, macrophytes or aquatic higher plants are subject to severe anthropogenic impacts. The dynamics in spatial distribution of macrophyte stands is indicative of serious alterations to limnological processes and as such is widely used, along with other indices, for assessing the ecological status of the water body under consideration.

There are some species of macrophytes which are either emergent or floating (i.e. "covering" the water surface), some others are submerged forming underwater stands. The latter case is particularly difficult for operational surveillance by means of visual observations or low-altitude aerial photography. Lidar sounding offers a solution to this problem.

Typical laser-induced emission spectrum of higher plants exhibits in the red to near-infrared spectral regions two bands of nearly equal intensity with maxima at about 685 and 740 nm (Fig. 3.35).

These arise from fluorescence pertaining to photosystems 2 and 1, respectively (for more details about the assumed mechanisms of emission at 740 nm see Bounkin *et al.* (1985). At the same time, the characteristic laser-induced fluorescence spectrum of phytoplankon displays only one strong band at 685 nm accompanied by a flattened inflexion at about 740 nm (Fig. 3.30), so the appearance of the above-mentioned band at 740 nm in the lidar responce signal should be ascribed to macrophytes.

This premiss constitutes a basis for detecting water areas encroached by higher aquatic plants: a strong response signal at $\lambda \geq 700$ nm can be unambiguously ascribed to macrophyte stands.

Furthermore, the intensity of the response signal at these wavelengths is an indication of the macrophyte type. Indeed, macrophytes "covering" the water surface should produce a response signal at $\lambda \approx 740$ nm comparable in amplitude with that at 680 nm, whereas a relatively low-level response signal at $\lambda \approx 740$ is likely due to submerged macrophyte stands (a decrease in the response signal in the red

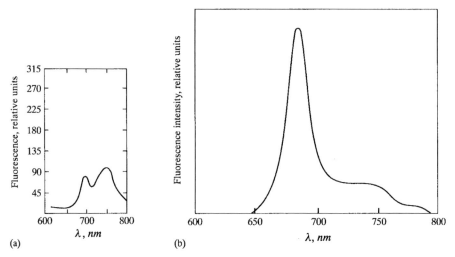

Fig. 3.35. Typical laser-induced fluorescence spectrum of for (a) soybean leaf (Chappell *et al.*, 1984) and (b) green algae (Browell, 1977).

and near infrared spectral regions is bound to dramatically increasE water absorption at $\lambda \geq 700$ nm, see § 3.1.1.1).

The process of differentiation between phytoplankton- and macrophyte-related response signals may be fully automated and, therefore, macrophyte stands mapping can be performed in near real time.

3.1.2.3.2   Lidar fluorosensing of oil films

Nowadays, oils are among very common pollutants affecting the ecology of inland water bodies. Since oils fluoresce in the blue region of the visible spectrum (see Burlamacchi *et al.*, 1983), the fluorescence lidar technique seems to be very feasible for detecting oil films on the surface of large lakes and water-storage reservoirs.

Burlamacchi *et al.* (1983) analyzed the efficiency of different laser wavelengths in inducing fluorescence of low-thickness oils films. Three different mixtures of lasing gases, KrF, XeCl and N_2 ($\lambda_{ex} = 249.5$ nm, 308 nm, 337.1 nm, respectively) were used while a fourth wavelength ($\lambda_{ex} = 420$ nm) was provided by a dye laser. Qatar marine (QM), Murban (MU), Es Sider (LI), Rhomashinskaya (RU) and Arabian light (LA) oils were investigated.

Fluorescence efficiencies of these types of oil are given in Table 3.2. Emission spectra of these five different crude-oil samples at four excitation wavelengths exhibited a surprising mutual conformity (Fig. 3.36 shows these spectra for $\lambda_{ex} = 337.1$ nm). Three emission bands are discernible in the obtained spectra: at ~400 nm, ~450 nm and ~520 nm. Their relative intensity varies with λ_{ex}. At low λ_{ex} bands at ~400 nm, ~450 nm are most pronounced (with the predominance of emission peaking at ~450 nm). When increasing λ_{ex}, the emission band at

Table 3.2. Fluorescence efficiency of five crude oil samples (Burlamacchi *et al.*, 1983).

Oil	Origin	$\lambda_{ex} = 249.5\,\text{nm}$	$\lambda_{ex} = 308\,\text{nm}$	$\lambda_{ex} = 337.1\,\text{nm}$	$\lambda_{ex} = 420\,\text{nm}$
Qatar Marine	Qatar	3	9	12	21
Murban	Abu Dhabi	5	18	21	55
Es Sider	Libya	4	12	13	18
Rhomashinskaya	Russia	2	9	7	12
Arabian light	Saudi Arabia	2	6	8	12

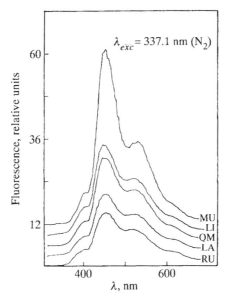

Fig. 3.36. Emission spectra of five crude oil samples at $\lambda_{ex} = 337.1\,\text{nm}$ (Burlamacchi *et al.*, 1983). Designation of curves is explained in the text.

$\sim$520 nm grows significantly (at $\lambda_{ex} = \sim$420 nm it even exceeds the 450-nm band) being accompanied by an inflexion at $\sim$620 nm.

Although fluorescence quantum yield is much higher at longer wavelengths, the available erxperimental data suggests (Burlamacchi *et al.*, 1983) that the overall efficiency for low-thickness films is the highest at $\lambda_{ex} = 308\,\text{nm}$.

The fluorescence intensity at emission peak wavelength depends on oil-film thickness ε: it nonlinearly increases with ε. Burlamacchi *et al.* (1983) assess that the fluorescence emitted by a 0.1 μm oil film is quite sufficient to record a spectrum at each laser shot. Since the water fluorescence becomes comparable with oil fluorescence for oil-film thickness in the range 0.01 μm, it can be perhaps considered as a lower limit.

Oil-film thickness assessments can be performed if the oil slick fluorescence is recorded at the wavelength of water Raman scattering. Choosing the laser excitation wavelength in such a manner that the oil-film quantum efficiency is very low compared with water Raman scattering. In this case, the actual contribution of oil-film fluorescence to the response signal at λ_R can be negligible. In the absence of an oil slick, Raman signal increases since there is no additional absorption due to oil. Given the precise value of the oil attenuation coefficient, film thickness can be regressed against the ratio of R-signals from a clean water area and an oil slick. Field measurements indicate that, within this approach, oil-thickness upper limit is about 4 μm: thicker films completely subdue the legitimate signal. Instead, microwave remote methods prove to be more efficient in quantifying the value of ε for newly spilled (and hence, generally thick) oil films.

An improved method of detecting oil slicks and quantifying their thickness was suggested by Chubarov et al (1995). The main advantage of this approach resides in the fact that it is not a "contrast-based" technique. Stemming from differential absorption concepts, it uses measurements of the ratio of bulk water Raman signals at two laser excitation wavelengths. Being inversely proportional to the difference of spectral values of the oil-film attenuation coefficient at two λ_R, this ratio can be related to oil-slick thickness, given that the relevant oil-film attenuation coefficient spectral values are known. The oil fluorescence signals obtained concurrently with water Raman signals for two excitation wavelengths can also be used for quantitative assessment of the oil type and its abundance on the water surface.

3.1.3   Atmospheric correction of satellite images

3.1.3.1   Atmospheric correction in the visual range

It was shown in Clarke et al. (1970) that chlorophyll concentration could be retrieved by airborne spectral measurements of upwelling radiance over the sea.

Following the ideas of that work, NASA launched the coastal zone color scanner (CZCS) on Nimbus 7 in 1978, which was specifically designed for color measurements of surface waters of the ocean (Gordon et al., 1980). The goal of the CZCS mission was to adjust the concept of measuring ocean color from space and to assess global-ocean primary-carbon productivity. It was a scanning radiometer that had four bands in the visible at 443, 520, 550 and 670 nm with bandwidths of 20 nm, one band in the near infrared at 750 nm with a bandwidth 100 nm for land and cloud detection, and an IR band for sea-surface-temperature measurements. It had high radiometric sensitivity and 0.825 km ground resolution at nadir, being oriented on global scale measurements.

After CZCS several sensors with improved water-color capability have been designed, some are now orbiting the Earth, and some are to be launched at the end of the century. They include sea-viewing wide-field-of-view sensor (SeaWiFS) on SeaStar, moderate resolution imaging spectroradiometer (MODIS) and multiangle imaging spectroradiometer (MISR) on EOS-AMI, medium resolution imaging spectrometer (MERIS) on ENVISAT, and global imager (GLI) on ADEOS2. Compared with CZCS, the new sensors have more spectral bands, reduced radiometric noise, and improved calibration capabilities. A number of improvements have been made to supply the atmospheric correction procedure with the necessary information: the additional spectral bands in the near infrared, where water can be considered black, have to help extrapolation of the atmospheric radiance to the blue and green channels, and some measurements have to improve determination of the aerosol model (Kaufman *et al.*, 1997).

Satellite water-color imagery is contaminated by atmospheric scattering and absorption along with surface reflection. This contamination must be eliminated to retrieve water-living radiance, which depends on biomass content in the water. The procedure is called atmospheric correction and it is strictly necessary to apply it, if any quantitative utilization of satellite imagery is to be done. The difference between atmospheric correction over land and over water must be kept in mind. In the latter case the process corrects not only atmospheric attenuation, but also surface effects.

In general, atmospheric correction of satellite data requires information that describes the variable atmospheric constituents influencing upwelling radiance: their optical properties and distribution in the atmosphere along with modeling of scattering and absorption processes. Atmospheric correction over water is the most complex retrieval problem, because nearly all the signal measured at the top of the atmosphere is composed of photons that have not interacted with the water body. In the blue spectral band, where phytoplankton pigments substantially absorb, typically 90% of the satellite radiance originates from the atmosphere. In the red, where phytoplankton fluoresces, the atmospheric part becomes 99.5–99.9% (Gordon, 1997). So, descriptions of atmospheric constituents as well as radiance measurements at satellites have to meet very rigorous requirements for accuracy.

The main contribution in solving the atmospheric correction problem over oceans was made by H.R. Gordon and his collaborators (see Gordon (1997) for main references). One of the basic parameters they use is the normalized water-leaving radiance $[L_w]_N$, which was defined by Gordon and Clark (1981) through

$$L_w(\lambda) = [L_w(\lambda)]_N \cos\theta_0 \exp\left[-\left(\frac{\tau_r(\lambda)}{2} + \tau_{oz}(\lambda)\right)\left(\frac{1}{\cos\theta_0}\right)\right] \quad (3.161)$$

where L_w is the radiance backscattered out of the water toward the zenith at a wavelength λ, $\lambda_r(\lambda)$, and $\tau_{oz}(\lambda)$ are the optical thicknesses of the atmosphere associated with Rayleigh (molecular) scattering and ozone absorption, respectively, θ_0 is the solar angle.

Reflectance ρ associated with radiance L is defined as $\pi L/F_0 \cos\theta_0$, where F_0 is the extraterrestrial solar irradiance, and θ_0 is the solar zenith angle. Reflectance is preferred because it may be possible to more accurately calibrate sensors in reflectance rather than radiance. $[L_w]_N$ can be converted to normalized water-leaving reflectance $[\rho_w]_N : [\rho_w]_N = \pi/F_0[L_w]_N$

$$\rho_w(\lambda) = [\rho_w(\lambda)]_N \exp\left[-\left(\frac{\tau_r(\lambda)}{2} + \tau_{oz}(\lambda)\right)\left(\frac{1}{\cos\theta_0}\right)\right] \equiv [\rho_w(\lambda)]_N t(\theta_0, \lambda) \quad (3.162)$$

where $t(\theta_0, \lambda)$ is an approximation to the diffuse transmittance of the atmosphere.

It was shown in Gordon *et al.*, 1988, that normalized water-leaving radiance at 443 nm and 550 nm depend on pigment concentration C, which is the sum of concentrations of chlorophyll a and its degradation product phaeophytin a for case-1 waters. Thus the normalized water-leaving radiance plays a central role in the application of water color imagery and atmospheric correction procedure becomes a critical factor in retrieving pigment concentration.

The radiance measured by a sensor at the top of the atmosphere (TOA) in a spectral band central at a wavelength λ_i, $L_i(\lambda_i)$ can be divided into four components:

$$L_t(\lambda_i) = L_{path}(\lambda_i) + T(\lambda_i)L_g(\lambda_i) + t(\lambda_i)L_{wc}(\lambda_i) + T(\lambda_i)L_w(\lambda_i) \tag{3.163}$$

where $L_{path}(\lambda_i)$ is the radiance generated along the optical path by scattering in the atmosphere and by specular reflection of atmospherically scattered light from the sea surface, $L_g(\lambda_i)$ is the contribution arising from specular reflection of direct sunlight from the sea surface (the effect of sun glitter), $L_{wc}(\lambda_i)$ is the contribution arising from direct and diffuse atmospheric radiance reflected from whitecaps on the sea surface, $L_w(\lambda_i)$ is water-leaving radiance, which has to be determined after atmospheric correction. T and t are the direct and diffuse transmittance of the atmosphere respectively. Diffuse transmittance is appropriate for water-leaving radiance and whitecap radiance as they have near uniform angular distribution. Direct transmittance is applicable when the angular distribution is approximately a Dirac delta function. As sun glitter is highly directional, its transmittance is approximated by direct transmittance. The latter is determined by

$$T(\theta_v, \lambda) = \exp\left[-\left(\tau_r(\lambda) + \tau_{oz}(\lambda) + \tau_a(\lambda)\right)\left(\frac{1}{\mu_v}\right)\right] \tag{3.164}$$

where $\mu_v = \cos\theta_v$, θ_v is the angle between the upwelling radiance and the upward normal at the TOA, and τ_r, T_a, and τ_{OZ} are respectively the rayleigh, aerosol and ozone optical thicknesses. In this equation absorption by water vapor has been ignored. Converting (3.163) to reflectance, we have

$$\rho_t(\lambda_i) = \rho_{path}(\lambda_i) + T(\lambda_i)\rho_g(\lambda_i) + t(\lambda_i)\rho_{wc}(\lambda_i) + T(\lambda_i)\rho_w(\lambda_i) \tag{3.165}$$

and we require an algorithm that provides accurate estimates of $\rho_{path}(\lambda_i)$, $T(\lambda_i)\rho_g(\lambda_i)$, $t(\lambda_i)\rho_{wc}(\lambda_i)$ and $t(\lambda_i)$. The sun glitter pattern, $T(\lambda_i)\rho_g(\lambda_i)$ is so large that imagery associated with it is virtually useless and must be discarded. Away from the sun glitter pattern, i.e. where values of $T(\lambda_i)\rho_g(\lambda_i)$ become negligibly small, the largest of the remaining terms, and most difficult to estimate, is $\rho_{path}(\lambda_i)$. This difficulty is principally due to atmospheric aerosols whose properties are highly variable in time and space.

Reflectance corresponding to atmospheric path radiance $\rho_{path}(\lambda_i)$ can be presented as a sum of three components (Gordon, 1997):

$$\rho_{path}(\lambda_i) = \rho_r(\lambda_i) + \rho_a(\lambda_i) + \rho_{ra}(\lambda_i) \tag{3.166}$$

where $\rho_r(\lambda_i)$ is reflectance resulting from multiple scattering by air molecules in the absence of aerosols, $\rho_a(\lambda_i)$ is reflectance resulting from multiple scattering in the absence of the air, and $\rho_{ra}(\lambda_i)$ is the interaction term between molecular and aerosol scattering (Deschamps *et al.*, 1983). ρ_r in (3.166) can be computed accurately and the problem is to evaluate the two other terms.

Critically important spectral bands for atmospheric correction over water are those with band centers $\lambda_i > 700\,\text{nm}$, i.e. bands in the near infrared (NIR). Virtually no light will exit the ocean in these bands because of strong absorption by liquid water, except in the most turbid coastal water, so measured radiance originates from scattering of solar irradiance by the atmosphere and by the sea surface. These bands can then be used to estimate the atmospheric effects by first assessing the contribution of the atmosphere in the NIR and then extrapolating it to the visible (Gordon, 1997).

It is useful to consider $\rho_{path}(\lambda_i)$ in the case of *single-scattering approximation*, which is valid when the optical thickness of the atmosphere is much less than 1. The last term in (3.166) ρ_{ra} is zero in this case and (3.166) reduces to

$$\rho_{path}(\lambda_i) = \rho_r(\lambda_i) + \rho_{as}(\lambda_i) \qquad (3.167)$$

where aerosol contribution ρ_{as} can be computed if aerosol optical depth, aerosol phase function, and aerosol single scattering albedo are known.

To provide atmospheric correction in this case it was suggested (Gordon, 1997) to measure the path reflectance at two bands in the NIR at λ_s and λ_l wavelengths, where subscript "s" stands for short and "l" for long wavelengths, e.g. for MODIS instrument it will be $\lambda_s = 750\,\text{nm}$ and $\lambda_l = 865\,\text{nm}$, for SeaWiFS $\lambda_s = 765\,\text{nm}$ and $\lambda_l = 865\,\text{nm}$, for MIRS $\lambda_s = 670\,\text{nm}$ and $\lambda_l = 865\,\text{nm}$, while for CZCS there were no NIR bands. Then, taking measurements outside the glitter pattern, assuming that $t(\lambda_i)\rho_{wc}(\lambda_i)$ has been provided and considering $\rho_w = 0$ at both wavelengths, ρ_{path} determined by (3.167) can be estimated from the measured ρ_t at λ_s and λ_l wavelengths. As ρ_r can be computed accurately, $\rho_{as}(\lambda_s)$ and $\rho_{as}(\lambda_l)$ are determined from the associated values of ρ_{path} at λ_s and λ_l. This allows estimation of the parameter $\varepsilon(\lambda_s, \lambda_l)$:

$$\varepsilon(\lambda_s, \lambda_l) \equiv \frac{\rho_{as}(\lambda_s)}{\rho_{as}(\lambda_l)}$$

If we can compute the value of $\varepsilon(\lambda_i, \lambda_l)$ for the wavelength λ_i from the value of $\varepsilon(\lambda_s, \lambda_l)$ this will yield $\rho_{as}(\lambda_i)$ which, when combined with $\rho_r(\lambda_i)$, provides the desired $\rho_{path}(\lambda_i)$:

$$\rho_{path}(\lambda_i) = \rho_r(\lambda_i) + \varepsilon(\lambda_i, \lambda_l)\rho_{as}(\lambda_l)$$

Clearly, the key to utilizing this procedure is estimation of $\varepsilon(\lambda_i, \lambda_l)$ from $\varepsilon(\lambda_s, \lambda_l)$ and it is important to gain some information about the possible spectral behavior of $\varepsilon(\lambda_i, \lambda_l)$. Investigations of that sort have been carried out in the work of Gordon and Wang (1994), where they compute $\varepsilon(\lambda_i, \lambda_l)$ for several aerosol models developed by Shettle and Fenn (1979). It has been shown that there should be strong variation in ε with aerosol properties and relative humidity (RH). Spectral variation in ε is due in large part to the spectral variation in aerosol phase function. Increase in particle size because of swelling with increasing RH reduces spectral variation in ε. There is an hypothesis that between the range of 412–865 nm, $\varepsilon(\lambda_i, \lambda_l)$ can be considered an exponential function of $\lambda_i - \lambda_l$ for the Shettle and Fenn (1979) models. Analysis of Gordon's (1997) data, when he considered other aerosol models, suggests that absorption-free models display $\varepsilon(\lambda_i, 865)$ dependence similar to Shettle and Fenn models. However, for models with strong absorption discrepancies occur.

The single-scattering approach is seen to work well only for sufficiently small optical depth and nonabsorbing aerosols, typically the case over the open ocean (Gordon, 1997). For situations having a strong continental influence, which are much more common for lakes than for oceans, aerosols are likely to be at least moderately absorbing. Also, τ_a can be significantly larger than the aerosol single-scattering limit in that case. Thus we have to consider a multiple-scattering approach.

Multiple-scattering effects have already been shown (Deschamps *et al.*, 1983; Gordon *et al.*, 1988) to be significant at the level of accuracy required for SeaWiFS and MODIS, that is $\Delta[\rho_w(443)] \approx 0.001\text{--}0.002$. In the case of multiple scattering the effect of atmospheric aerosol depends significantly on the aerosol model. We can modify the single-scattering algorithm (3.167), writing

$$\rho_a(\lambda_i) + \rho_{ra}(\lambda_i) = K[\lambda_i, \rho_{as}(\lambda_i)]\rho_{as}(\lambda_i)$$

where the dependence of K on $\rho_{as}(\lambda)$ represents the deviation of the $\rho_a(\lambda) + \rho_{ra}(\lambda)$ from $\rho_{as}(\lambda)$ relationship from linearity or the difference between the single-scattering and multiple-scattering effect on the measured radiance at TOA. K is nearly the same for the two NIR bands, but can be significantly different at 443 nm (Gordon, 1997).

The approach suggested by Gordon and Wang (1994) was to solve the radiative transfer equation for a set of N candidate models to provide what is essentially a set of lookup tables for $K[\lambda, \rho_{as}(\lambda)]$. As in the single-scattering algorithm, NIR bands are used to provide the aerosol model through

$$\varepsilon(\lambda_s, \lambda_l) = \frac{K[\lambda_l, \rho_{as}(\lambda_i)]\rho_a(\lambda_s) + \rho_{ra}(\lambda_s)}{K[\lambda_s, \rho_{as}(\lambda_s)]\rho_a(\lambda_l) + \rho_{ra}(\lambda_l)}$$

However, since the aerosol model is not known at this point, the ratio K is unknown. According to Gordon and Wang (1994), this ratio should not deviate significantly from unity, so they suggested computing the value $\varepsilon(\lambda_s, \lambda_l)$ using the formula

$$\varepsilon(\lambda_s, \lambda_l) = \frac{1}{N}\sum_{j=1}^{N}\varepsilon_j(\lambda_s, \lambda_l)$$

where $\varepsilon_j(\lambda_s, \lambda_l)$ is the value of $\varepsilon(\lambda_s, \lambda_l)$ derived from $\rho_a(\lambda_l) + \rho_{ra}(\lambda_l)$ and $\rho_a(\lambda_s) + \rho_{ra}(\lambda_s)$ by assuming that the ratio K for the jth aerosol model is correct. This procedure works reasonably well because the values of ε_j derived by using the individual models are all close to the correct value. The procedure has been further modified by recomputing a new average formed by dropping the two models with the largest values of $\varepsilon(\lambda_s, \lambda_l) - \varepsilon_j(\lambda_s, \lambda_l)$ and the two models with the most negative results. This procedure is carried out several times until the final value is computed using four models: two with $\varepsilon - \varepsilon_j < 0$ and two models with $\varepsilon - \varepsilon_j > 0$.

After the value for $\varepsilon(\lambda_s, \lambda_l)$ has been derived, the next task is to estimate $\varepsilon(\lambda_i, \lambda_l)$.

It is suggested to use the fact that the derived value of $\varepsilon(\lambda_s, \lambda_l)$ is bracketed by two of the N candidate aerosol models. It is assumed that $\varepsilon(\lambda_i, \lambda_l)$ falls between the same two aerosol models proportionately in the same manner as $\varepsilon(\lambda_s, \lambda_l)$. Finally, it is also assumed that $K[\lambda, \rho_{as}(\lambda)]$ falls between the same two values for these models in the same proportion as $\varepsilon(\lambda_s, \lambda_l)$. These assumptions help in solving the problem, but unfortunately they are not always true. However, to the extent that the actual aerosols are similar in their optical properties to the candidate models, the assumptions appear to be reasonably valid.

Tests of the suggested multiple-scattering algorithm (Gordon, 1997) have shown that it significantly improves retrieval of normalized water-leaving reflectance for all aerosol models that are going to represent the real atmospheric aerosol except those with high absorption. It has been shown that aerosol models which are going to represent the real atmospheric aerosol must have single- scattering albedos similar to those of actual aerosols. However, it is not easy to detect the presence of absorbing aerosols in the atmosphere and furthermore to control their optical properties. More observations (ground-based, airborne) are needed of the optical properties of absorbing aerosols. Combined use of data from different satellites might also improve the atmospheric correction procedure. It was also shown in the mentioned paper that the influence of the vertical structure of atmospheric aerosols cannot be neglected for absorbing aerosols. Use of satellite measurements sensitive to atmo- spheric profiles can help to solve the problem.

Thin cirrus clouds and stratospheric aerosols, remaining undetected in water- color imagery, may cause incorrect interpretation of satellite data. Gao and Kauf- man (1996) suggested using the new 1.38 μm infrared channel on MODIS, which enables identification of thin cirrus clouds and the assessment of stratospheric aerosol contribution.

The suggested atmospheric correction algorithm (Gordon, Wang, 1994; Gordon, 1997) requires a set of applicable aerosol models, which are usually based on the work of Shettle and Fenn (1979). However, these models have to be validated with emphasis on the variables that directly affect aerosol path radiance $\rho_a(\lambda) + \rho_{ra}(\lambda)$. Some ground-based and airborne measurements are definitely needed to develop realistic models for atmospheric correction purposes.

The presence of whitecaps (foam, streaks, underwater bubbles) at the ocean surface can considerably affect the satellite signal in the visible and near infrared, making it more difficult to estimate aerosol path radiance. The optical properties of whitecaps are not well known. Recent measurements (Frouin et al., 1996; Moore et al., 1996) indicate that whitecap reflectance decreases substantially in NIR, contrary to previous theoretical and laboratory studies. If confirmed, errors of an order of magnitude larger than the acceptable errors for biological applications are expected by neglecting spectral dependence of whitecap reflectance. More in situ measure- ments of whitecaps have to be done, to see if their properties depend on environ- mental factors (wind speed, ocean stratification, and others). The idea exists to develop methods of retrieval of whitecap reflectance based on MODIS and GLI measurements in NIR (Gordon, 1997).

Additional errors can occur in the atmospheric correction procedure due to residual sun glint effects and bidirectional characteristics of the water body reflectance. It was assumed in the presented algorithm that photons specularly reflected by the water surface are directly transmitted through the atmosphere. In the presence of aerosols some of the photons reflected by the glint might be scattered into the satellite viewing direction. Fraser *et al.* (1997) have suggested incorporating the glitter reflectance diagram into the atmospheric correction algorithm. On the other hand, upwelling radiance reflected by the water surface is not isotropic as shown by Morel, Gentilli (1993) and Morel *et al.*, 1995. The bidirectional characteristics of the water body reflectance affect the signal transmitted to the satellite but are ignored in current atmospheric algorithms. The dependence of bidirectional function parameters on water composition should be investigated and taken into account.

In conclusion, we can say that during recent years important steps have been taken to improve the atmospheric correction algorithm over water surface. Among the most important was reduction of radiometric noise, improvement of calibration capability, and inclusion of multiple-scattering effects, based on additional spectral bands of sensors on the newest satellites and the availability of new information on the optical properties of the water surface, and aerosols. The use of improved characteristics of the new satellite sensors (among them spatial and spectral resolution, viewing and polarization capability) must be extended for atmospheric correction purposes.

3.1.3.2   *Atmospheric correction in the infrared range*

Over the last 20 years the use of satellite data to obtain sea-surface temperatures (SST) and to provide global data sets on the temperature of waters all over the world has been gradually increasing. A great advance in the measurements of SST was made with the launch of the first advanced very-high-resolution radiometer (AVHRR) instrument on the polar-orbiting TIROS-N satellite. Later, the NOAA meteorological satellite with a five-channel AVHRR instrument on board was launched. At the present time a similar device works in space. It operates in five spectral intervals, visible: 0.58–0.68 µm, near-infrared: 0.725–1.10 µm and three infrared: 3.55–3.93 µm, 10.3–11.3 µm, 11.5–12.5 µm, the last two playing the main role in SST determination; spatial resolution is 1.1 km in the subsatellite point; data can be obtained for the same area twice a day. The launch of the along-track scanning radiometer (ATSR) on ERS satellites is the latest technological progress in satellite measurement of surface temperature. It receives signals with less noise and has applied a dual-view scanning system that allows atmospheric correction using different pathlength measurements as well as multiwavelength capability. SST images are available from the European Space Agency (ESA) as ATSR products.

The measurements of 4th and 5th, so-called "split-window", channels of the AVHRR instrument, centered at 10.8 and 11.9 µm, are usually utilized to obtain the temperature of water surface based on the differential absorption technique. However, in many regions this application is limited by frequent cloud cover over the surface, cloudiness being the main source of errors in water-surface temperature determination. A number of algorithms for cloud filtering have been suggested to remove cloud- contaminated pixels from satellite images. Since the properties of clouds at different levels are special, as a rule, each method for cloud filtering includes several steps based on various approaches to the analysis of satellite imagery. Among these there is the spectral approach, in which emphasis is laid on analysis and intercomparison of information from various spectral channels (McClain et al., 1985; Saunders and Kriebel, 1988), the analytical approach in which satellite data are analyzed for correspondence to some empirical model (Welch et al., 1988), and the textural approach in which all conclusions are based on the analysis of the satellite imagery texture (Key, 1990; Gallegos et al., 1993). Usually measurements from other channels of AVHHR are used in a cloud filtering procedure.

Infrared radiation emitted from the water surface is partially absorbed by the atmosphere and re-emitted at a colder temperature. The net effect for a satellite radiometer is that it observes a blackbody equivalent water-surface temperature which is different from, and usually colder than, the actual surface. In the infrared waveband of interest, between 3 and 12 mm, the principal absorbers are water vapor, ozone, and carbon dioxide, and only the last of these is appreciably constant and uniformly distributed through the atmosphere. The ozone content varies diurnally, being greater during the day when sunlight generates ozone by ultraviolet interaction with oxygen, seasonally and horizontally. Water-vapor content can vary significantly in horizontal distances of order 1000 km, and is generally much greater in warmer air, thus varying latitudinally and seasonally.

Attenuation by atmospheric water vapor is the largest error source in the computation of water-surface temperature using infrared satellite imagery after filtering out cloud contaminated pixels. The error can be up to 10 K in the Tropics. The well-known and widely used split-window algorithm uses a linear difference between measurements in two different thermal infrared satellite channels to correct for this atmospheric moisture effect by differential absorption in two bands (McMillin, 1975). Many authors (McMillin, 1975; Deschamps and Phulpin, 1980; McClain et al., 1985 and others) showed the possibility of using two nearby infrared channels to derive SST in terms of a linear combination of the two satellite brightness temperatures and suggested their own estimation of linear coefficients. In the present review we will show the most general approach to the problem to make it clear what assumptions are made for deriving each particular algorithm. Below we will mainly refer to SST measurements, because current algorithms are optimized for marine atmospheres. However, retrieval of lake-surface temperatures is of considerable interest for monitoring climatic change and for particular limnological studies as well. We cannot expect that the coefficients of atmospheric correction algorithms determined for open oceans or seas will work in the case of inland water-body measurements. All algorithms should be tuned (or checked at least) for lake temperature calculations.

Many papers have reported applications for the split-window approach summarized in Prata (1993), Barton (1995), Emery, Yu (1997). Some authors have investigated the problem of variation in split-window coefficients according to the atmospheric situation. Minnett (1990) has pointed out the improvement obtained in SST retrieval by a regional optimization of the split-window algorithm. Ottle, Vidal-Madjar (1992) have derived split-window algorithms for different classes of atmospheres using a large set of atmospheric situations and a radiative transfer model. Ulivieri *et al.* (1994) have taken in the dependence of algorithm coefficients on atmospheric parameters, including water-vapour amount and air temperature as parameters of the algorithm. Sobrino *et al.* (1993) pointed out that the split-window coefficients currently used are not adequate to account for all the atmospheric variability. They have studied the dependence of split-window coefficients on different atmospheric parameters such as water-vapor amount and some others, using 60 radio soundings extracted from the TOVS Initial Guess Retrieval database (TIGR) and the radiative transfer model LOWTRAN-7. As a result, they included the ratio of AVHRR channels 4 and 5 transmittances in their split-window coefficients. Walton (1988) followed another approach to take into account the atmospheric variability in split-window coefficients and developed a complex nonlinear algorithm, the Cross Product Sea Surface Temperature (CPSST). Coll *et al.* (1994), on the other hand, have also included water-vapor dependence of the split-window coefficients in a nonlinear quadratic algorithm. They introduced a quadratic dependence for the brightness temperature differences in their split-window formulation.

The radiance $I(\nu)$ passed through the atmosphere and measured at the satellite can be represented in a form (Harris and Mason, 1992)

$$I(\nu) = \varepsilon(\nu)\tau(\nu)B(\nu, T_s) + A_{D_\nu} + A_{R_\nu} \qquad (3.168)$$

where ν is a channel frequency, $B(\nu, T)$ is the Planck function and T_s is the surface temperature, ε_ν is emissivity of the surface, $\tau(\nu)$ is the transmittance of the total atmosphere and A_{D_ν} and A_{R_ν} refer to radiance components due to atmospheric emission observed directly and after reflection from the surface, respectively. These latter two terms, which are difficult to estimate from satellite measurements, are the main obstacles to accurate SST retrieval. However, they can be accounted for using the following approach.

Assuming that the atmosphere over the viewing area is horizontally homogeneous, a surface temperature increase of ΔT_s from one pixel to the next will lead to a variation in satellite received radiance. Mathematically, this is given by

$$\Delta I(\nu) = \varepsilon(\nu)\tau(\nu)\Delta B(\nu, T_s) \qquad (3.169)$$

Let $I(\nu) = B(\nu, T)$, where T is the so-called brightness temperature measured by the satellite sensor. So we have

$$\Delta B(\nu, T) = \varepsilon(\nu)\tau(\nu)\Delta B(\nu, T_s) \qquad (3.170)$$

As is commonly done, the Planck function $B(\nu, T_s)$ can be approximated by a Taylor expansion of the first order as

$$B(\nu, T_s) = \frac{\partial B(\nu, T)}{\partial T}(T_s - T) \qquad (3.171)$$

This approximation holds good over a small range of temperatures and a narrow wave number span (Yu and Barton, 1994). Following Deschamps and Phulpin (1980), we can have

$$\Delta B(\nu, T) = \varepsilon(\nu)\tau(\nu)\left[\Delta B(\nu, T) + (T_s - T)\Delta\left(\frac{\partial B(\nu, T)}{\partial T}\right) + \frac{\partial B(\nu, T)}{\partial T}(\Delta T_s - \Delta T)\right] \tag{3.172}$$

which can also be written as

$$\frac{\partial B(\nu, T)}{\partial T} = \varepsilon(\nu)\tau(\nu)\left[(T_s - T)\left(\frac{\partial^2 B(\nu, T)}{\partial T^2}\right) + \frac{\Delta T_s}{\Delta T}\frac{\partial B(\nu, T)}{\partial T}\right] \tag{3.173}$$

The Planck function can be expressed in the form

$$B(\nu, T) = \frac{C_1\nu^3}{e^{C_2\nu/T} - 1} \tag{3.174}$$

where constants C_1 and C_2 are $1.191\,0659 \times 10^{-5}$ mW and $1.438\,833$ K cm^{-1} respectively. The derivative of the Planck function is then

$$\frac{\partial B(\nu, T)}{\partial T} = \frac{C_2'}{T^2}\frac{e^{C_2'/T}}{e^{C_2'/T} - 1}B(\nu, T) \tag{3.175}$$

and

$$\frac{\partial^2 B(\nu, T)}{\partial T^2} = \left(\frac{C_2'}{T^2}\frac{e^{C_2'/T}}{e^{C_2'/T} - 1} - \frac{2}{T}\right)\frac{\partial B(\nu, T)}{\partial T} \tag{3.176}$$

where $C_2' = \nu C_2$. Replacing eq. (3.176) into (3.173), we have

$$\frac{\Delta T_s}{\Delta T} = \omega(\nu) - C(T)(T_s - T) \tag{3.177}$$

where

$$\omega(\nu) = \frac{1}{\varepsilon(\nu)\tau(\nu)}$$

and

$$C(T) = \frac{C_2'}{T^2}\frac{e^{C_2'/T}}{e^{C_2'/T} - 1} - \frac{2}{T}$$

Applying formula (3.177) for two infrared window channels, we get two similar relationships like (3.177) for channels 4 and 5 of the AVHRR. Dividing these two relations we then have a ratio of

$$R_{5/4} = \frac{\Delta T_5}{\Delta T_4} = \frac{\omega(\nu_4) - (T_s - T_4)C(T_4)}{\omega(\nu_5) - (T_s - T_5)C(T_5)} \tag{3.178}$$

Solving (3.178) for SST gives

$$T_s = \frac{\omega_4 - R_{5/4}\omega_5 + T_4 C_4 - T_5 R_{5/4} C_5}{C_4 - R_{5/4} C_5} \tag{3.179}$$

where $\omega_i = \omega(\nu_i)$, $C_i = C(T_i)$, $i = 4, 5$. In order to compare different algorithms which use the 2-channel temperature difference, eq. (3.179) can be written in the form

$$T_s = \frac{\omega_4 - R_{5/4}\omega_5}{C_4 - R_{5/4}C_5} + T_5 + \frac{C_4}{C_4 - R_{5/4}C_5}(T_4 - T_5) \tag{3.180}$$

The simplest form of this equation can be presented as the standard split-window algorithm

$$T_s = c + T_5 + \gamma(T_4 - T_5)$$

or

$$T_s = a_1 + a_2 T_4 + a_5 T_5 \tag{3.181}$$

where c, γ, a_1, a_2, and a_3 are constants, $a_1 = c$, $a_2 = \gamma$, $a_3 = 1 - \gamma$.

The coefficients of (3.181) have to be determined by regression either theoretically, by applying (3.168) for a set of atmospheres, or empirically, by correlating satellite data against *in situ* measurements. Numerous studies have been done based on the split-window technique (Deschamps, Phulpin, 1980; McClain *et al.*, 1985; Sobrino *et al.*, 1993; Harris, Mason, 1992; Yu, Emery, 1996). Different values suggested for coefficients of eq. (3.181) can be found in Barton (1995) and Emery, Yu (1997).

Some authors (Sobrino *et al.*, 1993; Harris, Mason, 1992) use radiative transfer model simulations to obtain coefficients c and γ and to suggest some improvements to the technique. Following this approach they control the response of split-window coefficients to the variations in input information (atmospheric transmittance, total water-vapor content, surface temperature, satellite zenith angle, etc.). They have found strong variability in coefficients due to particular atmospheric conditions (Sobrino *et al.*, 1993) and have improved the algorithm by incorporating information on transmittance into eq. (3.181) (Harris, Mason, 1992). Harris, Mason (1992) suggest the following improved form for the equation of the split-window algorithm

$$T_s = a_1 + a_2 \frac{T_4}{R_{5/4}} + a_3 \frac{T_5}{R_{5/4}} + a_4 T_4$$

where the parameter $R_{5/4}$ has been determined by (3.178) and has been found to be an almost linear function of the transmittance. The suggested approach is based on theoretical computations and the main difficulty in its application is the necessity of accurate knowledge of actual total water vapor in the atmosphere from some associated measurements.

Going back to eq. (3.180) and using the theoretical approximation suggested by Harris and Mason (1992) $\Delta T_j / \Delta T_i \approx \varepsilon_j \tau_j / \varepsilon_i \tau_i$ for the first term in (3.180) we have

$$\frac{\omega_4 - R_{5/4}\omega_5}{C_4 - R_{5/4}C_5} = \frac{\dfrac{1}{\varepsilon_4 \tau_4} - \dfrac{\Delta T_5}{\Delta T_4}\dfrac{1}{\varepsilon_5 \tau_5}}{C_4 - R_{5/4}C_5} \approx 0 \tag{3.182}$$

This means that the first term in (3.180) can be neglected (Yu, Barton, 1994) and eq. (3.180) can be simplified to

$$T_s = T_5 + \frac{C_4}{C_4 - R_{5/4}C_5}(T_4 - T_5) \tag{3.183}$$

In Yu, Barton (1994) it was suggested that eq. (3.183) be used as a base equation for the algorithm for SST retrieval.

A good understanding of the neglected term in eq. (3.182) is necessary for operational use of the last-mentioned algorithm. In practice, the term should be written only in terms of the ratio $R_{5/4}$, T_4 and T_5, because atmospheric transmittances are generally not available. This conversion can be achieved by the following method. With the use of theoretical calculations it was shown by Harris, Mason (1992) and Yu, Barton (1994) that there is a good linear relation between the ratio $R_{5/4}$ and ω_5/ω_5. So the ratio ω_5/ω_5 can be replaced as

$$\frac{\omega_4}{\omega_5} = d_0 + d_1 R_{5/4}$$

where d_0 and d_1 are constants. Therefore, the neglected term becomes

$$T_n = \omega_5 \frac{(d_1 - 1)R_{5/4} + d_0}{C_4 - R_{5/4}C_5}$$

In order to estimate the term easily, as well as to display an obvious physical meaning, a special value is introduced which is

$$R_0 = \frac{d_0}{1 - d_0}$$

Using this value the neglected term can be written as

$$T_n = \omega_5 \frac{(d_1 - 1)(R_{5/4} - R_0)}{C_4 - R_{5/4}C_5} \tag{3.184}$$

It can now be seen that R_0 is the value of the ratio $R_{5/4}$ when the neglected term is equal to zero. The sign of difference between R_0 and $R_{5/4}$ indicates whether the term is to be negative or positive. The magnitude of this difference will determine whether or not the term indeed can be neglected.

In Yu, Barton (1994) it was noted that parameters C_4 and C_5 have near-linear dependence with temperature in the range considered. This can be written as

$$C_i \cong A_i T_i + B_i$$

where A_i and B_i are constants, $i = 4, 5$. Using this replacement, algorithm (3.183) becomes

$$T_s = T_5 + \frac{A_4 T_4 + B_4}{A_4 T_4 - R_{5/4} A_5 T_5 + B_4 - R_{5/4} B_5} (T_4 - T_5)$$

The last formula is exactly the same as the CPSST algorithm, suggested by Walton (1988). It can be said that the CPSST algorithm is a simplification of algorithm (3.183).

Equation (3.183) can be treated as the most general algorithm for SST retrieval. The coefficient before temperature difference (γ in the split-window algorithm) is temperature dependent. The ratio of the spatial variation of brightness temperature in one channel to that in another, which is actually proportional to the ratio of the atmospheric transmittance in one channel to that in another, is used to determine the coefficient. This removes the requirement of using regression techniques for determining SST algorithm coefficients and proves its global applicability, which as claimed in Yu, Barton (1994) is the main advantage of the proposed technique. However, our experience indicates the very high sensitivity of the method to the selection of areas (squares) for which ratio $R_{5/4}$ is determined.

Once infrared radiometer data has been atmospherically corrected, there is no doubt about interpreting it in terms of a sea (lake) parameter. But it should be kept in mind that remote-sensing observations supply us only with sea *surface* temperature. Since IR radiation is emitted only by the layers of water molecules close to the surface, it is their temperature which characterizes the radiation and controls the observed brightness temperature. The actual thickness of the layer whose temperature is remotely sensed varies according to the wavelength of the radiation, but for the wavelengths of interest it is less than 0.1 mm (Robinson, 1994). The temperature measured from space is therefore a skin temperature, and its usefulness in contributing to limnology depends on being able to interpret it in terms of the underlying (so-called bulk) temperature of the lake.

This fact causes a fundamental problem with SST retrieval algorithms, which is that some of them require "calibration" with *in situ* (buoy or ship) temperature measurements in order to derive coefficients in the SST algorithms and others use atmospheric simulations for the same purpose. The fact is that satellite measurements represent water skin temperature, whereas a drifting buoy or ship measures bulk-water temperature. The difference between the two types of algorithms was studied in the work of Emery, Yu (1997). They demonstrated similarities in large-scale (>100 km) spatial patterns retrieved with different algorithms despite overall magnitude differences of SST. Features with smaller spatial scales happened to be different, with the skin SST fine structure unlike others derived with buoy-calibrated algorithms. This causes the need to carefully test SST algorithms for small water bodies before they can be applied.

3.2 PASSIVE AND ACTIVE MICROWAVE REMOTE SENGING

3.2.1 The physical basis for microwave diagnostics of lakes, reservoirs and watershed parameters

3.2.1.1 Brightness temperature of the lake–atmosphere system

The microwave emission of various natural formations and structures is the thermal emission of heated bodies. The physical mechanism for thermal emission consists in the conversion of internal thermal energy into the energy of the electromagnetic field, propagating beyond the heated body. This phenomenon depends on the physical state of the emitting body (solid body, liquid, gas) and is determined from its temperature and emission coefficient (emissivity), varying greatly in the microwave part of the frequency, depending on the state of the "emitting body–environment" interface.

The principal difficulty in thematic decoding of passive microwave information from sensors carried by aerospace platforms consists in a variety of factors differently affecting the radio-brightness temperature T_B of the Earth–atmosphere system. This circumstance, as well as an ambiguous retrieval algorithm which describes the interrelationship between the parameters of the state of the water surface and the land, have long hindered the use of microwave sounding techniques in meteorological and limnological observations.

The brightness temperature of lakes as a natural object $T_{B,\lambda}$ is physically the temperature of a blackbody (BB), whose brightness at a wavelength of sounding λ is equal to that observed. For the cases of sounding from a flying vehicle, research vessel (RV), or for model-simulation experiments, it is written as

$$T_{B,\lambda} = \Sigma T_S \exp\left(-\sec\Theta \int_0^H \alpha\, dz\right) + \sec\Theta \int_0^H T(z)\alpha \exp\left(-\sec\Theta \int_z^H \alpha\, dz\right)$$

$$+ \left\{(1-\varepsilon)\sec\Theta \int_0^\infty T(z)\alpha \exp\left(-\sec\Theta \int_0^z \alpha\, dz\right) dz\right\} \exp\left(-\sec\Theta \int_0^H \alpha\, dz\right)$$

$$(3.185)$$

where Σ is emissivity, T_S water (or land) surface temperature, $T(z)$ the air temperature at a height z; α the attenuation coefficient of the atmosphere;, Θ', φ' are the angular coordinates of the incident beam; Θ the viewing angle; φ the viewing azimuth; $r(\Theta_1, \varphi_1, \Theta', \varphi')$ the surface brightness coefficient; H the height at which the passive microwave sensor is located; z is the vertical coordinate.

Analysis of expression (3.185) shows, first of all, that the multi-factor nature of recorded $T_{B,\lambda}$ results in limited accuracy of retrieval of geophysical parameters. For example, for the case of microwave sounding of open water, ice cover and land, the problem is further hindered by the fact that emissivity of the lake Σ is determined not only by conditions of the water (ice) surface, but also by vertical distribution of temperature and structure of the upper layer of water (ice). The presence of slicks, liquid and solid fractions, air inclusions and pollutants also substantially changes the dielectic parameters both of water (ice) and land.

3.2.1.2 *Physical models of microwave temperature characteristics (emissivities) of water, ice and snow, macrophytes*

Fresh water

Both open water and ice are characterized by electronic, dipole- relaxation and elastic-dipole polarizations (Debye, 1957), which leads to a substantial dependence of their emitting properties on the frequency of the external electric field. This special feature, on the one hand, introduces limitations to the accuracy of solution of inverse problems in space oceanography and limnology, and on the other hand, it opens up prospects for spectral-polarization measurements and the search for optimal conditions to retrieve various geophysical parameters. In this connection, let us consider the spectral dependence of electrophysical parameters of water, ice and snow.

The electronic inertialess polarization of water and ice as polar (dipole) dielectrics is determined by shifting the cloud of atoms, molecules and ions under the influence of the external field and takes about 10^{-15} s to become stabilized. The dipole-relaxation polarization observed in dipole liquid and gaseous dielectrics consists in increasingly regulating the location of dipoles. The time for stabilization is about 10^{-10} s. The electric-dipole polarization is connected to elastic turning (deformations) of dipoles at small angles, when the dipoles cannot rotate freely.

The polarization of the polar liquid P, according to the new theory by Debye, is determined by the parameters of the medium and the field:

$$P(\omega) = \frac{\varepsilon - 1}{\varepsilon + 2} \frac{M}{\rho} = \frac{4\pi N}{3} \left[\alpha_0 + \frac{\mu^2}{3kT} + \frac{1}{1 + i\omega\tau} \right] \qquad (3.186)$$

where ε is the dielectric permeability; M is the molecular mass; ρ is the density of liquid; N is the Avogadro number; α_0 is the average polarization of molecules on the assumption that there is no interaction between the molecules; μ is the constant dipole moment; k is the Boltzmann constant, T is the absolute temperature; τ is the relaxation time; that is, the time needed for the dominating orientation of molecules, caused by the external field, to be replaced by a random distribution when the field is removed; $\omega = 2\pi v$ where v is the frequency of the external field.

The complex dielectric permeability of water in the microwave can be written as:

$$\varepsilon = \frac{\varepsilon_S - \varepsilon_0}{1 + i\dfrac{\lambda_S}{\lambda}} + \varepsilon_0 \qquad (3.187)$$

Here ε_r and ε_i are the real and imaginary parts of the dielectric permeability, written, respectively, as:

$$\varepsilon_r = n^2 - \chi^2 = \frac{\varepsilon_S - \varepsilon_0}{1 + \left(\dfrac{\lambda_S}{\lambda} \right)^2} + \varepsilon_0 \qquad (3.188)$$

$$\varepsilon_i = 2n\chi = \frac{(\varepsilon_S - \varepsilon_0)\dfrac{\lambda_S}{\lambda}}{1 + \left(\dfrac{\lambda_S}{\lambda} \right)^2} \qquad (3.189)$$

where $\lambda_S = 2\pi c\tau$ is the critical wavelength; ε_s is the statistical dielectric permeability; ε_0 is the optical dielectric constant, the sum of atomic and electronic polarizations; ε_0 is the emission wavelength.

The correctness of the Debye formulas for fresh water has been confirmed experimentally (Debye and Zuck, 1936; Frehlich, 1960). At the same time, ideas about the dielectric properties of sea and brackish water as an electrolytic solution of salts of Na, K, Mg, Ba, and other elements, have long been ambiguous. The reason for the degree of variability in the data by various experts is high values at radio frequencies of electric conductivity of the water solutions of electrolytes, as well as an ambiguity of theoretical assessments of the effect of these salts specific electric conductivity of solutions and the Debye relaxation absorption.

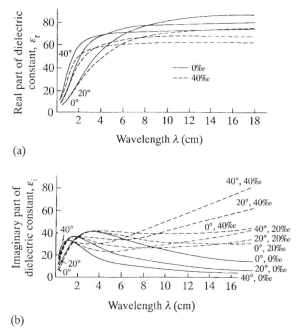

(a)

(b)

Fig. 3.37. (a). Real part of dielectric constant of water as function of wavelength, with temperature as the parameter). (b) Imaginary part of dielectric constant of water as function of wavelength, with temperature as the parameter.

A detailed analysis of these effects and their dependence on wavelengths has been made by Rabinovich and Melentyev (1970) and others. Without dwelling on description of assumed modifications of calculation schemes, we shall only indicate that laboratory experiments using an original technique to measure emissivity of sea-water and other liquids have verified the correctness of the Debye formulas for water solutions of electrolytes with an equivalent concentration of NaCl salts to several hundredths of the normal solution on the assumption of techniques given by Kondratyev *et al.* (1992) to take into account the dependence of ε_S, ε_0, and λ_S on the temperature of the solution and the concentration of salts suggested by Rabinovich and Melentyev (1970).

Analysis of calculation data (Fig. 3.37 a,b) shows that both ε_r and ε_i of fresh, brackish and sea water depend strongly on the temperature that the clearly expressed dependence on salt content of water increases with growing wavelength λ.

The substantial part of the dielectric permeability of natural waters ε_r increases with increasing λ, for fresh waters this increase manifests itself more clearly. The absolute value of ε_r for salt waters in the 0.3–18.0-cm wavelength region is less than that of ε_r for fresh waters at the same temperature. This difference manifests itself mostly at $\lambda > 2.0$ cm.

The variation in ε_i for fresh waters is determined by the Debye absorption, whose maximum value at a water temperature of 0°C lies near $\lambda = 3.0$ cm. With increasing temperature, this maximum shifts towards shorter wavelengths, the absolute value of ε_i being less in the whole wavelength interval under consideration. The appearance of salt in the water changes its electric properties in the region of the Debye absorption. A maximum of absorption, clearly expressed for each temperature, also remains for salt waters, but its absolute value decreases. With increasing wavelength, the effect of ion conductivity of the solution is shown, breaking the former temperature dependence of ε_i. The effect of ion conductivity in the interval of wavelengths longer than 1.5–3.0 cm (depending on the temperature of the solution) leads to an increase in absolute value of the imaginary part of dielectric permeability with increasing concentration of salts at any water temperature.

Results of model calculations of dielectric parameters and emissivity of fresh and mineralized water surfaces have been tabulated (Rabinovich, Melentyev, 1970); later this suggested calculation scheme was applied to decimeter and meter wavelengths by Raiser *et al.* (1975). The results of these calculations show further increase in dependence of e_i on salinity with growing wavelength, and are now used for microwave sounding of the mineral content of natural waters.

Fresh-water ice

While for negligibly elastic water the mechanism of dipole-relaxation polarization is not observed in the wavelength region 10^9–10^{10} Hz, in the case of ice this mechanism is active in the region 10^2–10^4 Hz (Eder, 1947). It is seen from Fig. 3.38a, b that the ε_r of fresh-water ice changes drastically, starting from acoustic frequencies. In the alternative field changes in polarization lag behind changes in the field, since shifting of changes in a solid dielectric cannot be instant.

The difference in the phase of oscillations of the external field and the total vector of dipole moments of the ice crystalline grid expressed in radians is called the angle of ice dielectric loss δ. It turns out here that the tangent of the angle of loss, for which the formula

$$\mathrm{tg}\,\delta = \frac{(\varepsilon_s - \varepsilon_0)\omega\tau}{\varepsilon_S + \varepsilon_0\omega\tau} \tag{3.190}$$

holds, has a clearly expressed maximum shifting with air temperature increasing towards higher frequencies.

At the same time, it should be borne in mind that the polarization theory has some substantial limitations to its applicability, which is connected to a simplified model representation of the behavior of dipole molecules in the external field. Therefore from the viewpoint of development of efficient techniques of remote sensing of the parameters of water, ice, and snow, it is important to develop a technique for observational studies of electric parameters of real natural structures and formations. Procedures to measure electrophysical properties of dielectrics have been successfully developed (Hippel, 1959).

Snow, and frozen soils

Principal techniques for measuring ε_r and ε_i of ice, snow, soils for negative temperatures were created by Bogorodsky and Oganesian (1987), because limitations

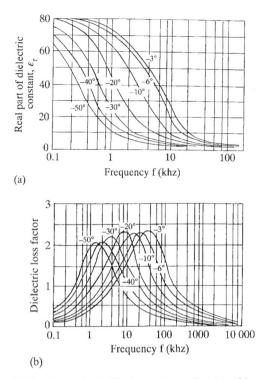

(a)

(b)

Fig. 3.38. (a) Real part of dielectric constant of fresh-water ice as function of frequency, with temperature as the parameter (Eder, 1947). (b). Imaginary part of dielectric constant of fresh-water ice as function of wavelength, with temperature as the parameter (Eder, 1947).

inherent in any measurement of dielectric properties of natural medium procedure must be taken into consideration when comparing the data of subsequent calculations and this comparison made carefully in the thematic interpretation of remotely sensed data. Note that the use of data of experimental studies of ice and snow electrophysical properties needs to take into account not only differences in the density and composition of samples as well as differences in the temperature and stress but also the technique and technology of microwave measurements of the dielectric properties (and their contrasts).

A detailed overview of the results of measurements for indisturbed ice has been made (Evans, 1965). The conclusion from this overview about the good agreement of data of measurements of ε_r of fresh-water ice is favorable for development of techniques for its remote sensing in the microwave. In the analysis of fixed values of ε_r we shall take data by Hippel (1959) as a reference point: at frequencies 3.0 and 10.0 GHz the ε_r of ice constitutes 3.20 and 3.17. Unfortunately, information about the frequency dependence of the tangent of the dielectric loss angle given by various authors is widely scattered.

Detailed tables of the values of real n and imaginary χ parts of the coefficient of refraction of fresh-water polycrystalline ice in the spectral interval from the UV to meter wavelengths have been published (Warren, 1984), who gives a summary of data for practically the whole microwave range, from 1.0 mm to 5.0 m. Ice temperature variability ranges from $-1°C$ to $-60°C$. As seen from Warren's data, ε_r depends weakly on temperature and varies from 3.16 to 3.20 as a function of wavelength and temperature, which agrees well with published data for fresh-water ice.

The tangent of the loss angle δ is, on the contrary, a strongly varying characteristic that depends on these parameters. The scattering of tg δ values for millimeter wavelengths is particularly substantial.

For practical purposes, a simple formula has been recommended:

$$\text{tg } \delta = A\, e^{-B|t_c|} + D \tag{3.191}$$

where A, B, D are the wavelength-dependent coefficients (Warren, 1984).

The ice and snow cover of seas and inland water basins have usually a heterogeneous structure. Therefore, to compare the data of field experiments with those of calculations, emitting properties should be calculated for various mixtures (Netushil, 1975; Basharinov et al., 1974).

The formula by Weiner (1910) can also be used to calculate the imaginary part of dielectric permeability of mixtures. However, its application only to real values of ε has been strictly justified.

Detailed experimental and theoretical studies of the variability of dielectric properties of dry and humid snow have been made (Bogorodsky et al., 1971). Variations in ε_r and ε_i of snow have been studied in the region of variability in the determining parameters: water equivalent 0–12.3% of the volume of ice sample; snow density varied from 0.09 to 0.42 g cm^{-3}; temperature from 0°C to $-15°C$; size of crystals from 0.5 to 1.5 mm. The measured data show that the dielectric properties of humid snow are closely connected with the dielectric properties of water and their variability with frequency. For dry snow the contribution of volume scattering takes place at a frequency of 37.0 GHz, mostly used in the experiment. The use of some empirical and theoretical parameterizations of the mixture models shows that both the Debye semi- empirical model and the mixture formula by Polder, Van Santen (1946) provide a good agreement of the measured data for the real part of the dielectric permeability of wet snow only when the shape of water inclusions turns out to be asymmetric and depends on snow water content.

Of interest are the results of studies on the effect of the shape of water inclusions in snow: new formations of water in snow are of the needle shape. Its transformation takes place at $m_v = 3\%$, when the inclusions start taking on a disc shape. These phenomena of transformations in the shape of water inclusions must show themselves with the polarization selection of microwave signal.

From the estimates of the effect of thermal-physical characteristics of snow on the microwave contrast of the snow-ice surface, it follows that the screening effect of snow depends much on its density (heat conductivity). Calculations confirmed by data of the IR survey show that the screening effect of snow cover is at a maximum when the snow is fresh. The calculations verified the correctness of the scheme chosen and showed the substantial contribution of scattering for fresh-water ice, for which, at short λ, the decrease in T_B due to scattering reaches almost 30 K, and confirm the prospects for spectral microwave sensing of the ice cover of inland water basins and continental ice. Calculations have shown that in relation to considerable absorption, the effective emitting layer of different types of ice is small: for white ice it is 10–15 cm, for nilas 1–2 cm. Calculations show that the presence of water film on the surface of nilas leads to a decrease in T_B of this ice: so, with small thicknesses of film (<5 mm) their T_B are even below the temperature of the open sea surface equal to 105 K, in the same conditions. It should also be stated that calculations for simple 3- layer models fail to simulate observational data (Kondratyev *et al.*, 1996).

It is seen from the data (Melentyev and Aleksandrov, 1991) that for fresh-water ice the character of angular dependence is determined by dielectric properties of all three media, and that the lake and river ice up to 50-cm thick is sufficiently transparent to the flux of water's own emission, which determines the character of the angular dependence of Σ. For extreme values in thickness of lake ice, 10–140 cm, the coefficient of polarization of the medium decreases, the character of the dependence of Σ on Θ both on vertical and horizontal polarizations is mainly determined by dielectric properties of the fresh-water ice itself. The revealed peculiarities of the emitting properties should be taken into account in thematic decoding: with scanning normal to the flight direction it is still possible to retrieve characteristics of the state of the 3- layer medium "water–ice–air". Figure 3.39 shows the values of Σ for different temperatures of ice. The increase in temperature of air and ice to $-10°C$ raises the Σ for all ice thicknesses and every sounding condition. The increase of the Σ on both polarizations shows itself stronger with ice thickness at a maximum.

Observational studies of emitting properties of real soils and water medium in its various states revealed interference effects (Tuchkov, 1968), which further confirm that the model of transfer of the emission of a flat-stratified isothermal structure with sharp interfaces between media has a physical sense. Calculations made for Σ of fresh-water ice should be supplemented by taking into account the presence of air bubbles in the fresh-water ice. The analysis of the lake ice cores shows that maximum ice porosity, for example, of Onega Lake constitutes 3%, the size of air bubbles is 1–2 mm. Calculations of the effective dielectric permeability of this ice were made using the Wiener equation in its complex form. Figure 3.40 shows the dependence on difference between emissivities of porous ice with the value of porosity 0.03, and monolithic ice on ice thickness.

At $\lambda = 10$ cm the air bubbles slightly reduce Σ (by 0.02 at most), and at longer wavelengths their effect practically does not show. At the same time, in the centimeter range the air bubbles vary substantially in their interference pattern, either increasing or decreasing Σ for different thicknesses of ice. At $\lambda = 0.8$ cm the $\Delta\Sigma$ reaches 0.175, at $\lambda = 2$ cm 0.08, and then with increasing wavelength it decreases more substantially. Naturally, the effect of air bubbles is at a maximum for thicker lake ice.

Macrophytes (physical models of the microwave emissivity)

Means of microwave remote sounding turn out to be extremely useful to retrieve some important hydrological and meteorological parameters determining the con-

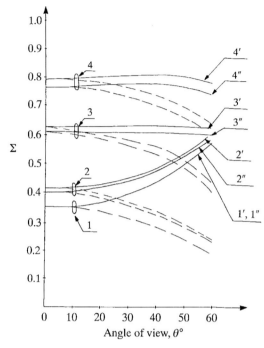

Fig. 3.39. Emissivity of system: fresh water–ice–atmosphere *versus* angle of view, $T'_{ice} = -10°$C, $T''_{ice} = -40°$C, $\lambda = 18.0$ cm; $H_{ice} = $ var(1—10 cm, 2—50 cm, 3—100 cm, 4—140 cm). Solid line—vertical polarization, dashed line—horizontal polarization.

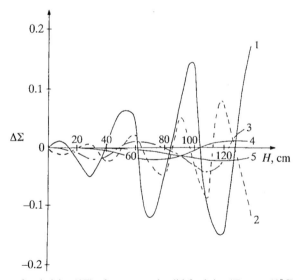

Fig. 3.40. Difference of emissivity ($\Delta\Sigma$) of porous and solid fresh ice ($T_{ice} = -10°$C).

dition and the quality of waters in the water body–watershed system as well as the land surface condition. However, these studies have been aimed so far at remote sounding of soil moisture and development of techniques of on-board estimation of water supplies in soils, since these problems are extremely urgent in agriculture in particular for yield forecast (Kondratyev *et al.*, 1992).

At the same time, the problems of using microwave means to assess the cyclicity of water in watershed territories, to study the characteristics of surface run-off—the important problems of land surface climatology, hydrology, limnology—have hardly been touched upon. A great prospect is microwave decoding to determine natural-economic structures and vegetation diagnostics—an assessment of phytomass, its seasonal dynamics, plant suppression; the respective studies have been undertaken both in Russia and abroad (Kondratyev *et al.*, 1977; Owe *et al.*, 1988).

There is an important possibility of using microwave techniques to assess anthropogenic effect on water bodies—for example, an increase of pollution and trophicity of inland water bodies leads either to the development or suppresion of higher algae, whose characteristics can be quantitatively evaluated using radiometric sensors (Kondratyev *et al.*, 1990a).

However, interpretation of brightness contrasts created by land surfaces is rather complicated. Emissivity of water surface can be calculated from dielectric constants of water, and land emissivity depends on dielectric constants of the substance and water almost always present in the substance, as well as the presence of air layers, inhomogeneities, and admixtures of other materials. Moreover, the emission of real surfaces is affected by relative size of roughness at the air–land interface.

The development of remote sounding of water quality has been so far reduced to studies of optical data of changes in concentration of phytoplankton chlorophyll as an indicator of the biological productivity of water bodies (Sid'ko *et al.*, 1984). At the same time, a change in water condition leads to variations in higher algae—either in their development or their suppression. These processes can be studied using spectral microwave techniques. Floristic diversity, degree of development, character of allocation of the plants in water bodies and over the shore are determined by climatic conditions, ecological conditions in the "water body–watershed" system, by different types of anthropogenic effect: the effect of industrial wastes, chemicalization of agriculture, construction of stock-breeding complexes, and others.

To assess the ability of microwave techniques to diagnose the condition of higher algae as an indicator of pollution and trophicity of water bodies, field simulation experiments have been carried out. A layer of reed (*Phragmites australis*) was arranged on a sheet of metal in three portions: 16, 16, and 17 kg; the respective contrasts of brightness temperatures were as follows: at $\lambda = 18.0$ cm—141.0; 21.0; 15.0 K ($T_{B\Sigma} = 177.0$ K); at $\lambda = 11.0$ cm—157.0; 20.4; 5.8 K ($T_{B\Sigma} = 183.2$ K).

Model calculations have also been made. Microwave emission of an unbroken water domain with the present dissolved organic matter, mineral suspensions, phytoplankton, also at maximum concentrations observed in natural lakes and reservoirs, varies negligibly. Estimates of the effect of these factors have shown that variations in the emission coefficient in the wavelength interval 0.8–60 cm show themselves only in the fourth figure of ε; that is, the variability of the state of the near-surface layer of the water body weakly affects its eigen-emission (with possible indirect influences on the change in environmental conditions left out of account). It is explained by the fact that even with concentrations of phytoplankton, dissolved organic matter, mineral and organic suspensions $30\text{--}40\,\text{mg/m}^3$, the dielectric permeability of water masses varies negligibly, since the dielectric permeability of the components determining this permeability have weights in proportion to their volume concentrations (Weiner, 1910).

To analyze the possibility of diagnosing higher algae, model calculations have been made for the 3-layer domain "water–macrophytes (on-water and above-water plants)–atmosphere" in different sounding conditions (wavelength, viewing angle, kind of polarization). The dielectric permeability of fresh water was estimated from the data (Rabinovich and Melentyev, 1970).

Both on-water and above-water plants are not a continuous medium but a living (raw) phytomass–air mixture. To calculate the dielectric permeability of two- and multi-component mixtures, various relationships have been developed. In all these relationships the dielectric permeability of the mixture ties in with experimental data and is determined with the dielectric permeability of the components and their volume concentrations taken into account. The authors used the Wiener equation in their calculations since, apart from the enumerated parameters, it also takes into account the form of inclusions—a parameter very important in calculations of plant emissivity.

In the case considered the Wiener equation in the complex form is written as:

$$\frac{\varepsilon_{cm} - 1}{\varepsilon_{cm} + 1} = \rho_p \left(\frac{\varepsilon_1 - 1}{\varepsilon_1 + u} \right) + (1 - \rho_p)\frac{\varepsilon_2 - 1}{\varepsilon_2 + u} \tag{3.192}$$

where ε_1, ε_2 and ε_{cm} are the complex dielectric permeability of plants, air, and their mixture, respectively; ρ_p is part of the volume occupied by vegetation; u is the relative parameter which describes the form of vegetation.

Considering that for air $\varepsilon_2 = 1 + j0$, the Wiener equation is re-written in the form similar to the equation for ice. With plants considered as cylinders with constituents of the complex dielectric permeability $\varepsilon_{rp}, \varepsilon_{ip}$ in the air, we transform:

$$(\varepsilon_{rcm} - 1)(\varepsilon_{rp} + u) - \varepsilon_{icm}\varepsilon_{ip} = \rho_p[(\varepsilon_{rcm} + u)(\varepsilon_{rp} - 1) - \varepsilon_{icm}\varepsilon_{ip}] \tag{3.193}$$

$$\varepsilon_{ip}(\varepsilon_{rcm} - 1) + \varepsilon_{icm}(\varepsilon_{rp} + u) = \rho_p[(\varepsilon_{rcm} + u)\varepsilon_{ip} + \varepsilon_{icm}(\varepsilon_{rp} - 1)] \tag{3.194}$$

The final expressions for determination of dielectric permeability of higher algae are as follows:

$$\left.\begin{array}{l}\varepsilon_{icm} = \dfrac{4p_p\varepsilon_{ip}}{(\varepsilon_{rp} + 1 - p_p\varepsilon_{rp} + p_p^2)^2 + [\varepsilon_{ip}(1 - p_p)]^2}; \\[3mm] \varepsilon_{rcm} = \dfrac{\varepsilon_{icm}(p_p\varepsilon_{rp} - p - \varepsilon_{rp} - 1)}{\varepsilon_{ip}(1 - p_p)}\end{array}\right\} \tag{3.195}$$

Figure 3.41 shows the dependence of emissivity Σ of the 3-layer domain on the

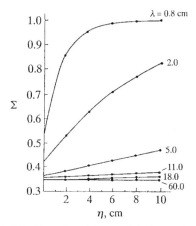

Fig. 3.41. Dependence of emissivity Σ on the thickness of the layer of macrophytes.

thickness of the vegetation layer ($h_{max} = 10$ cm) for the 0.8 to 60 cm wavelength interval.

Calculations are given for $\rho_p = 0.1$ (10% of the layer volume are occupied by vegetation; 90% by the air); $w = 0.383$ (the plant contains 38.3% moisture), $T_S = 0°C$, nadir viewing. The 0–1 cm layer of vegetation is attributed to the category of on-water vegetation: these are plants immersed in water, floating and rising above the water up to 1 cm. Air-aquatic vegetation more than 1 cm thick is classified as above-water vegetation.

As seen from Fig. 3.41, for such a domain, with any thickness of the vegetation layer, $\lambda = 2$ cm is the optimal wavelength of sounding, since in this case the value of $\Delta\Sigma$ is high and the linear relationship between Σ and h is sufficient. The 0.8-cm wavelength is more useful for separate diagnostics of on-water vegetation. At wavelengths longer than 2 cm the effect of vegetation on Σ decreases, at $\lambda \geq 11.0$ cm both on-water and above-water plants are practically transparent to microwaves.

Calculations of the angular dependence of the emission coefficient of on- water vegetation ($h = 1$ cm) in the wavelength region 0–60° with different values of w have shown that for wavelengths 11.0 cm and longer these characteristics practically repeat the angular dependence of Σ for the smooth water surface. At the same time, at $\lambda = 0.8 \cdots 2$ cm the angular dependence of Σ of on-water vegetation differs a lot from that of Σ of clean smooth water. An increase in volume moisture content of vegetation reduces the emissivity of the system.

Calculations of the effect of the relative volume content of the vegetation component ρ_p show that an increase in ρ_p from 0.01 to 0.1 leads to a considerable increase of emission in macrophytes, which enables assessment of the phytomass even from the data of 1-channel measurements at wavelengths shorter than 5 cm.

Numerous possibilities open up for polarization microwave sounding of water bodies due to considerable variations in the coefficients of polarization of the emission of the domain with changes in the content of the air in vegetation and the moisture in a leaf.

Possibilities of assessing the condition of high macrophytes are of interest for limnology. Calculations of the emission of rushes and reeds up to 300 cm high with different moisture content in the plants show that the effect of such macrophytes is evident even in the decimeter wavelength region.

Calculations of the emissivity of higher algae with condition parameters $\rho_p = 0.000\,55$; $w = 0.9$, $u = 1$, and $\rho_p = 0.001$, $w = 0.5$, $u = 1$ gave estimates of Σ close to those studied in model experiments in the town of Suisar$'$ (Lake Onega). The results of both calculations and experiments agree well. Thus, a series of field simulation experiments has made it possible to study in detail the emissivity of soil-vegetation cover. Various types of surfaces, different kinds of vegetation contribute differently to the measured brightness temperature. The complex experimental and theoretical estimates of variations in TB due to changes in phytomass and moisture content of vegetation have been used to decode the data of microwave measurements of the condition of lake eco-systems.

3.2.1.3   Characteristic of radio-brightness temperatures and retrieval algorithm development

In conditions of increasing anthropogenic effect on the environment the need arises for methods and means of qualitative assessment of the degree of ecological disturbance of both reservoirs and their watershed territories. Of special importance are multichannel remote microwave measurements of characteristics of the coastal zone of inland water bodies, river mouths, zones of watering and shallows (wet-lands). It is these areas that are most sensitive to changes in the eco- hydrological situation which forms in the "water body–watershed" system and sometimes, in the case of inland water bodies, covers vast territories. Until recently, areas of wetlands have remained inadequately studied because of difficulties arising in completing field observations and investigations. As shown by Kondratyev *et al.* (1990a), abrasive processes, variation in soil moisture, changes in the character of plant growth on the shore, floristic diversity, the degree of development (or suppression) of spatial allocation of vegetation communities over the littoral zone are reliably estimated with the help of microwave complexes carried by aerospace vehicles.

The remote studies of the interaction of the water body–watershed system using microwave radiometry foresee multi-level experiments and model calculations. An intercomparison of data of airborne microwave and key ground measurements have resulted in algorithms for converting brightness temperatures into volume soil moisture and height (phytomass) of higher algae. These algorithms enable micro-wave registrograms to be decoded and form the basis for landscape-ecological diagnostics, to carry out thematic decoding of the ecohydrological condition of Rybinsk Reservoir and geocomplexes of its shore areas using data of multichannel microwave measurements from low-flying aircraft (AN-2), to improve landscape-ecological diagnostics of the processes of interaction of water body–watershed system. This work is based on materials from the "INTERKOSMOS"–"Inland

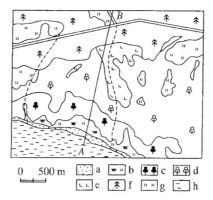

Fig. 3.42. Schematic map of the landscape of the watering zone of Rybinsk Reservoir 2 km south-west of the town of Cherkasovo: a—sandy shallow water, b—waterlogged sedge meadow on the peat–gley soil, c—long-moss birch forest on the peat–gley soil with sphagnum bogs in depressions, d—birch forest with bilberry growth on turf–podzol peaty soil, e—clearing with young pine and fir trees and drainage trenches, f—fir- and birch-grove with bilberry growth on turf–podzol soil, g—mixed-grass cereal meadow, h—water reservoir.

water bodies" experiment carried out at Rybinsk Reservoir in 1986–1989 (Kondratyev *et al.*, 1990b; Melentyev, Mel'nichenko, 1995).

The landscape structure of the watering zone of Rybinsk Reservoir has been formed under the influence of complicated differently directed processes. The naturally governed history of its development and anthropogenic impacts differentiate the region's landscapes. Particularly strong changes in landscape structure are connected to the appearance of Rybinsk Reservoir in the 1940s. The hydrological influence of the reservoir has led to a rise in level of groundwaters (GWL) in the coastal zone, to the formation of the belt structure of natural complexes. Figures 3.42–3.43 show landscape schematic maps of the watering zones of Rybinsk Reservoir drawn from data of aerophoto-survey in June 1989 for two characteristic landscape profiles: forest (Fig. 3.44) and meadow (Fig. 3.45). In both Figs. 3.42 and 3.43 the line AB shows flight paths of AN-2 aircraft carrying the microwave complex.

Specially organized model experiments to process the results of airborne microwave measurements, as well as comparison of data of the background measurements of soil moisture at the test site with data of phytomass assessment has suggested the following relationships of brightness temperature T_B of the surface–atmosphere system with soil moisture in the 0–10 cm layer (w_{0-10}) and the height of macrophytes growing in the coastal zone of the reservoir (h_M) (Kondratyev, Melentyev, 1991):

$$T_B = 281 - 1.28 w_{0-10} \qquad (3.196)$$

$$T_B = 107.54 - 9.43 H_M \qquad (3.197)$$

Data of microwave measurements have been processed considering these relationships as well as materials of the model experiment showing that an increment in phytomass m_M of air-water plants per 1000 g/m^2 gives a 5 K contrast in brightness temperature (the emissivity of reed *Phragmites australis* was studied).

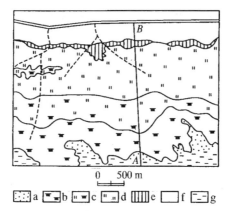

0   500 m

⬚a ▭b ▭c ▭d |||||e ▭f ▭g

Fig. 3.43. Schematic map of the landscape of the watering zone of Rybinsk Reservoir 3 km north-east of the town of Borok: a—sandy shallows, thinly overgrown, b—bog-grass meadow on peat–gley soil, c—mixed-grass–sedge meadow and sedge meadow on turf-meadow peaty soil, d—mixed-grass–cereal hummocky meadow on the turf- meadow soil, e—willow-bed with meadow-sweet in the near-slope depression, f—agrocenoses, g—water reservoir.

Fig. 3.44. Watering zone of Rybinsk Reservoir—aerophoto-survey of the forest landscape.

Figures 3.46 and 3.47 show registrograms of microwave survey at $\lambda = 18.0$ cm, as well as vertical profiles and boundaries of the extent of landscape profiles. It can be seen that the belt structure of the watering zone is strictly organized, it tells on the local radio-relief.

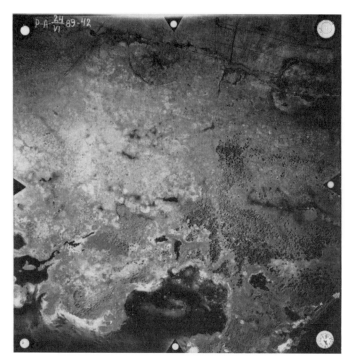

Fig. 3.45. Watering zone of Rybinsk Reservoir—aerophoto-survey of the meadow landscape.

Brightness temperatures corresponding to the belts of strong, moderate and weak watering, have different absolute values. The contrast ΔT_B for the belts of strong andweak watering constitutes ~100 K. Thus, the known geophysical parameter—the gradient of the factor of watering normal to the belts of watering—is characterized by a sharp decrease in T_B recorded within the belt of moderate watering. The values of soil moisture for different areas of the watering zone from microwave sounding data vary from 30–35% to a maximum approaching 100% moisture (the respective scale of conversion of T_B values into those of moisture is shown in Figs. 3.46 and 3.47, on the right). Parameter K_e is the ratio of the contrasts ΔT_B, characteristic of belts of weak T_{B1} and strong T_{B2} watering to the extent of the transitional zone of moderate watering L_m,

$$K_e = \frac{\Delta T_B}{L_m} = \frac{T_{B1} - T_{B2}}{L_m} \qquad (3.198)$$

enabling quantitative description of the dynamics of water exchange processes in the coastal zone of inland water bodies.

With GWL lowering towards the shore and soil moisture, measured by micro-wave radiometer, decreasing, a certain series of vegetation communities (biocenoses) and natural low-row complexes (facies) occur connected to processes taking place on the slopes and to features of watering processes. This circumstance makes it possible to use K_e not only for microwave indication of interpermeability of natural waters in

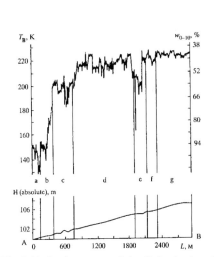

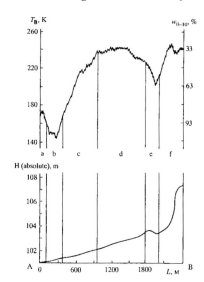

Fig. 3.46. Registrogram of the flight (top) and landscape profile (bottom) of the watering zone of Rybinsk Reservoir (2 km south-west of the town of Cherkasovo).

Fig. 3.47. Registrogram of the flight (top) and landscape profile (bottom) of the watering zone of Rybinsk Reservoir (3 km north-east of the the town of Borok).

the coastal zone but also for remote diagnostics of several phenomena connected to the interaction of the water body–watershed system and quantitative estimation of the eco-hydrological situation on a regional level.

The number of points processed to count the T_B on the forest profile constituted 127, on the meadow profile 76. Brightness temperature was taken from registrograms with an accuracy of 0.1 K; w_{0-10} and m_M were calculated with an accuracy of tenths of one per cent and g/m^2, respectively.

As demonstrated earlier, the condition of higher algae is the most important indicator of the waters of the inland water bodies and the degree of their ecological disturbance. For this purpose quantitative fixation of macrophyte phytomass enables use of the developed method to control the condition of natural waters.

From data of microwave measurements of the forest landscape profile near the town of Cherkasovo in the region of strong watering (zones a, b in Fig. 3.46) the ecological row is represented by the following belts of vegetation (quantitative estimates of geophysical parameters indicated above are given in parentheses): reed ($m_M = 1311$ g/m^2, $w_{0-10} = 100\%$); airy and bubbly sedge ($m_M = 1134$ g/m^2, $w_{0-10} = 90.2\%$) as well as black sedge and sedge with willow bushes ($m_M = 1311$ g/m^2, $w_{0-10} = 87.3\%$).

At the interface of strong and moderate watering there is a hydrophyll cereal–mixed-grass meadow, for which the following quantitative estimates were obtained using microwave radiometry: $m_M = 1615$ g/m^2, $w_{0-10} = 80.6\%$.

The belt of moderate watering is represented by three dominating natural complexes clearly reflected by brightness relief (zones c, d, e in Fig. 3.46). The first—a long-moss birch grove on peaty soil—is characterized by two levels of moisture. Over the thinned-out forest areas, according to microwave data, w_{0-10} constitutes 58.4–62.2% (~60.5% on average), over the open boggy areas covered with sphagnum moss 66.0–68.9% (higher values of moisture are typical of the length of aircraft flight paths located closer to the reservoir).

With transition to the next natural complex T_B smoothly increases, which corresponds to a gradual decrease in soil moisture and exchange of the long-moss grass-undergrowth layer for the bilberry one. The birch–bilberry layer on the turf–podzol peaty soil is characterized by several levels of moisture. Over typical sufficiently drained areas according to microwave data, w_{0-10} constitutes 47.7–48.7%; over less drained areas with an admixture of haircap moss in the grass cover 50.6–55.5%; over the open boggy areas of the forest covered by communities of black and turfy sedge, long manna grass, stunted willow-bed 57.3–62.2%.

The third natural complex in the belt of moderate watering formed in a boggy clearing of 15 years standing, overgrown with fir, birch and pine growth with the developed moss cover of haircap moss. For this belt, from microwave survey data, soil moisture fluctuates at three levels: at more drained sites with individual big birches and pines w_{0-10} constitutes 71.9–73.8%, at less drained 77.7–79.5%, in drainage trenches overgrown with moss 80.6–90.2%.

In the belt of weak watering, there are fir trees, birches and bilberries on turf–podzol soil. At typical forest sites soil moisture constitutes 52.6–55.5%, at low sites with horsetails, sedges and mosses in grass cover 57.3–58.4%

For the meadow landscape profile the course of changes in T_B (Fig. 3.47) clearly correlates with the course of the hypsometric curve on the profile AB, with changes in soil moisture and degree of hygrophilousness of vegetation. From microwave survey data, soil moisture from the water–shore line away from the shore decreases from 95 to 35%.

Shallows along this route are alternations of patches of sand, silt and various amphibian plants bordering the reservoir coast in the form of intermittent belts. The course of T_B on shallows (zone a in Fig. 3.47) is jagged due to special features of the underlying surface. Registrogram data enabled assessment of phytomass of amphibian vegetation: in the belt of *Sagittaria sagittifolia*, *Eleocharis palustris*, *Alisma plantago-aquatica* m_M is 654, in the belt of reed 1286–1527 g/m^2.

In the strong watering belt (zone b in Fig. 3.47) are marsh–grass meadows on peat–gley soil. Here the lowest values of T_B and, respectively, the highest values of w_{0-10} (85.3–95.0%) are observed by microwave on the profile. In some places the soil is without structure, oversaturated by 100%. Along this route, one belt is smoothly transformed into another. The marsh-grass vegetation is gradually replaced by the sedge meadows of the moderate watering belt and mixed-grass–sedge meadows located higher on the slope of the first ancient-lake terrace (zone c in Fig. 3.47). Here the moisture content also decreases smoothly (from 79.6–69.9% in the sedge meadows to 54.5–46.7% in the mixed-grass–sedge meadows). Local microdepressions are clearly fixed on the graph in the form of T_B decreases by 2–4 K, which corresponds to an increase in w_{0-10} by 1–3%.

In the weak watering belt are well-drained mixed-grass cereal meadows on the turf–meadow soil. From microwave sounding data soil moisture averages 30.2 to

Table 3.3. Retrieved estimates of soil moisture (in % total water capacity; dashes—
no data).

Watering	GWL, m	Height above the level of reservoir, m	w_{0-10}	
			In the meadow	In the forest
Strong	0.0–0.5	0.0–0.4	85.6–100.0	—
Moderate	0.5–1.3	0.4–1.5	46.7–69.9	47.7–67.0
Weak	1.0–1.8	1.5–3.0	30.2–41.6	52.6–58.4
Absent	>1.8	>3.0	28.5–35.2	—

37.2% increasing in microdepressions by 1–3%; in macrodepressions it reaches
41.6% (with average background value 32.5% characteristic of this zone).

The brightness relief of both forest and meadow landscape profiles clearly shows
T_B decreases typical of the near-terrace depressions stretched along the slope (zone e
in Figs. 3.46 and 3.47) with increased soil moisture. A depression on the meadow
profile is occupied by willows with meadow-sweet on turf peaty soil; the moisture
content increases from 41.0–48.7 to 57.3%. The value of w_{0-10} was at a maximum
(66.0%). The increase in moisture content in this case may be connected to two
counter-processes: subsoil flow from the direction of the second ancient-lake terrace
and watering from the direction of the reservoir.

On the slope and at the edge of the terrace located at a height of 5–6 m above the
level of the reservoir, outside the watering area, the soil moisture on the day of
survey constituted 28.5%, in depressions and drainage trenches it increased to
35.20%.

The data on soil moisture over the studied areas of the coastal zone of Rybinsk
Reservoir are given in Table 3.3 (airborne microwave measurements were taken on
22–25 June 1989; $\lambda = 18.0$ cm).

Thus, a combined analysis of microwave sounding materials, data of aerophoto-
survey, ground geo-botanic and landscape studies together with information of
model experiments has made it possible to develop the fundamentals of landscape-
ecological diagnostics about processes of interaction within the inland water body–
watershed system. Possibilities of the microwave method to obtain quantitative
estimates about the ecohydrological condition of the shores of lakes and reservoirs
have been vividly demonstrated by these complex studies.

3.2.2   Active microwave remote sensing of lakes

3.2.2.1   *Physical models of microwave energy scattering by natural water and land surfaces*

The amount of energy returned to radar is strongly influenced by surface roughness:
smooth water surfaces tend to reflect microwave energy away from the satellite or
aircraft SAR (synthetic-aperture radar) instrument (Stewart, 1985). Land surface

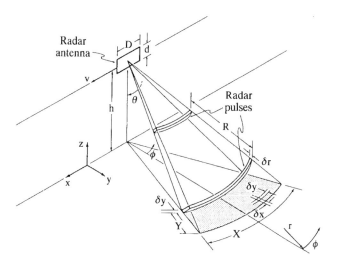

Fig. 3.48. Geometry of SAR observation of lakes (Stewart, 1985).

and ice cover tend to be rougher and reflect more energy back to the sensor. Radar backscatter is greater for slopes facing the radar than for slopes facing away from the sensor: this creates a "shaded relief" from which morphological information can be derived. SAR can distinguish textural differences created by the fresh-water surface (mesoscale features, currents, waves, and circulation), by fresh-water ice (development, extent, concentration), and by land topography (shallow water, wetlands, flooding, watershed) (Melentyev and Mel'nichenko, 1995). The amount of moisture in the soil or in (on) vegetation affects the amount of SAR backscatter which is represented as tonal variations of the signal.

According to Stewart (1985) the range resolution δy of SAR is determined by the length Δr of the transmitted radar pulse projected on the land, or lake surface:

$$\delta y = \Delta r(2 \sin \theta) = c\tau(2 \sin \theta) \tag{3.199}$$

where c is the velocity of light, τ the pulse duration, and θ the incident angle, measured from vertical. Geometry for describing SAR observation of lakes is demonstrated in Fig. 3.48.

The azimuth resolution depends on the precision with which the Doppler shift of the reflected signal can be determined. The radar transmits a train of coherent pulses, and then samples the reflected wave front at a series of points along the radar trajectory, the synthesized aperture. This results in the azimuthal resolution being independent of range and dependent only on size of the real antenna used by radar. In practice, λ, θ, r (range from satellite to the imaged area), v (satellite velocity), and Y (depth of area on surface illuminated by radar) are determined by the atmospheric windows available for space or airborne radar, the satellite's (aircraft's) orbit, the size of the scene to be viewed, and the processes to be observed at the lake surface. Together, these parameters determine the size of the antenna. For example, for

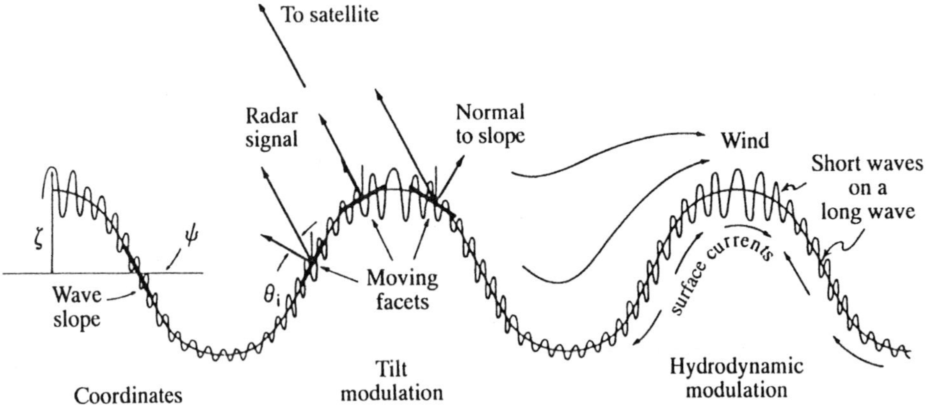

Fig. 3.49. Some processes influencing SAR observation of waves on the water surface (Stewart, 1985).

SeaSat $\lambda = 0.24$ m, $r = 850$ km, $\theta = 20°$, $c = 3 \times 10^8$ m/s, $Y = 100$ km. These yield $d < 2$ m, $D > 7$ m.

An important practical consideration in processing SAR images is the prodigious rate at which data are produced by such a radar, and the massive amount of data that must then be processed into images. SAR does not directly measure wave height but only the variation in reflectivity of the surface. Thus it is necessary to relate reflectivity to the properties of longer waves. Physical mechanisms have been investigate by many authors, and here the simplest form will be described. Figure 3.49 demonstrates some processes influencing SAR observations of waves on water surfaces. Large ocean waves (a) tilt the surface, (b) interact with short waves and distort the airflow over the surface (both processes modulate the height of the short waves that reflect the radar signal), and (c) large waves move the surface, thus distorting the SAR image of the surface and producing wavelike patterns in the image.

According to Stewart (1985), variations in reflectivity can be connected to the properties of long waves using the two-scale approximation to the water surface: with consideration of sea and lake surfaces covered by short waves of uniform height superposed on a long wave of the form:

$$\varsigma = A \cos(K_x x + K_y y - \Omega t) \qquad (3.200)$$

The long wave modulates reflectivity of the lake surface, and the radar cross-section can be written as:

$$\sigma_0 = \langle \sigma_0 \rangle + \delta \sigma_0 \qquad (3.201)$$

where $\langle \sigma_0 \rangle$ is the spatially-averaged cross-section per unit area of water surface and is the deviation induced by the long wave, A long-wave amplitude, x, y horizontal Cartesian coordinates, K_x, K_y components of long-wave number, Ω long-wave frequency, ς lake-surface elevation. Assuming radar reflectivity is linearly related to the properties of the long wave:

$$\delta\sigma_0 = \langle\sigma_0\rangle|R|A\cos(K_x x + K_y y - \Omega t + \Delta) \qquad (3.202)$$

$$\Delta = \tan^{-1}[\text{Im}(R)/\text{Re}(R)] \qquad (3.203)$$

where the complex quantity R is the modulation transfer function, Δ phase of R, $\delta\sigma_0$ deviation from $\langle\sigma_0\rangle$.

Variability in radar cross-section is usually attributed to three processes (Stewart, 1985): (a) surface tilting, (b) hydrodynamic interactions, (c) velocity bunching.

A glance at many images showed that the SAR did not always image waves when they were present on the surface. Because it is difficult to describe the influence of wave spectrum on SAR observations, especially of lakes, Ishimaru (1978) used theoretical and experimental study with controlling conditions of scattering by natural waters and ground surfaces. As demonstrated (Stewart, 1985), horizontally polarized radiation is more sensitive to wave slopes than is vertically polarized radiation; and range travelling waves (those with wave crests parallel to the radar velocity vector) produce a variation in reflectivity that is about 10 times larger than the variation produced by azimuth travelling waves (those with crests parallel to the radar beam).

Contrary to Stewart (1985) assumption, the water surface is not covered by short waves of uniform amplitude riding on long waves. Instead, the long waves modulate the amplitudes of short waves. Several processes contribute to this: long waves strain the surface, producing areas of convergence and divergence, and long waves distort the airflow over the surface, causing short waves to grow at various rates depending on the phase of the long wave and mean wind velocity. The influence of wave velocity on SAR image is shown in Fig. 3.50.

The radial component of the velocity of scattering areas on the water surface causes SAR to displace the position of these areas on the image of the surface. If the wave is not traveling in the range direction, then the periodic velocity field due to long waves can produce periodic bands of light and dark in the image, bands that resemble the image of a wave. This enables SAR to image waves travelling in the azimuth direction.

3.2.2.2 *Analysis of microwave signals scattered by surface roughness (radar signatures), results of retrieval of water and ice parameters*

Russia has a long history of development of methods and means of radar sounding: 1968 the launch of satellite carrying the 5-channel passive microwave complex, 1983 the SLR (side-looking radar)-carrying satellite "Okean", 1987 the SAR-carrying satellite "Almaz" with 12-m resolution, but has no system of operative distribution of satellite information for application study. At the present time "Okean" SLR imagery (radar wavelength $\lambda = 3$, 15 cm, resolution 1.5 km), ERS (Earth Research Satellite) and RADARSAT SAR data ($\lambda = 5$, 6 cm, resolution 20–30 m), Tupolev-134 NPO "MNIIP-Vega" (Moscow Research Institute of Technology-Vega) and Antonov-30 PINRO (Polar Research Institute of

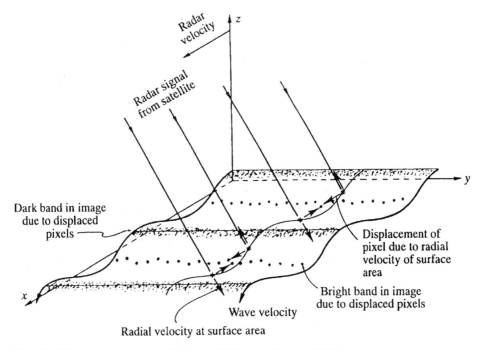

Fig. 3.50. The influence of wave velocity on SAR images (Stewart, 1985).

Marine Fisheries and Oceanography) airborne SAR instruments ($\lambda = 4$, 23, 70 and 250 cm, resolution 5–30 m) are the radar data currently used to support oceanography and limnology research. SAR measure the radar reflectivity of the underlying surface over areas 5 to 500 km with a resolution of 5–30 m. Sea-surface roughness, oil slicks, eddy currents, internal waves, wind under the water, sea-ice parameters can be studied by radar (Kondratyev *et al.*, 1996). Results of analyses of SAR data on lakes are more rare but they demonstrate that other water-surface phenomena as well are seen by SAR: internal waves (Naumenko *et al.*, 1992), eddies (Melentyev *et al.*, 1997), upwelling and downwelling (Filatov *et al.*, 1990), pollution (Melentyev *et al.*, 1998), fresh-water and sea ice of various type (Johannessen *et al.*, 1996). The results combine *in situ* and microwave remote-sensing studies of large lakes (both in the open-water season and wintertime conditions), and will be discussed more fully in Chapter 4.

The examples of thematic decoding of ERS-1&2 SAR images serve as the basis for analysis of radar signatures, for water surface parameter retrieval, for climatic and application study, for design and engineering activities on Arctic and sub-Arctic regions (Kondratyev *et al.*, 1995; Carsey, 1992). The visual distance and the operative all-weather character of SAR survey together with documented, thoroughly detailed natural phenomena testify to the prospects for the use of such imagery by geologists, b geophysicists and oil/gas-men.

REFERENCES

Adamenko V.N., Kondratyev K.Ya., Pozdnyakov D.V., and Chekhin L.P. 1991. *Radiative Regime and Optical Properties of Lakes*. Leningrad: Gidrometeoizdat., 300 pp. (in Russian)

Ahn Yu-Hwan, Bricaud A., Morel A. 1992. Light backscattering efficiency and related properties of some phytoplankters. *Deep-Sea Res.*, **39**, No. 11/12, 1835–1855.

Allali K., Annick Bricaud, Claustre Herve. 1997. Spatial variations in the chlorophyll- specific absorption coefficients of phytoplankton and photosymthetically active pigments in the equatorial Pascific. *J. Geophys. Res.*, **102**, No. C6, 12413–12423.

Atkinson P.M., Tatnall A.R.L. 1997. Neural networks in remote sensing. *Int. J. Remote Sensing*, **18**, No. 4. 699–709.

Austin R.W. 1980. Gulf of Mexico ocean-color surface-truth measurements. *Boundary-layer Meteorol.*, **18**, 269–285.

Babichenko S., Poryvkina L., Arikese V. 1993. Remote Sensing of Phytoplankton Using Laser-Induced Fluorescence. *Remote Sensing Environ.*, **45**, 43–50.

Baker K.S., Smith R.C. 1990. Irradiance transmittance through the air-water interface. In: *Ocean Optics X, Proc. SPIE Int. Soc. Opt. Eng.*, **1302**, 556–565.

Bannister T.T. 1990. Empirical equations relating scalar irradiance to a, b/a, and solar zenith angle. *Limnology, Oceanography*, **35**, 173–177.

Barton I.J. 1995. Satellite-derived sea surface temperatures: Current status. *J. Geophys. Res.*, **100**, 8777–8790.

Basharinov A.E., Gurvich A.S., Egorov S.T. 1974. *Radio-emission of the Earth as Planet*. Moscow: Nauka Press, 188 pp. (in Russian).

Berthon J.-F., Morel A. 1992. Validation of a spectral light-photosynthesis model and use of the model in conjunction with remotely sensed pigment observations. *Limnology, Oceanography*, **37**, No. 4, 781–796.

Bidigare R.R., Ondrusek M.E., Morrow J.H., Kiefer D.A. 1990. Bacterioplankton and solar light. In: *Ocean Optics X, Proc. SPIE Int. Soc. Opt. Eng.*, **1302**, 291–302.

Blasco D. 1975. Variations of the ratio in vivo-fluorescence/ chlorophyll-a and its application to oceanography. NASA TN F-16. Washington D.C.: NASA Pub., pp. 317.

Bogorodsky V.V., Gusev A.V., Khohlov G.P. 1971. *Freshwater Ice Physics*. Leningrad: Gidrometeoizdat., 227 pp. (in Russian).

Bogorodsky V.V., Oganesian A.G. 1987. *The Penetrating Radar Sounding of Sea Ice with the Digitally Processed Signals*. Leningrad: Gidrometeoizdat., 343 pp. (in Russian).

Bogorodsky, V.V., Kropotkin, M.A., Sheveleva, T.Yu. 1979a. Optical parameters of liquid water in IR, and influence of different factors. In: *Optical Remote Sensing of Oceans and Inland Water Bodies*. Novosibirsk: Nauka Press, pp. 227—234 (in Russian).

Bogorodsky, V.V., Martynova, E.A., Spitsyn, V. A. 1979b. IR emissivity of snow and ice cover. In: *Optical Remote Sensing of Oceans and Inland Water Bodies.* Novosibirsk: Nauka Press, pp. 235–241 (in Russian).

Bounkin A.F., Kasymova P.C., Mircamilov D.M. 1985. Investigation of specific features pertaining to fluorescence spectra of higher plants. In: *Problems of the Earth's Surface Lidar Sounding.* Tashkent: Nauka Press, pp. 72–89 (in Russian)

Bowers D.G., Harker G.E.L., Stepan B. 1996. Absorption spectra of inorganic particles in the Irish Sea and their relevance to remote sensing of chlorophyll. *Int. J. Remote Sensing,* **17**, No. 12, 2449–2460.

Bricaud A., Babin M., Morel A., Claustre H. 1995. Variability in the chlorophyll-specific absorption coefficients of natural phytoplankton: analysis and parameterization. *J. Geophys. Res.,* **100**, No. C7, pp. 13 321–13 332.

Bricaud A., Morel A. 1986. Light attenuation and scattering by phytoplanktonic cells: a theoretical modeling. *Appl. Opt.,* **25**, No. 4, pp. 571–580.

Bricaud A., Morel A., Prieur L. 1983. Optical efficiency factors of some phytoplankters. *Limnology, Oceanography,* **28**, No. 5, 816–832.

Bricaud A., Stamski D. 1990. Spectral absorption coefficients of living and nonalgal biogenous matter. *Limnology, Oceanography,* **35**, 562–585.

Bristow M., Nielsen D. 1981. Remote monitoring of organic carbon in surface waters. Las Vegas, Nevada: Pub. Environmental Monitoring Systems Laboratory, US Environmental Protection Agency, 35 pp.

Bristow M., Nielsen D., Bundy D., Furtek R. 1981. Use of water Raman emission to correct airborne laser fluorescensor data for effects of water optical attenuation. *Appl. Opt.,* **20**, No. 17, pp. 2889–2906.

Bristow M., Zimmermann R. 1989. Remote water quality monitoring with an airborne laser Fluorosensor. In: *Proceedings Seventh International Conference on Chemistry for Protection of the Environment, Catholic University of Lublin, Poland, 4–7 September 1989,* pp. 115–141.

Brooks A.S., Torke B.G. 1977. Vertical and seasonal distributions of chlorophyll a in Lake Michigan. *J. Fish. Res. Board Canada,* **34**, 2280–2287.

Browell E.V. 1977. Analysis of laser fluorosensor systems for remote algae detection and quantification. NASA Technical Note TN D-8447. Washington, DC: NASA Pub., 42 pp.

Brown C.A., Farmer F.H., Jarret O., Staton W.L. 1977. Laboratory studies of in vivo fluorescence of phytoplankton. *Proceedings Fourth Joint Conference on Sensing of Environmental Pollutants. New Orleans, Louisiana, 6–11 November 1977,* pp. 231–236.

Bukata R.P., Bruton J.E., Jerome J.H. 1983. Use of chromaticity in remote measurements of water quality. *Remote Sensing Environ.,* **13**, 161–177.

Bukata R.P., Jerome J.H., Bruton J. 1985. *Measurement of Water Quality Parameters.* CCIW Publ., 65 pp.

Bukata R.P., Jerome J.H., Bruton J.E. 1989. Determination of irradiation and primary production using a time-dependent attenuation coefficient. *J. Great Lakes Res.,* **15**, 327–338.

Bukata R.P., Jerome J.H., Bruton J.E., Jain S.C., and Zwick H.H. 1981. Optical water quality model of Lake Ontario. I. Determination of the optical cross sections of organic and inorganic particulates in Lake Ontario. *Appl. Opt.*,– **20**, 1696–1781.

Bukata R.P., Jerome J.H., Kondratyev K. Ya., Pozdnyakov D.V. 1991. Estimation of organic and inorganic matter in inland waters: optical cross sections of Lakes Ontario and Ladoga. *J. Great Lakes Res.*, **17**, 461–469.

Bukata R.P., Jerome J.H., Kondratyev K. Ya., Pozdnyakov D.V. 1995. *Optical Properties and Remote Sensing of Inland and Coastal Waters*. Boca Raton: CRC Press, 360 pp.

Burlamacchi P., Cecchi G., Mazzinghi P., Pantani L. 1983. Performance evaluation of UV sources for lidar fluorosensing of oil films. *Appl. Opt.*, **22**, No. 1, 48–53.

Campbell J.W., Esaias W.E. 1983. Basis for spectral curvature algorithms in remote sensing of chlorophyll. *Appl. Opt.*, **22**, 1084–1093.

Carsey F.D. (Ed.) 1992. *Microwave Remote Sensing of Sea Ice*. Washington, DC: American Geophysical Union (AGU) Monograph, 462 pp.

Chappell E.W., Wood F.M., McMurtrey L.E. 1984. Laser-induced fluorescence of green plants: I. A technique for the remote detection of plant stress and species differentiation. *Appl. Opt.*, **23**, No. 1, 134–138.

Chubarov V.C., Deydan T.A., Fadeev V.V., Petrov P.S. 1995. New remote laser method for measurement of oil slick thickness on water surface. In: *Proceedings Third Thematic Conference on Remote Sensing for Marine and Coastal Environments, Seattle, Washington, D.C., 18–20 September 1995*. Ann Arbor: ERIM Publ., pp. I-433–I- 439.

Clarke G.L., Ewing G.C., Lorenzen C.J. 1970. Spectra of backscattered light from the sea obtained from aircraft as a measurement of chlorophyll concentration, *Science*, **167**, 1119–1121.

Coble P..G., Brophy M.M. 1996. Investigation of the geochemistry of dissolved organic matter in coastal waters using optical properties. In: *Ocean Optics XII, Proc. SPIE Int. Soc. Opt. Eng.*, **494**, 56–61.

Coble P.G. 1994. Investigation of the geochemistry of dissolved organic matter in coastal waters using optical properties. In: *Ocean Optics XII, Proc. SPIE Int. Soc. Opt. Eng.*, **1302**, 291–302.

Coll C., Casellers V., Sobrino J.A., Valor E. 1994. On the atmospheric dependence of the split-window equation for land surface temperature. *Int. J. Remote Sensing*, **15**, 105-122.

Commission Internationale de l'Eclairage (CIE) 1957. *Vocabulaire International de l'Eclairage*. Paris: C.I.E. Pub. No. 1, 2nd edn., 136 pp.

Committee on Colorimetry. 1963. *The Science of Color*. Washington, DC: Optical Society of America, 6th edn., 385 pp.

Debye P. 1957. *Polar Molecules*. New York: Dover, 90 pp.

Debye P., Zuck G. 1936. *Theory of the Electric Properties of Molecules*. Moscow: ONTI Press, 144 pp. (in Russian).

Dedieu G., Deschamps P.Y., Kekk Y.H. 1987. Satellite estimation of solar irradiance at the surface of the Earth and of surface albedo using a physical model applied to Meteosat data. *J. Clim. Appl. Meteorol.*, **26**, 79-87.

Deschamps P.Y., Herman M., Tanré D. 1983. Modeling of the atmospheric effects and its application to the remote sensing of ocean color. *Appl. Opt.*, **22**, 3751–3758.

Deschamps P.Y., Phulpin T. 1980. Atmospheric correction of infrared measurements of sea surface temperature using channels at 3.7, 11, and 12 mcm. *Boundary Layer Meteorol.*, **18**, 131–143.

Doerffer R., Fisher J. 1994. Concentrations of chlorophyll, suspended matter, and gelbstoff in case II waters derived from satellite coastal zone color scanner data with inverse modeling methods. *J. Geophys. Res.*, **99**, No. C4, 7457–7466.

Doubovick O.V., Ostchepkov S.L., Lapenock T.V. 1994. Iterational/regularizing method of solution to nonlinear inverse problems and its application to interpreting water leaving radiance spectra. *Proc. Russian Acad. Sci.. Ocean Atmos. Phys.*, **30**, No. 1, 106–113 (in Russian).

Eder F.X. 1947. Das elektrishe verhalten von Erde (anommale dispersion und absorbtion. *Annalen der Physik*, **1**, No. 78, 6.

Emery W.J., Yu Y. Satellite sea surface temperature patterns. *Int. J. Remote Sensing*, **18**, 323-334.

Estep L. 1992. Estimations of bottom reflectance spectra. *Int. J. Remote Sensing*, **13**, 393–397.

Estep L. 1994. Bottom influence on the estimation of chlorophyll concentration in water from remotely sensed data. *Int. J. Remote Sensing*, **15**, No. 1, 205–214.

Evans S. 1965. Dielectric properties of ice and snow. *J. Glaciol.*,**2**, No. 42, 773–792.

Fadeev V.V., Chekaliuk A.M., Choubarov V.V. 1982. Nonlinear lidar fluorometry of compound organic matter. *Proceedings USSR Acad. Sci.*, **262**, No. 2, 338–342 (in Russian).

Fadeev V.V., Klyshko D.N., Roubin L.B. 1979. Analysis of water components making use of laser fluorometry and Raman scattering. In: *Optical Methods of Studying Oceans and Inland Waters*. Novosibirsk: Nauka Press, pp. 87–98 (in Russian).

Fee E.J. 1969. A numerical model for the estimation of photosynthetic production, integrated over time and depth in natural water. *Limnology, Oceanography*, **14**, 906–911.

Filatov N.N., Beletsky D.V., Zaytsev L.V. 1990. *Variety of Hydro-physical Fields of the Onega Lake*. Petrozavodsk: Publishing House of the Karelian Scientific Center of Russian Academy of Sciences, pp. 113 (in Russian).

Fisher J., Doerffer R. 1987. An inverse technique for remote detection of suspended matter, phytoplankton and yellow substance from CZCS measurements. *Adv. Space Res.*, **7**, No. 2, 21–26.

Fisher J., Heinemann T., Preusker R. 1997. Retrieval of aerosol properties from MERIS and SeaWiFS measurements. *J. Geophys. Res.*, **102**, 12 862–12 881.

Fisher J., Kronfeld U. 1990. Sun-stimulated chlorophyll fluorescence. 1: Influence of oceanic properties. *Int. J. Remote Sensing*, **11**, No. 12, 2125–2147.

Fraser R.S., Mattoo S., Yeh E.-N., McClain C.R. 1997. Algorithm for atmospheric and glint corrections of satellite measurements of ocean pigment. *J. Geophys.Res.*, **102**, 17 107–17 118.

Frehlich G. 1960. *Theory of Dielectrics*. Moscow: Foreign Liter. Publ. (in Russian).

Frouin R., Swindling M., Deschamps P.Y. 1996. Spectral reflectance of sea foam in the visible and near infrared: In situ measurements and remote sensing applications. *J. Geophys. Res.*, **101**, 14 361–14 371.

Gallegos S.C., Hawkins J.D., Cheng C.F. 1993. A new automated method of cloud masking for AVHRR full-resolution data over the ocean. *J. Geophys. Res.*, **98**, 8505–8516.

Gallie E.A., Murtha P.A. 1992. Specific absorption and backscattering spectra for suspended minerals and chlorophyll a in Chilko Lake, British Columbia. *Remote Sensing Environ.*, **39**, 103–118.

Gallie E.A., Murtha P.A. 1993. A modification of chromaticity analysis to separate the effects of water quality parameters. *Remote Sensing Environ.*, 44, 47–65.

Gao B.C., Kaufman Y.J. 1996. Selection of the 1.375 μm MODIS channel for remote sensing of cirrus clouds and stratospheric aerosols from space. *J. Atmos. Sci.*, **52**, 4231–4237.

Gautier C., Diak G., Masse S. 1980. A simple physical model to estimate incident solar radiation at the surface from GOES satellite data. *J. Appl. Meteorol.*, **19**, 1005–1012.

Ge Yu., Voss K.J., Gordon H.R.1995. In situ measurements of inelastic light scattering in Monterey Bay using solar Fraunhofer lines. *J. Geophys. Res.*, **100**, No. C7, 13 227–13 236.

Gershun A. 1936. On the photometry of turbid media. *Trudy Gos. Okeanogr. Inst.*, **11**, 99–125 (in Russian).

Gordon H.R. 1979. Diffuse reflectance of the ocean: the theory of its augmentation by chlorophyll a fluorescence at 685 nm. *Appl. Opt.*, **18**, No. 8, 1161–1166.

Gordon H.R. 1988. Ocean color remote sensing systems: Radiometric requirements. In: *Recent Advances in Sensors, Radiometry, and Data Processing for Remote Sensing. Proc. SPIE Int. Soc. Opt. Eng.*, **924**, 151–167.

Gordon H.R. 1989. Dependence of the diffuse reflectance of natural waters on the sun angle. *Limnology, Oceanography*, **34**, No 8, 1484–1489.

Gordon H.R. 1997. Atmospheric correction of ocean color imagery in the Earth Observing System era. *J. Geophys. Res.*, **102**, 17 081–17 106.

Gordon H.R., Brown O.B. 1974. Influence of bottom depth and albedo on the diffuse reflectance of a flat homogeneous ocean. *Appl. Opt.*, **13**, No. 9, 2153–2159.

Gordon H.R., Brown O.B., Evans R.H., Brown J.W., Smith R.C., Baker K.S., Clark D.K. 1988. A semianalytic radiance model of ocean color. *J. Geophys. Res.*, **93**, 10 909–10 924.

Gordon H.R., Brown O.B., Jacobs M.M. 1975. Computed relationships between the inherent and apparent optical properties of a flat homogeneous ocean. *Appl. Opt.*, **14**, No. 2, 417-427.

Gordon H.R., Clark D.K. 1981. Clear water radiances for atmospheric correction of coastal zone color scanner imagery. *Appl. Opt.*, **20**, 4175–4180.

Gordon H.R., Clark D.K., Mueller J.L., Hovis W.A. 1980. Phytoplankton pigments derived from the Nimbus-7 CZCS: Initial comparisons with surface measurements. *Science*, **210**, 63–66.

Gordon H.R., McCluney W.R. 1975. Estimation of the depth of sunlight penetration in the sea for remote sensing. *Appl. Opt.*, **14**, 413–416.

Gordon H.R., Morel A.Y. 1983. *Remote Assessment of Ocean Colour for Interpretation of Satellite Imagery*. New York: Springer-Verlag, 115 pp.

Gordon H.R., Smith R.C., Zaneveld J.R.V. 1984. Introduction to ocean optics. In: *Ocean Optics VII*, Blizard M.A. (ed.), *Proc. SPIE Int. Soc. Opt. Eng.*, **489**, 2–41.

Gordon H.R., Wang M. 1994. Retrieval of water-leaving radiance and aerosol optical thickness over the oceans with SeaWiFS: A preliminary algorithm, *Appl. Opt.*, **33**, 443–452.

Gower J.F.R., Borstad G.A. 1993. Use of imaging spectroscopy to map solar-stimulated chlorophyll fluorescence, red tides and submerged vegetation. In: *Proceedings 16th Canadian Symposium on Remote Sensing*. Ottawa: Canada Remote Sensing Society, pp. 95–98. Green S.A., Blough N.V. 1994. Optical absorption and fluorescence properties of chromophoric dissolved organic matter in natural waters. *Limnology, Oceanography*, **39**, No. 8, pp. 1903–1916.

Grew G.W. 1981. Real-time test of MOCS algorithm during Superflux 1980. In: Campbell J.W., Thomas J.P. *The Chesapeake Bay Plume Study: Superflux 1980*. Hampton, VA: Langley Research Center, pp. 121–128.

Hakvoort J.H.M. 1994. *Absorption of Light by Surface Water*. The Netherlands: Delft University Press, 145 p.

Harris A.R., Mason I.M. 1992. An extension to the split-window technique giving improved atmospheric correction and total water vapor. *Int. J. Remote Sensing*, **13**, 881–892.

Haves S.K., Carder K.L., Harvey G.R.1992. Quantum fluorescence efficiencies of fulvic and humic acids: Effects on ocean colour and fluorometric detection. In: *Ocean Optics XI*, *Proc. SPIE Int. Soc. Opt. Eng.*, **1750**, 212–223.

Hippel A.R. 1959. *Dielectrics and Their Application*. Moscow: Gosenergoizdat., 586 pp. (in Russian).

Hoepffner N., Sathyendranath S. 1993. Determination of major groups of phytoplankton pigments from the absorption spectra of total particulate matter. *J. Geophys. Res.*, **98**, No. C12, 22 789–22 803.

Hoge F., Swift R. 1983. Airborne dual laser excitation and mapping phytoplankton in a Gulf Stream warm core ring. *Appl. Opt.*, **22**, 2272–2281.

Hoge F.E., Berry R.E., Swift R.N. 1986. Active-passive ocean color measurement. I. Instrumentation. *Appl. Opt.*, **25**, No. 1, 39–47.

Hoge F.E., Lyon P.E. 1996. Satellite retrieval of inherent optical properties by linear matrix inversion of oceanic radiance models: an analysis of model and radiance measurement errors. *J. Geophys. Res.*, **101**, No. C7, 16 631–16 648.

Hoge F.E., Swift R.N. 1986. Chlorophyll pigment concentration using spectral curvature algorithms: an evaluation of present and proposed satellite ocean color sensor bands. *Appl. Opt.*, **25**, No. 20, 3677–3682.

Houghton W. *et al.* 1983. Field investigations of technique for remote laser sensing of oceanographic parameters. *Rem. Sens. Environ.*, **13**, 17–32.

IGBP. 1988. The International Geosphere-Biosphere Programme: A study of global change. IGBP Rep. No. 4. Paris: IGBP, 200 pp.

Ishimaru A. 1978. *Wave Propagation and Scattering in Random Media*, 2 Vols. New York: Academic Press, pp. 272–336.

Jadhav D.B. 1987. On the method of detecting phytoplankton pigments by the property of their fluorescence. In: *Proceedings IEEE International Geoscience and Remote Sensing Symposium IGARSS '87, University of Michigan, USA, 18–21 May 1987*, pp.1–3.

Jassby A.D., Platt T. 1976. Mathematical formulation of the relationship between photosynthesis and light for phytoplankton. *Limnology, Oceanography*, **21**, 540–547.

Jerlov N.G. 1976. *Marine Optics*, Elsevier Oceanography Series 14. Amsterdam: Elsevier, 321 pp.

Jerome J.H., Bukata R.P., Bruton J.E. 1988. Utilizing the components of vector irradiance to estimate the scalar irradiance in natural waters. *Appl. Opt.*, **27**, 4012–4018.

Jerome J.H., Bukata R.P., Bruton J.E. 1990. Determination of available subsurface light for photochemical and photobiological activity. *J. Great Lakes Res.*, **16**, No. 3, 436–443.

Jerome J.H., Bukata R.P., Miller J.R. 1996. Remote sensing reflectance and its relationship to optical properties of natural waters. *Int. J. Remote Sensing*, **17**, No. 16, pp. 3135–3155.

Jerome J.H., Bukata R.P., Whitfield P.H., Rousseau N. 1994. Colours of Natural Waters: 1. Factors controlling the dominant wavelength. *Northwest Science*, **68**, No. 1, 43–52.

JGOFS. 1988. Report of the first session of the SCOR committee for JGOFS. *Sci. Comm. Ocean. Res.* Stockholm: IGBP, 50 pp.

Johannessen O.M., Sandven S., Pettersson L.H., Melentyev V.V. 1996. Real-time ice monitoring of the inland navigable waterway: Finnish Gulf–Lake Ladoga–Lake Onega—White Sea using ERS-1 satellite radar images. In: *Proceedings of the International Conference on Inland and Maritime Navigation and Coastal Problems of East European Countries*. Gdansk: Technical University, Vol. 1, pp. 191–201.

Karnaoukhov B.N. 1988. *Spectral Analysis of Cells in Prospective of Ecology and Environment Protection (Cellular Biomonitoring)*. Poushchino, Russia: Scientific Center for Biological Studies, pp. 125 (in Russian).

Kaufman, Tanre D., Gordon H.R., Nakajima T., Lenoble J., Frouin R., Grassl H., Herman B.M., King M.D., Teillet P.M. 1997. Passive remote sensing of tropospheric aerosol and atmospheric correction for aerosol effect. *J. Geophys. Res.*, **102**, 16 815–16 830.

Key J. 1990. Cloud cover analysis with arctic Advanced Very High Resolution Radiometer data. *J. Geophys. Res.*, **95**, 7661-7675.

Kiefer D.A., Mitchel B. Greg. 1983. A simple, steady state description of phytoplankton growth based on absorption cross section and quantum efficiency. *Limnology, Oceanography*, **28**, No. 4, pp. 770–776.

Kim A.W., Weber S.H., Wilson M.C. 1977. Assessment of laser beam effective attenuation in marine waters. *Appl. Opt.*, **16**, 345–362.

Kirk J.T.O. 1983. *Light and Photosynthesis in Aquatic Ecosystems*. Melbourne, Australia: Cambridge University Press, 405 pp.

Kirk J.T.O. 1984. Dependence on relationship between inherent and apparent optical properties of water on solar altitude. *Limnology, Oceanography*, **29**, No. 2, 350–356.

Kirk J.T.O. 1986. *Light and Photosynthesis in Aquatic Ecosystems*. London: Cambridge University Press, 239 pp.

Kirk J.T.O. 1991. Volume scattering function, average cosines, and the underwater light field. *Limnology, Oceanography*, **36**, 455–467.

Kondratyev K.Ya. 1987. *Global Climate and its Changes*. Leningrad: Nauka Press, 213 pp.

Kondratyev K.Ya. 1992. *Global Climate*. St. Petersburg: Nauka Press, 359 pp. (in Russian).

Kondratyev K.Ya., Buznikov, A.A., Pozdnyakov D.V 1972. Optical remote sensing of water pollution and phitoplankton. *Water Resources*, **3**, 65–76 (in Russian).

Kondratyev K.Ya., Guitelson A., Chevko V. 1988. Fulvic and humic acids absorption in the visible spectrum. *Dokl. Akad. SSSR*, **293**, 837–841.

Kondratyev K.Ya., Johannessen O.M., Melentyev V.V. 1996. *High Altitude Climate and Remote Sensing*. Chichester, UK: Wiley/Praxis, 200 pp.

Kondratyev K.Ya., Melentyev V.V. 1991. Modern situation on the small rivers of USSR and ways for their using, saving and re-construction. Leningrad: Gidrometeoizdat. *Proceedings GGI*, **2**, 76 pp. (in Russian).

Kondratyev K.Ya., Melentyev V.V. Bobylev L.P., Johannessen O.M., Pettersson L.H., Sandven S. 1995. Real-time ice monitoring of the Russian Arctic with using ERS-1 SAR. In: *Proceedings of the International Conference on Russian Arctic Offshore (RAO-95)*. St. Petersburg: Technical University, Vol. 1, pp. 141–142.

Kondratyev K.Ya., Melentyev V.V., Kriulkov V.A. 1990a. Microwave remote sensing of natural-economic structures (watershed of River Sit' as example). *Water Resources*, **2**, 178–184 (in Russian).

Kondratyev K.Ya., Melentyev V.V., Mishev D.N. *et al.* 1990b. Comprehensive study of watershed, wetlands and shallow waters with using microwave (Rybinsk Reservoir as example). *Earth Obs. Remote Sensing*, **3**, 136–138 (in Russian).

Kondratyev K.Ya., Melentyev V.V., Nazarkin V.A. 1992. *Remote Sensing of Water Areas and Watersheds (Microwave Methods)*. St. Petersburg: Gidrometeoizdat., 248 pp. (in Russian).

Kondratyev K.Ya., Melentyev V.V., Rabinovich Yu.I., Shulgina E.M 1977. Remote sensing of water resources and soil moisture. *Meteorology and hydrology*, **6**, 78–89 (in Russian).

Kondratyev K.Ya., Moskalenko N.K., Pozdnyakov D.V. 1983. *Atmospheric Aerosol*. Leningrad: Gidrometeoizdat., 230 pp. (in Russian).

Kondratyev K.Ya., Pozdnyakov D.V. 1990. *Optical Properties of Natural Waters and Remote Sensing of Phytoplankton*. Leningrad: Nauka Press, pp. 190 (in Russian).

Kondratyev K.Ya., Pozdnyakov D.V. 1995.Some aspects of optical remote sensing applications in contemporary limnological studies. *Water Resources*, **21**, No. 5, 563–574.

Kondratyev K.Ya., Pozdnyakov D.V., Isakov V.Yu. 1990c. *Radiation and Hydro-optical Experiments on Lakes*. Leningrad: Nauka Press, 115 pp. (in Russian).

Kondratyev K.Ya., Pozdnyakov D.V., Pettersson L.H. 1998. Water quality remote sensing in the visible spectrum. *Int. J. Remote Sensing*, **19**, No. 5, 957–981.

Kopelevich O.V., Rodionov V.V., Stoupakova T.P. 1987. Effect of bacteria on optical characteristics of ocean water. *Oceanology*, **27**, No. 6, 921–926 (English translation).

Kropotkin, M.A., Buznikov, A.A. 1974. Evaluation of the IR contrasts of sea and fresh water for retrieving of water parameters. *Water Resources*, **6**, 98–03 (in Russian).

Lee Z.P., Carder K.L., Marra J., Steward R.G., Perry M.J. 1996a. Estimating primary production at depth from remote sensing. *Appl. Opt.*, **35**, No. 3, 463–474.

Lee Z.P., Carder K.L., Peacock T.G., Davis C.O., Mueller J. 1996b. Method to derive ocean absorption coefficients from remote sensing reflectance. *Appl. Opt.*, **35**, 453–462.

Maritorena S., Morel A., Gentili B. 1994. Diffuse reflectance of oceanic shallow waters: influence of water depth and bottom albedo. *Limnology, Oceanography*, **39**, No. 7, 1689–1703.

Marquardt D.W. 1963. An algorithm for least-squares estimation of non-linear parameters. *J. Int. Soc. Appl. Math.*, **11**, No. 2, 36–48.

Marshall B.R., Smith R.C. 1990. Raman scattering and in-water ocean optical properties. *Appl. Opt.*, **29**, No. 1, 71–84.

McClain C.R., Yen Eueng-nan. 1994. CSZC bio-optical algorithm comparison. Case studies for the SeaWiFS calibration and validation. NASA Technical Memo. 104566-13. Greenbelt, Maryland: NASA, 52 pp.

McClain E.P., Pichel W.G., Walton C.C. 1985. Comparative performance of AVHRR- based multichannel sea surface temperatures. *J. Geophys. Res.*, **90**, 11 587–11 601.

McMillin L.M. 1975. Estimation of sea surface temperatures from two infrared window measurements with different absorptions. *J. Geophys. Res.*, **80**, 5113–5117.

Melentyev V.V., Aleksandrov V.Yu. 1991. Model calculations of microwave emissivity of fresh-water ice and frozen soils. *Proc. AARI*, Issue 421, 138–146 (in Russian).

Melentyev V.V., Johannessen O.M., Kondratyev K.Ya., Tikhomirov A.I., Worontsov A.M., Bobylev L.P. 1998. ERS-1&2 SAR as space tracelogy tool. *Rep. Russian Acad. Sci.*, **6**, 117–124 (in Russian).

Melentyev V.V., Mel′nichenko I.G. 1995. Landscape and ecological microwave diagnostics of the system waterbody-watershed accommodation. *Waters Resources*,**22**, No. 1, 110–114 (in Russian).

Melentyev V.V., Tikhomirov A.I., Kondratyev K.Ya., Johannessen O.M., Sandven S., Pettersson L.H. 1997. Study of the freeze-up phase change on large temperate-zone inland water bodies and possibilities of their microwave diagnostics. *Earth Obs. Remote Sensing*, No. 3, 61–72 (in Russian).

Minnett P.J. 1990. The regional optimization of infrared measurements of sea surface temperature from space. *J. Geophys. Res.*, **95**, 13 497–13 510.

Mitchell B.G., Kiefer D.A. 1986. Chloprophyll-a specific absorption and fluorescence excitation spectra for light-limited phytoplankton. *Deep-Sea Res.*, **35**, No. 5, 639–663.

Mitchell B.G., Kiefer D.A. 1988.Variability in pigment specific particulate fluorescence and absorption spectra in the northeastern Pacific Ocean. *Deep-Sea Res.*, **35**, No. 5, 665–689.

Mittenzwey K.-H., Sinn G., Roof N., Harsdorf D. 1997. An improved lidar method for monitoring surface waters: experiments in the laboratory. *Int. J. Rem. Sensing*, **18**, No. 11, 2271–2276.

Moon P., Spencer D. 1942. Illumination from a non-uniform sky. *Illumination Engineering*, **37**, 717–726.

Moore K.D., Voss K.J., Gordon H.R. 1997. Diffuse reflectance in the open ocean and their contribution to water-leaving radiance. In: *Ocean Optics XIII, Proc. SPIE Int. Soc. Opt. Eng.*, in press.

Morel A. 1980. Optical properties of pure water and sea water. In: Jerlov N.G. and Steeman Nielsen (eds.), *Optical Aspects of Oceanography*. London: Academic Press, pp. 1–24.

Morel A. 1988. Optical modelling of the upper ocean in relation to its biogeneous matter content (case I waters). *J. Geophys. Res.*, **93**, No. C9, 10 749–10 768.

Morel A., Berthon J.-F. 1989. Surface pigments, algal biomass profiles, and potential production of the euphotic layer: Relationships reinvestigated in view of remote sensing applications. *Limnology, Oceanography*, **34**, No. 8 1545–1562.

Morel A., Gentilli B. 1993. Diffuse reflectance of oceanic waters. II: Bidirectional aspects. *Appl. Opt.*, **32**, 6864–6879.

Morel A., Gentini B. 1991. Diffuse reflectance of oceanic waters: its dependence on sun angle as influenced by the molecular scattering contribution. *Appl. Opt.*, **30**, No. 30, 4427–4438.

Morel A., Prieur L. 1977. Analysis of variations in ocean water. *Limnology, Oceanography*, **22**, 709–722.

Morel A., Voss K., Gentilli B. 1995. Bidirectional reflectance of oceanic waters: A comparison of modeled and measured upward radiance fields. *J. Geophys. Res.*, **100**, 13 143–13 150.

Myrmehl C., Morel A. 1996. Accounting for the marine bidirectionality when processing remotely sensed ocean colour data. In: Blizard M.A. (ed.), *Ocean Optics X, Proc. SPIE Int. Soc. Opt. Eng.*, **494**, 12–21.

Naumenko M.A., Etkin V.S., Litovchenko K.T. 1992. Investigation of radar signatures of lake surface with the "Kosmos -1870" ("Almaz-0") SAR. *Proceedings International Geoscience and Remote Sensing Symposium (IGARS '92)*. Houston, Vol. 2, pp. 1774–1776.

Nelder J.A., Mead R. 1965. A simplex method for function minimization. *Comput. J.*, **7**, 308–313.

Netushil A.V. 1975. Models of electric fields in heterogenic media of irregular structures. *Electricity*, **10**, 1–8 (in Russian).

Nieke B., Vincent W.F., Therriault J.C., Legendre L., Berton J.F., Condal A. 1997. Use of a ship-borne fluorosensor for remote sensing of chlorophyll-a in a coastal environment. *Rem. Sens. Environ.*, **60**, No. 2, 140–152.

Noever D.A., Matsos H.C., Cronise R.J. 1994. Ocean-atmosphere CO_2 exchange: an accessible lab simulation for considering biological effects. *Climatic Change*, **27**, 299–320.

Ocean Optics. Vol. 1. 1983. Physical ocean optics. Moscow: Nauka Press, 372 pp. (in Russian).

Ottle, Vidal-Madjar. 1992. Estimation of land surface temperature with NOAA-9 data. *Remote Sensing Environ.*, **40**, 27–41.

Owe M., Chang A., Colus R.E. 1988. Estimating surface soil moisture from satellite derived vegetation index. *Remote Sensing Environ.*, **24**, No. 2, 331–346.

Paramonov L.E. 1995. A theoretical analysis of algae absorption optical spectra. *Oceanology*, **35**, No. 5, 719–724 (in Russian).

Petrova N.A. 1990. *Phytoplankton Successions with the Anthropogenic Eutrophication of Large Lakes*. Leningrad: Nauka Press, 200 pp. (in Russian).

Petzold T.J. 1972. Volume scattering functions for selected ocean waters. San Diego: Scripps Institute of Oceanography. Ref. 72-28, 79 pp.

Philpot W. 1987. Radiative transfer in stratified waters: a single-scattering approximation for irradiance. *Appl. Opt.*, **26**, 4123–4132.

Platt T. 1969. The concept of energy efficiency in primary production. *Limnology, Oceanography*, **14**, 653–659.

Platt T. 1986. Primary production of the ocean water column as a function of surface light intensity: Algorithms for remote sensing. *Deep-Sea Res.*, **33**, 149–163.

Platt T., Caverhill C., Sathyendranath S. 1991. Basin-scale estimates of oceanic primary production by remote sensing: The North Atlantic. *J. Geophys. Res.*, **96**, 15 147–15 159.

Platt T., Gallegoes C.L., Harrison. 1980. Photoinhibition of photosynthesis in natural assemlages of marin plankton. *J. Marine Res.*, **38**, 687–701.

Platt T., Sathyendranath S. 1988. Oceanic primary production: estimation by remote sensing at local and regional scales. *Science*, **241**, 1613–1620.

Polder D., Van Santon J.H. 1946. The effective permeability of mixtures of solids. *Physica*, **12**, 257–271.

Poole L.R., Esaias W.E. 1982. Water Raman normalization of airborne fluorosensor measurements: a computer mode study. *Appl. Opt.*, **21**, No. 20, 3756–3761.

Poole L.R., Esaias W.E. 1983. Influence of suspended inorganic sediment on airborne fluorosensor measurements. *Appl. Opt.*, **22**, No. 3, 380–381.

Porto S. 1966. Angular dependence and depolarization ratio of the Raman effect. *J. Opt. Soc. Am.*, **56**, 1585–1589.

Pozdnyakov D.V., Kondratyev K.Ya. 1997a. Remote sensing of natural waters in the visible spectrum. III. Some special issues. *Studying the Earth from Space*, **3**, 3–18 (in Russian).

Pozdnyakov D.V., Kondratyev K.Ya. 1997b. Remote Sensing of Natural Waters in the Visible Spectrum. II. Ways of Reverse Problem Solution. *Earth Res. Space*, **2**, 1–27 (in Russian).

Pozdnyakov D.V., Kondratyev K.Ya., Bukata R.P., Jerome J.H. 1998a. Numerical modelling of natural water colour: implications for remote sensing and limnological studies. *Int. J. Remote Sensing*, **19**, No. 10, 1913–1932.

Pozdnyakov D.V., Kondratyev K.Ya., Pettersson L.H. 1998b. Remote sensing assessment of phytoplankton primary productivity: methodological aspects. *Earth Studies from Space*, **4**, 61–82.

Pozdnyakov D.V., Lyaskovsky A.V. 1999. Comparison analysis of water quality retrieval algorithms for case II waters. *Earch Obs. Rem. Sens.*, in press.

Prata A.J. 1993. Land surface temperatures derived from the AVHRR and ATSR. *J. Geophys. Res.*, **98**, 16 689–16 702.

Preisendorfer P.W. 1976. *Hydrologic Pptics*, Vol. 6. Springfield, VA: NTIS, 376 pp.

Preisendorfer R.W. 1961. Application of radiative transfer theory to light measurements in the sea. Paris: Monogr. Int. Union Geod. Geophys., Vol. 10, pp. 11–30.

Present State of the Ladoga Lake Ecosystem. 1987. Leningrad: Nauka Press, 210 pp. (in Russian).

Prieur L., Morel A. 1971. Etude théorique du régime asymptotic: Relations entre charactéristiques optiques et coefficient d'extinction relatif à la pénétration de la lumier du jour. *Cah. Oceanogr.*, **23**, 35–48.

Prieur L., Sathyendranath S. 1981. An optical classification of coastal and oceanic waters based on the specific spectral absorption curves of phytoplankton pigments, dissolved organic matter, and other particulate materials. *Limnology, Oceanography*, **26**, 671–689.

Rabinovich Yu.I, Melentyev V.V. 1970. The effect of temperature and salinity on the emission of smooth water surface in the centimeter wavelength region. *Proc. MGO*, Issue 235, 61–71 (in Russian).

Raiser V.Yu., Sharkov E.A., Etkin V.S. 1975. The effect of temperature and salinity on the emission of smooth water surface in the decimeter and meter wavelength regions. *Izv. AN SSSR, FAO*, **11**, No. 6, 652–656 (in Russian).

Report on the International Workshop on Continental Shelf Fluxes of Carbon, Nitrogen and Phosphorus. 1996 Hall J., Smith S.V., Boudreau P.R. (eds.), LOICZ/R&S/96-9. Texel: LOICZ, 50 pp.

Robinson I.S. 1994. *Satellite Oceanography*. Chichester: Praxis Publ. Ltd.

Roesler C.S., Perry M.J. 1995. In situ phytoplankton absorption, fluorescence emission, and particulate backscattering spectra determined from reflectance. *J. Geophys. Res.*, **100**, No. C7, 13 279–13 294.

Roesler C.S., Perry M.J., Carder K. 1989. Modelling in situ phytoplankton absorption from total absorption spectra in productive inland marine water. *Limnology, Oceanography*, **34**, 1510–1523.

Rusk, A.N., Williams, D., Querry, M.R. 1971. Optical constants of water in the infrared. *J. Opt. Soc. Am.*, **61**, No. 7, 895–903.

Sakshaug E., Slagstad D. 1991. Light and productivity of phytoplankton in polar marine ecosystems: a physiological view. *Polar Res.*, **10**, No.1, 69–85.

Sakshaug E., Slagstad D. 1992. Sea ice and wind: effects on primary productivity in the Barents Sea. *Atmosphere-Ocean*, **30**, No. 4, 579–591.

Sathyendranath S. 1995. Personal communication.

Sathyendranath S., Morel A. 1983. Light emerging from the sea—interpretation and uses in remote sensing. In: Cracknell A.P. (ed.), *Remote Sensing Applications in Marine Science and Technology*. London: D. Reidel, pp. 323–357.

Sathyendranath S., Platt T. 1997. Analytic model of ocean color. *Appl. Opt.*, **36**, No. 12, 2620–2629.

Saunders R.W., Kriebel K.T. 1988. An improved method for detecting clear sky and cloudy radiances from AVHRR data. *Int. J. Remote Sensing*, **9**, 123–150.

Schiller H., Doerffer R. 1999. Neural network for emulation of an inverse model—operational derivation of case-ii water properties from MERIS data. *Int. J. Remote Sensing*, in press.

SeaWiFS. 1987. Report of the joint EOSAT/NASA SeaWiFS working group. Washington: NASA/EOSC, 91 pp.

Shettle E.P., Fenn R.W. 1979. Models of the lower atmosphere and the effects of humidity variations on their optical properties, Report AFGL-TR-790214. Hanscom, MA: Air Force Goephys. Lab., Hanscom Air Force Base.

Shifrin K.S. 1983. *Introduction to the Optics of the Ocean*. Leningrad: Gidrometeoizdat., pp. 278 (in Russian).

Sid'ko F.Ya., Sherstyankin P.P., Aponasenko A.D 1984. Optical remote sensing of phytoplankton of the Baykal Lake. *Earth Obs. Remote Sensing*, **5**, 11–15 (in Russian).

Smith R.C., Baker K.S. 1981. Optical properties of the clearest natural waters (200—800 nm). *Appl. Opt.*, **20**, No. 2, 177–184.

SNNS, Stuttgart Neural Network Simulator. 1995. *User Manual*, Version 3.1. University of Stuttgart, Institute for parallel and distributed performance systems (anonymous ftp. Informatik.uni-stuttgart.de (129.69.211.2).

Sobrino, Li Z.L., Stoll M.P. 1993. Impact of the atmospheric transmittance and total water vapor content in the algorithms for estimating sea surface temperatures. *IEEE Trans. Geosci. Remote Sensing*, **31**, 946–952.

Sosik H.M., Mitchell B.G. 1995.Light absorption by phytoplankton, photosynthetic pigments and detritus in the California Current System. *Deep-Sea Res.*, **42**, No. 10, 1717–1728.

Spitzer D., Dirks R.W.J. 1987. Bottom influence on the reflectance of the sea. *Int. J. Remote Sensing.*, **8**, No. 3, 279–290.

Spitzer D., Wernand A.D. 1981. In situ measurements of absorption spectra in the sea. *Deap-Sea Res., Part A*, **28A**, 165–174.

Stavn R.H., Weiderman A.D. 1992. Raman scattering in ocean optics: quantitative assessment of internal radiant emission. *Apl. Opt.*, **31**, No. 9, 1294–1303.

Stewart R.H. 1985. *Methods of Satellite Oceanography*. Berkeley, CA: University of California Press, 360 pp.

Stramski D., Kiefer D.A. 1990. Optical properties of marine bacteria. *Proceedings of the Society of Photo. Opt. Instrum. Eng., Ocean Optics X*, 1302, 250–268.

Sturm B. 1983. Selected topics of coastal zone color scanner (CZCS) data evaluation. In: Cracknell A.P. (ed.), *Remote Sensing Applications in Marine Science and Technology*. London: D. Reidel, pp. 137–167.

Sturm B. 1993. CZCS processing algorithms. In: *Ocean Colour: Theory and Applications in a Decade of CZCS Experience*. Netherlands: ESA, pp. 95–116.

Sugihara S., Kishino M., Okami N. 1984. Contribution of Raman scattering to upward irradiance in the sea. *J. Oceanogr. Soc. Japan*, **40**, 397–403.

Talling J.F. 1957. The phytoplankton population as a compound photosynthetic system. *New Phytol.*, **56**, 287–295.

Tassan S. 1994. Local algorithms using SeaWiFS data for the retrieval of phytoplankton pigments, suspended sediment, and yellow substance in coastal waters. *Appl. Opt.*, **33**, No. 12, 2369–2378.

Tikhonov A.N., Arsonin V.Ya. 1979. *Methods of Solution to Incorrect Problems*. Moscow: Nauka Press, 180 pp. (in Russian).

Tuchkov L.T. 1968. *Natural Noise Emissions at Radio Frequencies*. Moscov: Soviet Radio, 161 pp. (in Russian).

Ulivieri C., Castronuovo M.M., Francioni R., Cardillo A. 1994. A split window algorithm for estimating land surface temperature from satellites. *Adv. Space Res.*, **14**, 59–65.

Ulloa O., Sathyendranath S. 1992. Light Scattering by marine heterotrophic bacteria. *J. Geophys. Res.*, **97**, No. C6, 9619–9629.

Viljanen M., Rumyantsev V., Slepukhina T., Simola H. 1996. Ecological state of Lake Ladoga. In: *Karelia and St. Petersburg*. Jyväskylä: Joensuu University Press, pp. 107–128.

Vodacek A., Green S.A., Blough N.V. 1994. An experimental model of the solar-stimulated fluorescence of chromophoric dissolved organic matter. *Limnology, Oceanography*, **39**, No. 1, pp. 1–11.

Vodacek A., Philpot W.D. 1987. Environmental effects of laser-induced fluorescence spectra of natural waters. *Remote Sens. Environ.*, **21**, No. 1, 83–95.

Vollenweider R.A. 1966. Calculation models for estimating day rates phytoplankton photosynthetic primary production. In: Goldman C. (ed.), *Primary Productivity in Aquatic Environments*. Berkeley: University California Press, pp. 425–457.

Walton C.C. 1988. Nonlinear multichannel algorithms for estimating sea surface temperature with AVHRR satellite data. *J. Appl. Meteorol.*, **27**, 115–124.

Warren S.C. 1984. Optical constants of ice from the ultraviolet to the microwave. *Appl. Optics*, **23**, No. 8, 1206–1225.

Waters K.J. 1995. Effects of Raman scattering on water-leaving radiance. *J. Geophys. Res.*, **100**, No. C7, 13 151–13 161.

Weaver E.C., Wrigley R. 1994. Factors affecting the identification of phytoplankton groups by means of remote sensing. NASA Technical Memo. 108799. Moffet Field. CA: NASA, 117 pp.

Webb W.L., Newton M., Starr D. 1974. Carbon dioxide exchange of Alnus rubra: A mathematical model. *Oecologia*, **17**, 281–291.

Weiner O. 1910. Zur theorie der refraktionkonstanten. *Berichts über die Verhandlungen der königlich sächsischen Gesellschaft der Wissenschaften zu Leipzig-Mathematischphysikalische Klassen*, **62**, No. 5, 256–268.

Welch R.M., Sengupta S.K., Chen D.W. 1988. Cloud field classification based upon high spatial resolution textural features. Grey level co-occurrence matrix approach. *J. Geophys. Res.*, **93**, 12 663–12 681.

Whitlock C.H., Poole L.R., Ursy J.W. 1981. Comparison of reflectance with backscatter for turbid waters. *Appl. Opt.*, **20**, 517–522.

Whitte W.G., Whitlock C.H., Harriss R.C. 1982. Influence of dissolved organic materials on turbid water optical properties and remote sensing reflectance. *J. Geophys. Res.*, **87**, 441–446.

Yacobi Y.Z., Gitelson A., Mayo M. 1995. Remote sensing of chlorophyll in Lake Kinneret using high-spectral-resolution radiometer and Landsat TM: spectral features of reflectance and algorithm development. *J. Plankton Res.*, **17**, No. 11, 2155–2173.

Yu Y., Barton I.J. 1994. A nonregression-coefficients method of sea surface temperature retrieval from space. *Int. J. Remote Sensing*, **15**, 1189–1206.

Yu Y., Emery W.J. 1996. Satellite derived sea surface temperature variability in the Western Tropical Pacific Ocean, 1992–1993. *Remote Sens. Environ.*, **58**, 299–310.

Zaneveld J.R., Kitchen J.C. 1993. Vertical structure of productivity and its vertical integration as derived from remotely sensed observations. *Limnology, Oceanography*, **38**, No. 7, 1384–1393.

Zaneveld J.R.V. 1995. A theoretical derivation of the dependence of the remotely sensed reflectance of the ocean on the inherent optical properties. *J. Geophys. Res.*, **10**, No. C7, 13 135–13 142.

Zepp R.G., Schlotzhauer P.F. 1981. Comparison of photochemical behavior of various humic substances in water. III. Spectroscopic properties of humic substances. *Chemosphere*, **10**, 479–486.

4

Combined *in situ* and remote sensing of inland water bodies of the moderate climatic zone

4.1 HYDRODYNAMICS OF LARGE DEEP LAKES: RESULTS OF THE COMPLEX STUDIES

4.1.1 Remote studies of major thermohydrodynamic processes and events in lakes, estuaries and coastal zones

Investigations of lake thermohydrodynamic processes by remote sensing relies on measurements of indirect indicators residing either *on the surface* or *in the upper layers* of basins. As discussed in the previous chapter, remote sensing can be effected both in passive and active modes using a very wide range of wavelengths in the visible, infrared and microwave spectral regions.

One of the examples of successful utilization of purely surficial features as indicators of lake thermohydrodynamic phenomena dates back to 1972 when Bukata and McColl exploited mirror-reflected solar illumination (sun-glint) patterns recorded by aerial photography over Lake Ontario. It has generally been considered that sun-glint areas should be discarded from analyses since sun-glint completely obscures such subsurface features as lake-bed topography, and spatial distributions in submerged aquatic vegetation and suspended minerals. However, closer evaluation of sun-glint on a water body surface reveals a distinct, reproducible pattern across the sun-glint area. The authors not unduly conjectured that under suitable conditions reflections from a smooth air–water interface may be readily delineated from reflections from a more wind- roughened surface. The occurrence of smooth areas can be prompted by surfactants of various kinds which are known to inhibit development of wind-induced roughness. The authors found that the intensity of reflected radiation is maximum in the center of the sun-glint area, and a higher energy return within this maximum arises from a smooth rather than a wind-roughened surface, whereas at the outskirts of the sun-glint, a smooth surface displays a lower energy return (i.e. appears darker) than a diffuse surface. Consequently, at some intermediate angle of solar zenith, a reversal in relative intensities originating from smooth/wind-roughened tonal signatures should be observable. Comparing mosaics of the northern shore of Lake Ontario obtained under conditions of sun-glint with relevant thermal scans of the same region, Bukata and McColl assigned the revealed sun-glint areas (which proved to be restricted to the coastal zone) to thermal upwelling emerged along the shoreline.

A careful evaluation of sun-glint areas in photography over the Niagara plume, river mouth effluents, coastal sediment transport, etc., further substantiated the surmise that sun-glint could be used as a tool to identify and delineate dynamic features if, indeed, the source of such features is thermal in nature. An important requirement for sun-glint usability appears to be that spatial variations in the thermal time-gradient themselves change in time; i.e. there must be a change in slope of the plot dT/dt against distance. Such requirements could be met during the onset and decay phases of upwelling events, and may or may not be fulfilled in long-lived phenomena once the upwelling is established in time.

Later, Bukata et al. (1975) used ERTS-1 Band 5 (600–700 nm) MSS digital data to assess the water transport in the Point Pelee and Rondeau areas (Lake Erie). The higher the reflectance values in this channel, the more turbid the water mass. Spatial/temporal distributions of captured signal with the wavelengths specified above revealed several important features in the patterns of sediment transport in Lake Erie: the presence of suspended sediment load which is considerably more pronounced on the eastern shoreline of Point Pelee than on the western shoreline, a circular vortex structure located about 2.6 km off the Point and ~1.1 km in radius, a series of Archimedes spiral flow lines encompassing the vortex core. Based on spaceborne data, and considering prevailing bottom and surface current patterns characteristic of this lake, the authors reconstructed the mechanisms of geophysical accretion to the Point Pelee and Rondeau areas, and determined paths of bifurcated sediment transport to both locations. Interestingly, because the vortex set up in close vicinity to Point Pelee (as mentioned above), propagation of the Point Pelee tip into the lake follows a spiral pattern, providing a visual endorsement of transport mechanisms conceived from ERTS-1 data.

The idea of using remotely captured radiance at wavelengths in the red part of the spectrum as a (non-linear) function of suspended matter concentration for studying hydrodynamic features in inland waters was further developed by Bukata et al. (1988) and virtually shared by Fisher et al. (1991). Fisher et al. conducted their field experiment in the upper Elbe estuary. The area comprised a section of the Elbe River with a depth of 10–15 m and the mouth of a shallow side arm with a depth of about 2–3 m. The analyzed images revealed within the shallow region several areas with high concentrations of suspended matter separated by sharp fronts moving into the middle of the main stream. This finding proved to be of particular interest, since it suggests that the suspended matter front can move against the ebb current.

Vernal thermal fronts are common features in large northern temperate lakes (see Chapter 2): through a combination of solar warming and spring runoff, fronts form in coastal regions separating the thermally stratified inshore region where water-surface temperatures are in excess of 4°C from the isothermal or inversely stratified offshore region where water surface temperatures are under 4°C. Fronts move offshore until the 4°C isotherm disappears across the lake. So, water surface temperature is a meaningful parameter for delineating the temporal and spatial dynamics of thermal fronts. Thermal-front remote-sensing issues will be discussed at greater length in § 3.2 whereas in this section we are going to invoke this subject in relation to accompanying variations in hydro-optical fields.

Sherstyankin (1993) analyzed optical field patterns in conjunction with front dynamics in Lake Baikal and found that spatial distributions and temporal variations in the *water attenuation coefficient* could be closely related to the processes of frontogenesis. This work has been further substantiated by Bolgrien *et al.* (1995). It was demonstrated that using NOAA AVHRR, and *in situ* temperature and chlorophyll fluorescence data, features such as ice cover, thermal fronts, and the dispersion of river water could be retrievable from a series of space images. Based on analyses of AVHRR and Space Shuttle data, these workers have created a complete schema of the seasonal water-temperature cycle in Lake Baikal.

Analyzing sequential images from AVHRR band 2 and band 4 as well as photographs from Space Shuttle, Bolgrien *et al.* (1995) managed to identify areas covered with different types of ice and trace the progress in ice-out in different parts of the lake. In summer, retrieved thermal field distributions throughout the lake could be confidently related to areas of surface nutrient depletion and phytoplankton productivity. The AVHRR image analyses performed in conjunction with shipborne transects revealed upwelling events which resulted both from the upward displacement of the thermocline by internal waves and the wind-driven horizontal displacement of the epilimnion. The short Siberian summer was evident in the decrease in water surface temperature from August to September. Detected from space, thermal stratification was found to last from early July until early November. In a similar way to vernal warming, autumnal cooling as deduced from AVHRR band 4 appears to commence nearshore and progress offshore. The upper layer of the lake becomes isothermal in November and begins to freeze in January.

The construction of a flood-protection dike in the Neva River Bay in the immediate vicinity of St. Petersburg, Russia has given rise to serious concern about the ecological status of this estuarine water body. Many environmentalists ascribed the obvious and well-documented worsening of the public-health situation in the bay over the last few years to an adverse influence produced by the dike on the bay-water chemism and the indigenous biota *status quo*. Cutting the bay off from the Gulf of Finland, the dike was believed to seriously affect natural hydrodynamic processes in this enclosed water body, and thus bring about dramatic alterations to the ecology of this aquatic environment.

Victorov (1996) analyzed multispectral images of the Neva River Bay (the eastern margins of the Gulf of Finland) from the Russian satellite Meteor-30, as well as from MSU-SK (with average ground resolution of $175 \times 200 \, \text{m}$) and MSU-E (with average ground resolution of $30 \times 45 \, \text{m}$). A limited number of satellite photographic images were also used in his study. Some new features have been revealed: previously undetected inflows into the bay (which are believed to be small rivulets forming the Neva River delta). Patterns of suspended matter spreading across the bay have been established from qualitative analyses of satellite data and later on quantitative analyses. Six zones of different water transparency levels and associated suspended minerals loading have been evidenced throughout the bay: water transparency ranged from less than $1\,m^{-1}$ down to less than $0.2\,m^{-1}$ which corresponded to concentrations of suspended minerals less than $10\,mg/l$ and greater than $60\,mg/l$.

Satellite photographic images have been utilized to assess the influence of the flood- prevention dike on the Neva River Bay environment. Analyzing the characteristic patterns of spatial distributions of suspended matter as retrieved from spaceborne photographs taken before (about 1984) and during dike construction, Victorov has revealed serious dike-driven changes to the bay's hydrodynamics. A fine structure of suspended matter along the northern coastline of the bay in the vicinity of the dike exhibited remarkable patterns of eddies in the form of "mushrooms" which were not characteristic of bay-water movements prior to dike construction. Dike construction resulted in pronounced changes to water exchange processes between the bay and the Gulf of Finland: the coastal jet-streams reaching the dike are forced to turn back. It leads to formation of large-scale whirls and deposition of suspended matter within the bay area, *viz.* in front of the dike and along the bay coastline. A combined analysis of high spatial-resolution satellite images and relevant *in situ* data enabled the author to reveal and interpret previously unknown features occurring at frontal zones separating different water masses in the bay and in the eastern part of the Gulf of Finland. MSU-E images were also analyzed to derive water-level oscillations in the Neva River delta as well in the Neva River Bay for detecting outflow currents in the flood-protection dike's northern passage.

Recently, Oppenheimer (1997) reported on successful experience of utilizing the Landsat Thematic Mapper (TM) facility to study thermohydrodynamics of crater lakes. Although crater lakes *per se* could hardly be considered as objectives of traditional limnological studies, the concepts laid down at the basis of the author's research are arguably applicable as well to common inland water bodies. The thermodynamics of crater lakes are governed by various heat and mass inflows and outflows. Heat is lost by evaporative, sensible heat, and radiative fluxes from the surface, and by seepage and overflow. When there is a net heat loss through the water surface, the temperature of the surface skin (T_s) is lower, by a few degrees, or fractions of a degree, than the bulk water temperature (T_b). As discussed earlier in Chapter 3, the temperature difference (ΔT_{b-s}) is a function of net heat transfer (q_{net}) and skin thickness which depends on wind speed (u), drag coefficient (C_d), and kinematic viscosity (v_w) and thermal diffusivity (k) of the water. The author investigated through concurrent infrared sensing and *in situ* water temperature measurements the fluctuations of ΔT_{b-s} under variable meteorological conditions over a crater lake in Indonesia. It was found in particular that ΔT_{b-s} varies inversely with air temperature and nearly linearly with wind speed (from *ca* $1.1°C$ at $u < 1\,m/s$ to *ca.* $2.3°C$ at $u = 7\,m/s$). Based on these findings, it was shown that TM band 6 can be used for measuring and mapping water-surface temperatures which further can be employed for remotely assessing the heat fluxes at the air–water surface interface. However, it should be taken into account that heat budget estimates require additional assumptions or measurements of humidity, wind speed, solar flux.

Oppenheimer (1996) described two methods for using Landsat TM band 6 data to achieve this goal.

The other reason why this work by Oppenheimer is discussed in the present section resides in the fact that Oppenheimer discovered a distinct correlation between thermal and water colour fields in crater lakes. In § 3.1.1.3 we discussed the potentials of water colour characteristics as a tool to infer characteristic features of ecology dynamics in natural water bodies.

Earlier, Panin, Karetnikov (1991) have also studied energy exchange at the air–water interface in such northern temperate lakes as Ladoga and Onega during the periods of natural warming and cooling. Their data suggest that under conditions of microconvection the difference ΔT_{b-s} generally does not exceed 0.6–0.8°C, and is conducive to establishing warm or cold water surface skins. A stable density stratification in the near-surface atmospheric layer can entail much higher values of ΔT_{b-s} (3–5°C). The formation of cold surface skins invariably results in drastic changes in albedo of such areas which can easily be detected by optical remote sensors.

Taking into account the above findings, Kondratyev *et al.* (1996a) used atmosphere-corrected NOAA IR data to obtain adequate water-surface temperatures for mapping 2-D spatial distributions of some salient thermohydrodynamic features throughout Lake Ladoga. Numerical modelling results proved to be indicative of existence during stratification periods on time scales of 2 to 5 days (depending on concrete meteorological conditions) of a reliable correlation ($r = 0.7$–0.9) between the water-surface temperature (WST) and bulk water (within a 20–30 m layer) temperature (Filatov *et al.*, 1990). It makes it possible to interpret more confidently observed heterogeneous features of WST distribution in large lakes. Based on this assumption and employing *in situ* and NOAA data, Kondratyev *et al.* (1996a) have investigated Lake Ladoga regions with high thermal contrasts, especially frontal and upwelling zones.

The temporal/spatial evolution of the frontal zone in spring is of considerable ecological significance: while being close to the shore line, near-shore waters are pretty isolated from pelagic waters and may develop high levels of pollution due to intense inflow and runoff supply. Interestingly, the characteristic frontal zone width varies from 1 to 6 km depending on horizontal water-temperature gradients and is comparable with the Rossby baroclinic radius of deformation (Naumenko *et al.*, 1991; Naumenko, Karetnikov, 1993).

Investigating spatial/seasonal variations in Lake Ladoga's water turbidity, Roumyantsev (1987) has shown that frontal zones are characterized not only by drastic changes in water temperature and movements but also by pronounced horizontal gradients in water transparency (turbidity). During the period of vernal warming, turbid waters are confined to the littoral zone which is generally influenced by river discharge and runoff and separated from the rest of the water body by the frontal zone. The pelagic zone at this time of year remains rather optically homogeneous. In summer, as the thermal front moves offshore, the zone with spatially uniform hydro-optical properties gradually shrinks and retreats to the deep, central part of the lake until it finally submerges beneath the water surface.

These frontogenesis-driven spatial/temporal changes in water turbidity are bound to affect the upwelling radiance providing a clear indication of thermal front development.

The formation and development of thermal fronts in a stratified deep lake affect many hydrological processes and are at the basis of such a phenomenon as generation of internal waves. Bukata *et al.* (1978) were amongst the first who successfully utilized the Landsat-1 band 4 (0.5–0.6 µm) radiance data collected over Lake Superior. The spaceborne image showed a lake-wide pattern of regularly spaced alternating rows of high and low energy return from the lake. The apparent wavelength of the pattern (11–15 km) together with simultaneous observations of the thermal structure of the lake suggested that remotely detected thermoclinic vertical displacements (occurring in conjunction with upwelling along the north shore) were the result of the long-crested internal standing-wave pattern which is the 19th nodal transverse internal seiche and has a period very close to that of the semi-diurnal tidal-producing force. Time series concurrent measurements seemed to corroborate this assignment. Interestingly, the analyzed Landsat-1 images in bands 6 (0.7–0.8 µm) and 7 (0.8–1.1 µm) suggested that near-surface concentrations of chlorophyll were aligned with the standing-wave pattern: possibly it was a result of transporting phytoplankton to the surface by vertical oscillations occurring in the mixed layer. It implies that internal wave activity can perhaps be detected as well by analyzing remotely sensed patterns of chlorophyll spatial distributions. In § 4.3.1, we will further discuss the feasibility of remotely studying phytoplankton chlorophyll spatial/temporal distributions for investigating the many phenomena related to lake dynamics.

Now, returning to "non-chlorophyllous" indicators of water movements, it should be emphasized that it is not only optical wavelengths that have proved to be efficient in lake thermohydrodynamic studies. Kondratyev *et al.* (1992) have explored *microwave* spectral-region feasibility in studying lacustrine internal-wave events.

4.1.2 Surface roughness inhomogeneity as indicator of water mass dynamics: spatial variations in areas with subdued roughness driven by water density

The presently observed substantial variations in the environmental condition have been determined by both the effect of various external factors and internal variability in climatic systems themselves. The study of distribution of water masses in anthropogenically affected inland water bodies, revealing climatically significant natural processes, and their correct separation from external forces are the most important problems of limnology today. This problem is complicated: even well-developed climate theory cannot reliably distinguish between natural and anthropogenic factors responsible for global climate warming taking place during the last century. In K.Ya. Kondratyev's opinion (Kondratyev, 1987) the impact of external, anthropogenic factors has been considerably overestimated, whereas the scale of possible manifestations of natural factors has been played down.

The development of indicators of environmental dynamics are of great importance here. Characteristics of the global cycles of substances and energy (water, carbon, nitrogen, sulphur, etc.) can serve as examples of indicators of the condition and dynamics of the global environment and natural resources.

The condition of lacustrine thermal regimes is an integral indicator of thermics and watering of large territories up to the size of continents. The time of onset and duration of hydrological seasons in lakes depend not only on morphological and morphometric features of composite lake hollows (configuration of the bowl, volume and area, their average depth), they are considerably determined by features of the meteorological regime, by regional climatic conditions. The thermo-dynamic and ice regime of inland water bodies, due to their sensitivity to variable climate and weather conditions, is a particularly important climatic parameter, since it is the spatial and temporal variability in lake water-temperature that serves as an indicator of regional distribution of the heat and moisture regime in the Earth–atmosphere system.

The formation of water masses in deep fresh-water lakes in the moderate climatic zone is mainly connected with accumulation and expenditure of heat, water body morphology and determined by physical properties of water. The spatial variations in SST and in the condition of the water–atmosphere interface (including variations in its roughness due to external forcing of natural and anthropogenic factors, such as variations in near-water wind speed, pollution, etc.), changes in water quality, ice condition and structure, leading to additional changes in water surface roughness, dielectric constants of water and ice as well as characteristics of reflection and scattering of the sensing signal, open up opportunities to develop techniques for remote diagnostics of inland water bodies in the visible, IR and microwave spectral regions.

Particularly sharp is the contrast between water surface roughness and water mass condition in the region of the thermal bar (TB), characteristic of the second phase of springtime heating (Fig. 4.1) and autumn cooling of inland water bodies and dividing the lake into thermally active and thermally inert (cold) areas (TAA, TIA).

In the period of summer heating and during the first phase of autumn cooling the lake water mass is vertically divided by a temperature-leap layer (metalimnion) into epilimnion and hypolimnion.

The thermal bar as a frontal interface of large deep lakes of the moderate climatic zone differs from those of the ocean (Fedorov, Kuz′mina, 1977). It appears when water masses pass the temperature of their maximum density (4°C) as a result of non- uniform warming of water thickness in spring and cooling in autumn. The springtime TB shows itself as a frontal zone, where water temperature in the lake, due to convergence of warm coastal water and cool water in the pelagic area, reaches 4°C. The intrusion and densening of mixed waters lead to their vertical circulation, creating the TB front bordering the lake coastline. During maturation the TB front gradually shifts towards the growing depth. The speed of shift turns out to be directly proportional to the heat flux reaching the water and inversely proportional to the lake bottom slope. The location of the TB front fixed at the moment of thermal survey marks the location of similar depths in the lake: the TB is located farther from the water-shore line in regions with a small slope and closer to it in regions with a steep bottom (Tikhomirov, 1982). The TB is a tracer for state and

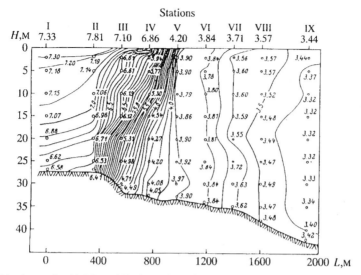

Fig. 4.1. This shows the 30 May 1959 thermal section of the south-western part of Lake Ladoga.

variability of water masses, their evolution and dynamics, and the possibility of fixing the TB position with the help of spectral remote sensors can serve as the basis for developing techniques for spectral diagnostics of inland water bodies. Using visual, infrared or microwave sensors that measure the differences in temperature, the difference in water–air interface parameters (roughness), or the difference in dielectric constants allows the water parameters in the regions of TAA, TIA, and TB to be controlled.

Figure 4.1 shows the 30 May 1959 thermal section of the south-western part of Lake Ladoga. Measurements were taken after a strong storm with prevailing north-easterlies; however, at the moment of survey there was clear sunny weather with wind speed 2.7 m/s and less, the TB line was marked by a foam strip stretching from north-east to south-east (Fig. 4.2).

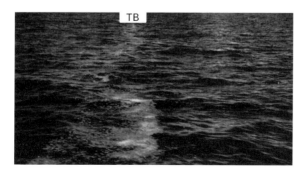

Fig. 4.2. Thermal bar on Lake Ladoga and its manifestation on the water surface, 30 May 1959; RV photographic survey.

The possibility to fix the location of the TB thin structure in the lake and associated processes and phenomena from thermal survey data in the IR, where the distribution of radiative temperature over Lake Ladoga's surface is clearly seen in the region where TB waters mix. The airborne survey with IR thermovisor "Vulkan", 15 June 1985 shows dark tone of signal corresponding to cold water, bright tone to warm water. According to *in situ* data of RV, TIA and TAA water masses have different temperature and mineralization (conductivity), and the diameter of the rings decreases the further they get from the TB frontal zone to the TIA.

Let us briefly discuss the problem of the existence of TB in the salty waters of sea gulfs and bays connected to runoff from large rivers in the north-west of Russia (Neva, Severnaya Dvina, etc.) as well as with runoff from the Great Siberian rivers (the seas of the arctic shelf). For many years this problem was discussed by the scientific community in Russia, because an instrumental fixation of this phenomenon would make it possible to specify ideas about spatial distribution and temporal variability in the water masses of these water bodies. Unfortunately, there are no published data confirming the presence of TB in sea waters subject to intensive systematic or periodic freshening with fresh water; we can only quote information from a member of staff of the Oceanographic Institute (Vinogradov, 1987) about detection of the TB front in the Gulf of Finland in the Baltic Sea. For this purpose, paramount importance is given to data of synchronous photographic and radar survey of the location of the autumn TB in the Yenisei Gulf of the Kara Sea. The 26 September 1997 survey from the nuclear icebreaker "Yamal" registered a frontal zone pinpointing the 10-m isobath in the region of Dikson Bay. The frontal zone differed in its special stability and ability to recover from multiple criss-crossing by the escorting motor boat breaking the local condition of the slick, but in 3–4 minutes it recovered after the disturbance (Fig. 4.3).

The contrast in TB zone revealed from data of spectral remote surveys opens up opportunities to study distribution of water masses in lakes and features of their

Fig. 4.3. Thermal bar on brackish water, 73°37.080′N/80°021.737′E, Yenisei Gulf, 26 September 1997; RV photographic survey, wind velocity 6 m/s, direction 260°.

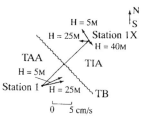

Fig. 4.4. Thermal bar on Lake Ladoga, and currents in the region of TB, horizons 5, 25, and 40 m, 30 May 1959; RV *in situ* data.

thermal regimes, and also to study circulation processes in the lake. So, for example, an idea about direction and speed of dense currents connected with the existence of TB in large inland water bodies can be obtained from the 30 May 1959 survey of Lake Ladoga (Figures 4.1 and 4.4). Figure 4.4 presents the results of measurements of currents at depths 5, 25, and 40 m by deep-water spinner at a distance of 1 km from TB in the regions of TAA and TIA.

Remote *in situ* monitoring of the SST on large lakes and reservoirs is the best means of studying water bodies: knowledge of water-body surface temperature enables one to judge processes at the water–atmosphere interface, the distribution of heat from the depth and dynamics of waters, to assess the intensity of the associated biological and biochemical processes, conditions of existence of hydroflora and hydrofauna. The temperature of the surface layer serves as an indirect indicator of the quality of lake waters. Water temperature is one of the most important and most informative parameters of condition of natural waters. In this connection, lakes have been classified by thermal regime from the viewpoint of using and the development of remote-sensing techniques.

The physico-chemical and biological properties of lake water masses (and hence their electro- and radio-physical parameters) are determined by properties of incoming river and slope runoff, the duration of their transformation in the lake governed by the character and intensity of external (both natural and anthropogenic) forcing on the water body.

As mentioned above, existing classifications of lakes do not consider the distribution of natural waters of water bodies themselves by their physico-chemical, electrical and radio-physical properties, nor amounts of suspensions, kinds and production of hydrobionts. Existing facilities make it difficult to study these properties and water quality of water bodies themselves: the indicated parameters of water masses are non-uniformly distributed over the basin and require a volumetric expression of the point where the water was sampled and its properties were measured. As shown below, the use of spectral sensors partially solves the problem of representativity of *in situ* point-to-point measurements.

From the viewpoint of contemporary limnology, a lake is a single functioning natural object with the properties of spontaneous ordering and auto-regulation being naturally and anthropogenically forced upon it (Kondratyev *et al.*, 1992). The physico-chemical and biological systems of lakes closely correlate and can be described by a number of parameters in their various combinations and by some optical and radio-physical parameters.

The structures of water objects are volumetric, 3-dimensional and should be spatially considered in three coordinates: in the horizontal plane (along and across) and with respect to the depth of the lake.

The shape of the hollow of large lakes is complicated and, in contrast to small lakes, cannot be reduced to a regular geometrical figure when studying the thermo-dynamic regime (Khomskis, 1969). Until recently the morphometric description of large inland water bodies has been carried out only by horizontal morphometric sections: the area of the horizons, relationship of volumes of water mass located between them. However, for remote diagnostics of lake ecosystems and comparative assessment of the condition and distribution of the elements of limnetic structures over the water body, their volumetric 3-dimensional description and referencing of vertical and horizontal variability in morphometric characteristics are needed.

4.2   THERMAL REGIME OF LAKES: RESULTS OF COMPLEX STUDIES

4.2.1   Thermal structures of the lakes: transformation in the annual cycle, specific to regional features

The recent intensive development of techniques and methods for remote sensing has made it possible for space earth science and space oceanography to become independent scientific directions. For a long time the use of spectral airborne and satellite-borne sensors to study lakes, rivers, and reservoirs has not reached scales of their possible applications. In many respects this is explained by the relatively high cost of aerospace information, the apparent incommensurability of expenditures on space-based studies of inland water bodies compared with the cost of expeditional work on lakes using traditional *in situ* measuring instruments. Lagging behind of satellite limnological programs has also been caused by the limitations of spatial resolution of space-borne means (limitations were particularly substantial at initial stages of development of remote-sensing techniques). Despite the known advantages of all-weather use of instruments and considerable depths of sounded layer, radar (active and passive) sensing, whose local resolution initially constituted kilometers and even tens of kilometers, seemed to be ineffective for use in limnology.

Systematic studies of inland water bodies started in the 1930–1940s, when the problem of global fresh-water supplies became extremely urgent because of growth of industry, agriculture, and increasing population. The authors believe that inadequate knowledge of the physical processes taking place in fresh waters, noted by all investigators (for understandable "technical" reasons inland water basins cannot be efficiently studied in winter or in autumn–winter, winter–spring transi-tional periods when working on ice is particularly dangerous), has been considerably extended by the use of space-borne radar developed intensively in the 1990s. Considerable progress in radar sensing has been made after launching space vehicles

ERS-1 and 2 (1991 and 1995, respectively) and RADARSAT (1996) carrying synthetic aperture radar (SAR) with local resolution 20–30 m. Radar developers state that in the very near future the resolution of space-borne radar will reach 5–6 and even 2–3 m (with the spatial swaths 20–30 and 50 km preserved).

Complex *in situ* remote studies, in which the authors took part, covered the principal lakes of the moderate climatic zone of Russia: Ladoga, Onega, Teletskoye, Sevan, Baikal, Taimyr, and some basic reservoirs of the Volga and Dnieper Cascades, as well as the Great American lakes. A combination of regional climatic conditions was coordinated with limnoclimate, latitudinal and altitudinal location of lakes, scale of water bodies, their area, volume and shape. The results of sub-satellite measurements, whose data will be given below, serve as the basis for development and improvement of techniques for thematic decoding of satellite information.

Of course, space-borne survey gives a considerably generalized picture of water-body condition, remote-sensing data are correctly decoded only with adequate knowledge of the regime features of the water body under study, possible conditions on the lake at the moment of survey, meteorological conditions in the preceding period. Also, knowledge of the physical principles of remote sensing and understanding ways of their practical application is of great importance. Thermal classification of lakes by length of hydrological seasons, thermal structure and hydro-physical processes should be developed by embracing remote-sensing techniques.

Complex aerospace studies of the thermodynamic and ice regime of large lakes of the moderate zone have produced theoretical and model calculations (Kondratyev *et al.*, 1992, Tikhomirov, 1982) showing similarities and differences in the genetic relationship of phenomena and processes taking place in lakes and lake-wise basins. For this reason they have been typified and classified, and this is considered a rational way to study related objects. The most complete characteristic of lakes can be obtained only from complex evaluation of all processes subsequently taking place in lakes. Classifications based on complex analysis of the temperature regime, taking into account the morphology and internal water exchange most convenient for remote diagnostics of lakes, take a special place in limnology.

As mentioned above, Forel (1895) was the first to classify lakes by their thermal regime, he took as a criterion the value of the temperature of maximum water density $4°C$, bearing in mind that, within the annual thermal cycle, exceeding the temperature of maximum water density is followed by free convection and total vertical mixing of lake waters. Forel laid the basis for classification of fresh-water lakes on their thermal regime and internal water exchange, clearly dividing them by geographic zones. Three types of lakes were selected: tropical-zone lakes with annual temperatures above $4°C$; moderate-zone lakes with waters exceeding $4°C$ every spring and autumn (dynamic lakes); polar lakes with the annual temperature below $4°C$. Later on, Forel classification was repeatedly specified and improved.

There are numerous publications on the modification of classifications of dimictic lakes, whose authors to substantiate thermal classifications have introduced different values of water temperature in the epilimnion in their relationships with the temperature of other layers of the lake as well as with the total volume, its individual parts and differences. However, note should be taken that due to inadequate stability of these relationships, variations in temperature and volume of epilimnion waters in different seasons, their year-to-year variability, the use of these classifications to develop remote-sensing techniques is not promising.

The authors have used a classification of thermal structures and hydro-physical processes of lakes; their variability in different hydrological seasons was suggested by Tikhomirov (1982); Melentyev *et al.* (1997). This classification is based on temperature distribution, the TB location, and the thermal structure of certain types of lakes that retain their principal features with varying temperatures in different layers and their year-to-year changing lengths. With the possibility of fixing manifestations of the TB on lake surfaces from remote-sensing data, this classification is considered most suitable for developing methods and means of remote diagnostics of inland water bodies.

According to our methodology, the lakes of the moderate climatic zone are subdivided into three classes (two basic classes and one intermediate).

Hypothermic lakes are characterized by the presence of both TAA and TIA, long duration of the TB in spring and in autumn, the presence of a clearly cut layer of a temperature leap in the summer. These are large and deep lakes in which springtime water heating (and, respectively, cooling in the autumn) on both sides of the TB follow laws of their own (Fig. 4.5).

Epithermic lakes have no conditions for TB formation. This is explained by sub-ice water warmed by the heat accumulated in bottom sediments. Ice-free water temperature in these lakes reaches 4°C. These lakes are shallow, the horizontal and vertical thermal stratification of their water masses in the ice- free period is weakly expressed; in summer they are characterized by simultaneous heating and cooling of total water masses.

Metathermic lakes (intermediate class) combine in different degrees the properties of hypo- and epithermic lakes. Multi-reach lakes with their separate gulfs and bays

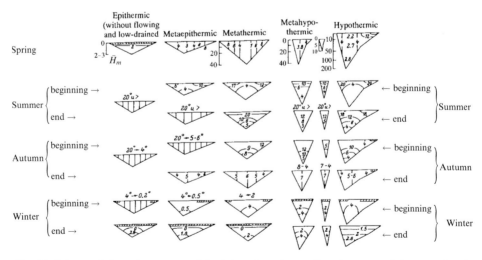

Fig. 4.5. Scheme of thermal structure of the lakes of moderate climatic zone, and its annual cycle.

are usually characterized by different morphometric parameters and thermal features, so that any part of the lake can be individually attributed to a certain thermal class on the "radio-portrait" of the lake.

Classification of moderate-zone lakes by thermal conditions, based on formation and difference of thermal structures with their specified lifetimes in different hydrological seasons and taking account of special locations of TB and morphometric parameters of the hollows, has been used by authors for thematic decoding of aerospace sensing data in the IR and microwave spectral regions.

In this classification of lakes, among other basic indications determining their belonging to a certain class, the following have been used: duration of hydrological seasons, presence in lakes of certain thermal structures, location of the TB as a converter of thermal structures in transitional periods in spring and in autumn. Only in lakes with TBs has a complete spectrum of thermal structures and hydro-physical processes in their annual cycle been detected. The classification selects direct and inverse thermal stratification, as well as homothermy when the water temperature in the lake is the same both horizontally and vertically. In the winter period, an *inverse thermal stratification* is observed in the thermal structure of lakes, when water temperature at the bottom remains near 4°C, and near the surface and beneath ice it is about 0°C. In the warm season, when water temperature increases from the bottom to the surface, a *direct thermal stratification* forms. In the period of warming there can appear a dichotomy with minimum temperature at some depth, when free convection has no time to equalize temperatures of the warm upper layer and the cooler lower layer, that exists with an inverse thermal stratification. In summer, there are usually three clearly expressed vertical thermal layers in the lake: upper—the warmest, with temperature reducing negligibly—*epilimnion*; middle—*metalimnion* or a layer of temperature leap characterized by temperature drop with depth (sometimes down to 10°C per meter of depth); and lower—a relatively cool layer—*hypolimnion* characterized by a smooth and negligible decrease of temperature with depth.

The basic water mass of large deep lakes is concentrated in hypolimnion; these lakes refer to the *hypothermic* class. Shallow small and moderate lakes, whose water mass is epilimnion for most of the year, are attributed to the *epithermic class*. The remaining lakes refer to the *metathermic* class. Among the enumerated classes of lakes can be found intermediate sub-classes: epimetathermic, metaepithermic, metahypothermic, and hypometathermic. Introduction of these sub-classes turns out to be useful when describing specific features of the condition of water masses of any water body from satellite observational data.

The annual thermal cycle of water bodies of inland moderate zones is divided into hydrological periods (seasons): hydrological spring, summer, autumn, and winter differing in duration and not coinciding in time with the calendar seasons. According to Zaykov (1955), the moment when the process of water heating persistently starts dominating the process of cooling is taken as the onset of spring. The moment when cooling starts prevailing over the process of heating is considered to be the end of the spring and the beginning of the autumn period; the end of the spring and autumn periods is connected to the time when water masses of the lake exceed the temperature of maximum density. With such an approach to the division of the annual thermal cycle, the moments of onset and duration of hydrological seasons of th moderate-zone lakes are determined not only by the season and their latitudinal location but also (and considerably) by morphometric data. This means that in lakes of the same region with different morphometric characteristics the hydrological seasons will begin at different times. Lakes with certain morphometric characteristics, differing in time of onset and duration of hydrological seasons, turn out to be similar with respect to heat accumulation and thermal structure of water masses. This creates an opportunity to solve an inverse problem: fixing the character of variations in SST using satellite sensing data, the condition of the water mass itself can be retrieved and the morphometric characteristics of inland water bodies can be indirectly assessed.

In transitional periods (spring and autumn) the water temperature in the lake happens to be rather homogeneous but never everywhere similar. In some cases existing small differences in water temperature can be ignored (within 0.1°C) and assess this condition as approaching homothermy. Nevertheless, in order to analyze from aerospace survey data, which of the dynamic features are potentially inherent to any thermal structure of water masses, due to which forces it is maintained, and how one thermal structure is replaced by another, it is very important to know differences in water temperature not only in the horizontal but also in the vertical.

By analogy with summertime direct and wintertime inverse thermal stratifications, it is possible to divide thermal inhomogeneity of lake waters in transitional periods into the vertical isothermy with springtime inhomogeneity (*springtime horizontal spatial thermal inhomogeneity*) when the temperature of water decreases from vertical to vertical towards the center of the lake with growing depth, and *the autumn horizontal spatial thermal inhomogeneity*, with vertical-to-vertical increase in water temperature when the depth grows. Below we shall use the terms "spring" (direct) and "autumn" (inverse) horizontal thermal inhomogeneity of lake water mass, meaning vertical isothermy (Fig. 4.6).

The beginning of the hydrological spring depends weakly on weather conditions. It is determined by astronomical factors: sun elevation and daytime duration; that is, the amount of heat permeating the water mass and the condition of water surface (the presence of ice and ice cover of lakes). Every year, on lakes of the same class times of onset of the spring heating period turn out to be sufficiently close. On large

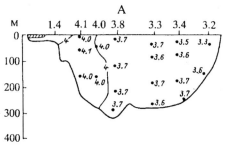

Fig. 4.6. Water temperature structure of the autumn cooling period in Siberia, Altay Mounts—Lake Teletskoye, 29–30 November 1978: very strong winter (Selegey and Selegey, 1978).

lakes that freeze later, snow cover on ice is either thin or completely missing. In spring, more heat penetrates the ice than the environment. Covered with almost snow-free ice, hypothermic lakes are the first to be heated in spring. The Sun's heat in March hardly reaches the water masses of lakes with snow-covered ice because of the high albedo of ice. Such lakes in the north-western region of Russia enter the period of the springtime heating in April, and those farther to the north in May.

The duration of the period of spring heating is determined not only by intensity of the heat flux to the water basin but also by the amount of heat needed to warm the water mass from its temperature minimum to the temperature of its maximum density. The temperature minimum of freezing lakes depends on their average depth. In shallow lakes with an average depth of 4–5 m the water mass in winter is markedly warmed due to heat accumulated in bottom sediments during the warm period of the year. Clearly, when the average depth of the lake increases, the amount of such heat decreases, and in the period of cooling the warming of water due to the heat supply manifests itself inversely. So, considering that deep lakes become covered with ice later and that it takes a longer time for their water masses to cool, and that they are practically devoid of warming due to bottom sediments, the temperature minimum ($\bar{T}_b$) of deeper lakes should be lower. However, this is valid only for lakes with average depths 25–30 m. Freezing lakes with average depths exceeding 25 m have deep-water regions in which the water on horizons more than 50 m deep does not cool below 1°C, and at the depth 100 m the water temperature is 2–2.5°C. Due to the higher temperature of deep-water regions the temperature minimum of such deep lakes turns out to be higher (Figs. 4.7 and 4.8).

In other words, the shallower the average depth of the lake, the less heat from outside is needed for TB to form and manifest itself on the water body surface, and for the water to reach the temperature of maximum density, especially in shallow lakes markedly warmed by bottom sediments. Contrarily, deep-water lakes need more heat and time to warm to the temperature of maximum density, especially when TIA is isolated from TAA by TB.

When the whole water thickness of the basin reaches the temperature of maximum density, the hydrological summer sets in. On shallow lakes this season sets in from the moment they are completely free from ice (in the north-west of Russia in late April–early May), with the period of spring heating practically going

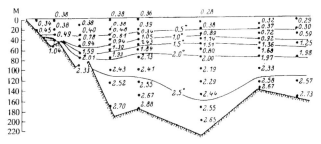

Fig. 4.7. Horizontal isothermy with inverse thermal stratification at the first phase of springtime heating at Lake Ladoga, contact measurements, 24–25 April 1900: 1899/1900 winter was strong (Witting, 1929).

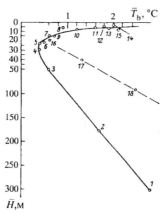

Fig. 4.8. Minimum temperature of water mass ($\bar{T}_b$) as a function of mean depth of lake (H) at the beginning of hydrological spring. Freezing lakes: 1—Baikal, 2—Teletskoye, 3—Ladoga, 4—Onega, 5–8—Bays of Lake Onega, 9—Ladm-ozero, Tut-ozero, 10—Kosm- ozero, 11—Yan-ozero, 12—Lake Kubenskoye, 13—Lake Beloye, 14—Lache, 15—Lake Glubokoye, 16—Great Sevan. Non-freezing lakes: 17—Small Sevan, 18—Michigan.

on under the ice. The end of summer on these lakes is determined by the time the heat supply of the water mass reaches a maximum in late June–first half of July. This heat supply is maintained for some time varying little depending on weather conditions. Heat reaching the lake is compensated for by heat expenditure on evaporation, surface–atmosphere eddy heat exchange, and efficient emission. From the second half of July the shallow lakes start to cool with the beginning of the period of autumn cooling.

The period of summer heating on deep non-freezing lakes starts in May (Lake Michigan, pre-reconstructed Lake Sevan), and on those covered with ice—approximately when the summer period on shallow lakes ends; that is, in the second half of July. So, on shallow lakes the period of spring heating lasts for several days, and on deep lakes up to 4 months. On all other lakes the duration of the period of spring heating lies within the time limits indicated, being determined by all the factors enumerated above, but mainly by morphometric characteristics of lake hollows. The hydrological summer on deep lakes in the moderate climatic zone is the shortest season. It lasts for 40–50 days and ends in late August–early September. The hydrological autumn on these lakes lasts for 3 to 4 months: on deep lakes it ends in late December–early January, and on shallow lakes in mid-to-late October. Winter is the longest period on shallow lakes at about 5 months; on deep lakes it does not exceed 3 months.

Hypothermic lakes have deep hollows with average depth more than 25–30 m, the area of their water-table reaches several hundred square kilometers and more. These morphometric characteristics determine the special features of their thermal regime.

The beginning of the period of spring heating on Lake Ladoga is 15 March, Lakes Michigan and Onega 15 April, the end of spring heating on: Lake Michigan is 5 May, Lake Baikal 10 August. So, its duration varies from 40 days to 4 months. The period of spring heating is divided into two phases. The first phase on freezing lakes starts with the freeze-up. At this time the thermal condition of lakes of the class considered is characterized by inverse thermal stratification and horizontally homogeneous temperature of lake water shown in Figure 4.7. As the lake becomes free of ice, the pattern of the distribution of isotherms in coastal regions is replaced by springtime (direct) horizontal thermal inhomogeneity (Figure 4.9).

In non-freezing lakes (Lake Michigan) vertical isothermy with inverse horizontal inhomogeneity (the water temperature increases with depth), having appeared in autumn, is replaced as in freezing lakes by direct (spring) isothermy. The second phase of spring heating starts from the moment of TB appearance which divides lakes into TAA and TIA. At some time of the early TB the whole lake preserves vertical isothermy with spatial thermal inhomogeneity, and up to the end of the spring period in TIA with water temperature below 4°C. In TAA, as the TB shifts towards deep waters, a thermocline appears in the coastal regions of the lake (Figs. 4.1). Cyclonic circulation originating in TAA simultaneously with TB develops further. When water reaches the temperature of maximum density, there are no longer conditions in deep waters for the existence of TB. A dome of dense water forms here, with warm water above it. The clear-cut thermal limnetic layers appear.

In the early hydrological summer the surface temperature difference over the area of hypothermic lakes reaches a maximum. Cyclonic circulation reaches its maximum. The layer of hypolimnion is the powerhouse. The hydrological summer lasts for 40–80 days. By the end of summer the temperature-leap layer becomes somewhat smeared, and closed drifting circulations appear in some regions of the lake. In coastal regions, cyclonic circulation connected to the slope of isotherms remains in the upper half of the metalimnion. The dome of dense water drops down to 70–100 m. The above-dome layer of hypolimnion with water temperature from 4 to 6°C is most powerful. In late August–September, when heat accumulated by the lake reaches its maximum, the period of autumn cooling begins. This period is divided into two phases. During the first phase the temperature-leap layer drops, the power of epilimnion grows, and water temperature reduces somewhat and equalizes throughout the lake. At the end of the first phase the dense water dome only remains near the bottom at extreme depths. In coastal regions horizontal thermal inhomogeneity appears with water temperatures between 8–12°C, preceding the appearance of the TB. Lakes get to the second phase of autumn cooling in late October–early November, when the TB divides the lake. From this moment on the water in TAA cools very rapidly ice appears in coastal regions, and in the remaining

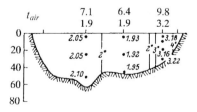

Fig. 4.9. Water temperature of Lake Onega, second phase of springtime heating, 06.06.1966.

part of the lake the thermal structure of water masses is expressed by vertical isothermy with inverse horizontal thermal inhomogeneity. The deep-water areas of the lake preserve temperatures above 4°C for a long time. On these lakes the TB disappears in late November–December. The period of winter cooling sets in. In the deep-water area the dense water dome is covered with outlying cold water. When the dome drops, its warm water reaching coastal regions delays the onset of freeze-up. Freeze-up covers the lake concentrically, the deep-water area remains ice-free for a long time. In the period of freezing, under the influence of convective mixing and weak horizontal currents, the water density and its thermal regime are regulated, forming an inverse thermal stratification with the horizontal isotherm.

During heating, lake water-surface temperature is lower than air temperature, and in the period of cooling higher.

Before and during the freeze-up a very thick layer of water cools considerably: near the lower ice edge the water temperature is approximately 0°C, gradually increasing until at depths 30–40 m it nearly reaches 5°C. As a result of strong cooling of the upper layer, the annual heat supply of lakes with average depth 25–30 m is at a minimum.

All fresh-water lakes of moderate zones, except for hypothermic ones, freeze-up completely. Hypothermic lakes are subdivided into freezing, partially covered with ice, and remaining open in winter. Among them, the lakes annually covered with ice are Onega, Baikal; completely or partially covered with ice depending on winter severity are Ladoga, Teletskoye, pre-reconstructed Lake Sevan; remain ice-free – Lake Huron and others.

Attributing a water body to a certain class from aerosurvey data, it should be borne in mind that, according to the classification considered, only single-reach lakes can completely correspond to a certain type of lakes. Multi-reach lakes with their gulfs and bays can differ considerably in their thermal regime. For example, Lake Onega on the whole should be referred to the hypothermic type. However, the principal reach of the lake and its multiple bays and gulfs can be attributed to all the classes of the suggested thermal classification of lakes.

Metathermic lakes are very numerous and diverse in their morphometric characteristics. The individual properties of these lakes range widely; some of them in some hydrological seasons temporarily acquire the features of lakes of other classes—hypo- and epithermic lakes. Metathermic lakes are subdivided into three groups: metathermic proper, metaepithermic, and metahypothermic. In some cases, within the metathermic class two subgroups can additionally be selected: epimetathermic and hypometathermic lakes. The average depth of metathermic lakes is 25 m. However, this parameter as well as area can also be characteristic of lakes of other classes. To decide which class a lake belongs to is determined by strength of flow, relationship between lake area and its plan configuration, the height and forest coverage of shores, and morphological characteristics of their hollows.

Metathermic lakes proper have average depth 8–25 m and they freeze. The hollows of these lakes are rather flat, but have deep-water regions. The areas of such lakes can reach tens and thousands of square kilometers. In metathermic lakes proper, during freeze-up the water is warmed by heat accumulated in bottom sediments.

However, this warming together with heat flux permeating the ice is insufficient for water masses to be heated to the temperature of maximum density at the moment of freedom from ice. In spring, the TB appears in such lakes and exists in different metathermic lakes from several days to several weeks. When the TB disappears in TIA, a dome-shaped water body (hypolimnion) with maximum density temperature forms in pelagic zones. During the summer hypolimnion transforms into metalimnion. The speed of this transformation differs in some years because of different weather conditions. It is characteristic of metathermic lakes proper that, by the end of hydrological summer, the metalimnion is located at the bottom. The hypolimnion, a dome of dense water with temperature below 4°C, and the above-dome layer with temperature 4–6°C are absent. With the beginning of the first phase of autumn cooling the layer of metalimnion is destroyed and replaced by horizontal thermal inhomogeneity. The second phase of autumn cooling proceeds in the presence of TB and ends with its disappearance. In winter the water mass near the bottom is warmed by heat accumulated in summer and autumn by bottom sediments.

Metaepithermic lakes with average depth 3–8 m have a flat hollow and thick layer of silt; they are drained or strongly drained; they freeze with ice reaching the ground in coastal regions as a result of lowering water level in winter; they are either round or asymmetric on plan and open to wind-driven mixing.

The thermal regime of spring heating of this group of lakes is very complicated. In the sub-ice period water masses in these lakes are warmed by heat accumulated by bottom sediments; in low-drained regions it can be followed by inversion of density and temperature near the bottom. Before the beginning of spring heating, thaw waters with temperature *ca.* 0°C start flowing into the lakes of this group; with thaw water inflow the level of water in the lake starts rising. Still under the ice, cold thaw waters start forcing out considerable amounts of warmer water heated in winter by bottom sediments. Solar heat penetrating the ice is insufficient to warm thaw water up to the pre-spring temperature of the lake water mass, so that heat supply of bottom sediments runs out at the same time as thaw water replaces winter waters. When the lake becomes ice-free, the thaw waters replacing winter waters have temperatures much below 4°C, and the temperature of the water from rivers and coastal slopes is above 4°C. For several days the TB exists at the interface of these two water masses. In the days following this, the period of summer heating sets in with spatial horizontal thermal inhomogeneity, and then due to waters mixing the temperature in the lake equalizes. In May, due to inflow of river water the heat supply differential in lakes of this group is greater compared with subsequent months. In the period of summer heating these lakes lose a considerable portion of their heat supply through runoff. In other respects the thermal regime of lakes of this group in summer coincides with that of epithermic lakes. In autumn, with a drastic fall in temperature, a short-lived TB can appear in lakes of the metaepithermic group with an area of several square kilometers. In winter, with lowered water level, ice in these lakes settles on the ground over a large area, and transfer of heat takes place with the waters of in- and outflowing rivers.

Metahypothermic lakes are diverse in area, with average depths down to 25 m and have either cone- or trench-shaped deeply cut-in hollows, and are sheltered from the wind. The lakes of this group are much smaller hypothermic lakes. By the beginning of the spring heating the warming of the bottom layer is weakly expressed. In the second phase of spring heating, depending on morphological characteristics, lakes of this group differ in the proximity of water temperature to that of maximum density, as well as in the volume and life-time of TAA and TIA in them. With the steep slopes of the hollow bottom, there is no TB in such lakes, and in this case the whole water mass of the lake simultaneously reaches the temperature of maximum density. During the period of hydrological summer and in early autumn lakes of this group differ in their properties from other groups of metathermic lakes by a clear and persistent division of water masses into epi-, meta-, and hypolimnions. In early summer, in pelagic regions, the water mass with temperature of maximum density constitutes a powerful core. However, later on its upper boundary slowly disappears and a stable hypolimnion forms. The layer of temperature leap (metalimnion) becomes clearly expressed and seriously inhibits mixing. So in lakes of this group the temperature of epilimnion is not below the surface temperature of metathermic lakes. In the pre-freeze-up period, with strong winds and low air temperatures, the whole water thickness of the lake cools, and water temperature at maximum depths drops to 2°C. In winter, water temperature rises here only above that part of the basin which the epilimnion bordered on in the summer.

Epithermic lakes are widespread, they are shallow, with average depths down to 3 m. They differ in that their hollows are most often flat, with weakly expressed relief and several depressions in the bottom. The areas of these lakes vary widely, they are either round or asymmetric and open to wind-driven mixing. By the beginning of spring heating the temperature of the bottom layer is close to 4°C and sometimes higher due to warming of water by the heat accumulated by bottom sediments in the open period. Solar heat mainly reaches the water after snow on top of the ice melts. The period of spring heating takes place under the ice, in the absence of TB. During break-up of ice the lake water masses are in a condition close to homothermy. With lake areas reaching tens of square kilometers, springtime horizontal thermal inhomogeneity can exist in early hydrological summer. Being warmed in calm weather, the water thickness of these lakes can be slightly stratified in the vertical. Over the whole open period water masses are represented by epilimnion, and for this reason lakes of this class are called epithermic.

In late June–early July the heat supply of water masses is at a maximum, and this marks the end of the period of summer heating. At the end of hydrological summer, heat supply of bottom sediments constitutes 40–60% of total heat supply of the water body. Accumulation of considerable amounts of heat depends on water thickness of lakes during summer being represented by one strongly heated layer, the epilimnion, the temperature of which even in very cold summer months exceeds 16°C, and in very warm months 25°C. Having reached a maximum, the heat supply in water masses remains almost constant for some time, varying little. Heat coming to the lake in this period turns out to be almost balanced by heat loss. In conditions close to homothermy the water mass follows air temperature oscillations, with the temperature of water higher than air temperature during the whole open period except for May.

The uniform cooling of the water thickness of these lakes does not create conditions for the autumn TB. In the period of intensive autumn and winter cooling the thermal condition of water mass with a large water-table area can be expressed by weak inverse horizontal isothermy. After freeze-up, water of the bottom layer is intensively heated, a clear inverse thermal stratification is established, replaced in some lakes by the stable temperature inversion of density with the bottom layer warmed above 4°C and water temperature directly beneath the ice about 0°C.

4.2.2   Retrieving lake temperatures: numerical modeling and results of sub-satellite experiments

Let us consider analytical results of field studies of the variability in some parameters of lakes as applied to microwave aerospace survey and evaluate the efficiency and accuracy of retrieval algorithms. The scale of averaging of passive microwave data is determined, on the one hand, by the level of development of the antenna system carried by aerospace vehicles and, on the other hand, by users' requirements for limnological information. So, for example, it was assumed that local spatial resolution when passive-microwave sensing from aircraft (average altitude of sensing $\bar{H} = 5.0\,\mathrm{km}$) for working wavelengths 0.8–18.0 cm averaged 100, 200, 500, and 1000 m, and when sensing from the orbital station ($\bar{H} = 300\,\mathrm{km}$) and meteorological satellite ($\bar{H} = 600\,\mathrm{km}$) was 10–30 km.

Our studies of spatial and temporal variability in characteristics of lake conditions have been complex, within the framework of several sub-satellite national and international experiments on Lakes Ladoga, Onega, Ilmen', Beloye, Kubenskoye, at the Rybinsk and other Volga reservoirs, at lakes in Bulgaria and Poland (Kondratyev *et al.*, 1988, 1990, 1992; Kondratyev, Melentyev, 1991; Melentyev and Mel'nichenko, 1995). Standard measurements were taken according to an extended program which foresaw quickened (including every hour and every 15 minutes) observations, as well as additional *in situ* and remote measurements of water temperature, biota parameters, characteristics of roughness, foam, near-water wind speed, precipitation; in some cases new meteorological, limnological and spectral instruments and sensors were developed for this purpose.

Methodology of experimental studies from ships foresaw research vessel (RV) routes crossing different water masses and also cycles of long-term "point" measurements at stationary test sites were organized. The choice of test sites was determined by the need to study remote-sensor sensitivity to background variability in meteorological elements and re-arrangements of hydrological and meteorological fields.

An example of analysis of the spatial variability of temperature of inland freshwater bodies was made from the data of the complex experiment carried out in the summer of 1984 on Lake Onega, and from materials of *in situ* investigations over the period 1956–1983. Figure 4.10 demonstrates the results of RV measurements by towing thermoelement, mercury thermometers immersed down to 35 cm, as well as IR and microwave radiometers. Every temperature value from the data of remote sensors refers to a 8-minute interval of averaging.

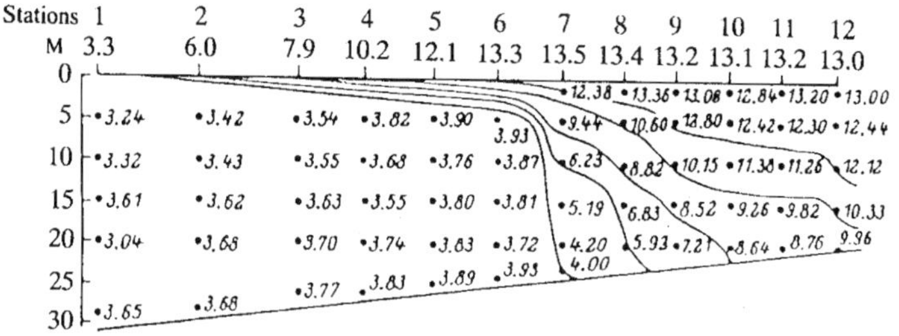

Stations 1   2   3   4   5   6   7   8   9   10   11   12
M 3.3   6.0   7.9   10.2   12.1   13.3   13.5   13.4   13.2   13.1 13.2   13.0

0 — •12.38 •13.36 •13.06 •12.84 •13.20 •13.00

5   •3.24 •3.42 •3.54 •3.82 •3.90 3.93 •9.44 •10.60 •13.80 •12.42 •12.30 •12.44

10 •3.32 •3.43 •3.55 •3.68 •3.76 •3.87 •6.23 •8.82 •10.15 •11.30 •11.26 •12.12

15 •3.61 •3.62 •3.63 •3.55 •3.80 •3.81 •5.19 •6.83 •8.52 •9.26 •9.82 •10.33

20 •3.04 •3.65 •3.70 •3.74 •3.63 •3.72 •4.20 •5.93 •7.21 •8.64 •8.76 9.96

25   •3.77 •3.83 •3.89 3.93 4.00

30 •3.65 •3.68 •3.77 •3.83 •3.89

Fig. 4.10. Water temperature of Lake Ladoga—observations perpendicular to the line of thermal bar, 31 May 1979.

By its thermal regime, Lake Onega is a moderate-zone lake and the temperature of this lake and its variations depend on the character of preceding and accompanying atmospheric circulation. From the data of measurements taken on 22 July 1984 at Velikaya Bay–Voznesenye section covered practically the whole Onega basin. During the survey spatial horizontal variability in lake-surface temperature was smooth in most cases, which was explained by intensive wind-driven mixing of water masses ending at the time of RV survey in thermobar destruction. Instant values of temperature measured in the open area were within 14.6–18.2°C; when approaching the source of the Svir' River near the town of Voznesenye, an outflow of cool bottom waters from the lake was recorded with temperature minimum of 13.6°C. From IR radiometer data, the temperature MSD (mean square deviation) for a 100-m stretch was ±0.05°C. In the open part of the lake, variations in temperature contrast of adjacent 100-m stretches constituting 0.2 and 0.25°C were recorded in some cases. Maximum difference in the temperatures of adjacent non-mixed surface layers recorded by IR and microwave radiometer constituted, respectively, 1.4 and 1.8°C (Fig. 4.11)—a temperature leap was observed in the zone where lake and river-water masses mix.

Microwave radiometers carried by the RV give higher values of water-surface temperature variations along the 100-m stretch (±0.1°C on average; ±0.2 and ±0.3°C at a maximum) compared with IR radiometers. This is explained by methodical ways of measurements, the effect of the ship's hull, the absence of a gyrostabilizing platform, the effect of solar and atmospheric radiation, and hence greater errors in temperature retrieval, but mostly by a greater thickness of the layer forming the emission. The

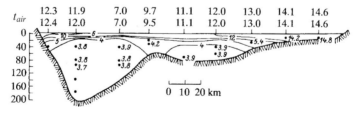

Fig. 4.11. Water temperature of Lake Ladoga, middle of hydrological summer, 1962.

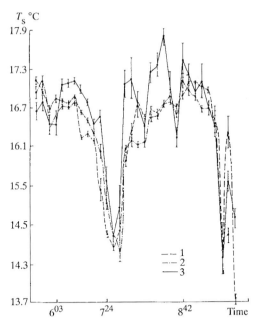

Fig. 4.12. Results of IR (1) and microwave (2—$\lambda = 13.5$ cm; 3—$\lambda = 18.1$ cm) survey of Lake Onega; 22 July 1984.

thermal regime of the lake is well described by data of point-to-point microwave measurements averaged over 100-, 200-, 500-, and 1000-m lengths of the route: the 22 July 1984 measurements of temperature anomalies in the southern part of Lake Onega are a 1.8°C decrease in temperature at a length of 2.5 km and a 2.5°C increase at a length of 3.5 km. Averaging of surface temperatures of the lake obtained from the spaceborne diagram-array instrument with spot diameter of 10.0 km enables only qualitative information about changes in the temperature regime of the basins to be obtained.

Before discussing the problem, note one important fact seen from the data comparing Fig. 4.12.

Analyzing the results of airborne or satellite measurements using a wide directional diagram it should be borne in mind that smearing the boundaries of fixed contrasts of brightness temperatures and dragging of fronts of the processes under study are possible. Otherwise, one can get an impression of an earlier detection of temperature anomalies compared with IR survey and *in situ* observations.

As discussed above the thermal structure of large lakes in periods of spring heating and autumn cooling is rather complicated and should be divided into two phases: before and after formation of the thermal bar. Figure 4.13 shows the distribution of water temperature over the surface layer of Lake Ladoga in the early period of spring heating, as well as the respective values of brightness temperature at $\lambda = 18.0$ cm, $\theta = 60°$, the regime of scanning, vertical and horizontal polarization (Fig. 4.13b, c), calculated using 7 June 1976 data of aerological sounding (HMS [Hydrometeorological Station] Sortavala).

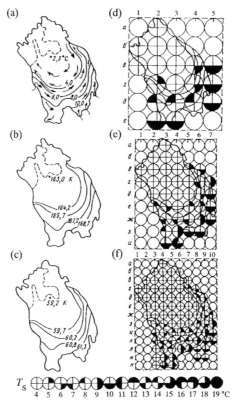

Fig. 4.13. Water-surface temperature (a) and results of the numerical experiment on Lake Ladoga temperature retrieval: (b), (c) ideal instrument, $\lambda = 18\,\text{cm}$, $\theta = 60°$, vertical and horizontal polarization, correspondingly; (d)–(f) scanning regime with different antenna system $\theta_{0.5} = 5.0°$, $3.5°$, and $1.8°$, correspondingly; 7 June 1976.

Figure 4.13e, f shows the pattern of distribution of water temperatures for the same date retrieved from the data of the numerical experiment with different widths of the directional diagram. As seen from these data, improvement of antenna system parameters by development of instruments having directional diagram width of $\theta_{0.5} = 1.8°$ and respective local resolution of 10.5 km markedly improves the detailed picture of the phenomena (Fig. 4.13); however, at $\theta_{0.5} = 3.5°$ (Fig. 4.13e) the possibility not only to divide TAA and TIA but also to trace the thermal spatial inhomogeneity of water masses in spring is still preserved. Studies of temporal variability in lake temperatures based on long-term "point" observations were carried out in different physico-geographical regions, and the most representative database was obtained by thermistor sensors immersed at a depth of 1.0 cm, IR and microwave radiometers within the framework of the "INTERKOSMOS—Inland waterbodies" experiment (Kondratyev *et al.*, 1988).

Figure 4.14 illustrates the results of calculations of the finite-difference increment in measured brightness temperature $\Delta T_B / \Delta T_s$ for the summer model of the cloud-free atmosphere (water vapour content in the air $w = 22.20\,\text{mm}$).

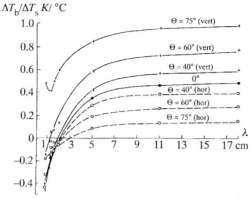

Fig. 4.14. Dependence of the finite-difference increment of the contrast $\Delta T_B/\Delta T_s$ for fresh-water masses on the angle of view at vertical (v) and horizontal (h) polarization; satellite; wind velocity $v = 0\,\mathrm{m/s}$; cloud-free.

The parameter $\Delta T_B/\Delta T_s$ is an informative but strongly variable characteristic of water-body conditions: the value of contrast $\Delta T_B/\Delta T_s$ when sensing in a free atmosphere depends on wavelength, viewing angle and polarization. As seen from Fig. 4.14, for fresh-water temperature retrieval by sensing from any platform, wavelengths λ longer than 8 cm should be used, where contrast $\Delta T_B/\Delta T_s$ is at a maximum and practically constant at the selected angle of view and polarization.

The problem of choosing optimal conditions for temperature retrieval is also solved based on the data in Fig. 4.14: measurements should be taken at angles other than nadir, at vertical polarization. At horizontal polarization, not only a decrease in emissivity takes place but also a decrease in contrast $\Delta T_B/\Delta T_s$, which can be used in solving geophysical problems where it is desirable to eliminate the effect of variations in water-surface thermodynamic temperature on microwave emission. The characteristic mentioned above—weak dependence of emissivity Σ and brightness temperatures on temperature near $\lambda = 2.0\,\mathrm{cm}$—also remains which can also be used in solving problems of satellite meteorology. As seen from our calculations (Kondratyev *et al.*, 1992), a change in angle of view as well as variations in atmospheric and water surface conditions can lead to shifting of the line on the graph where the sign of the finite-difference increment of $\Delta T_B/\Delta T_s$ changes. Note the high value of negative contrast $\Delta T_B/\Delta T_s$ in the short-wave range as well as a non-linear increment (decrease) in contrast $\Delta T_B/\Delta T_s$ with the changing angle of view, growing particularly at angles exceeding 60°. The main recommendation is the need to provide a constant angle (due to the spatial stabilization of the platform) and to eliminate the effects of depolarization of signals (provision for the constant angle at meeting the surface studied by constructing special antenna systems and signal processing (Voronin *et al.*, 1986).

The data in Fig. 4.14 demonstrate the values of finite-difference increments in $\Delta T_B/\Delta T_s$ which are fixed by the antenna of an ideal radiometer carried by different platforms. There is no substantial advantage for mapping the temperature of lakes and reservoirs at optimal λ and θ of lowering the height of the radiometer mounting in case of a cloud-free atmosphere—the contrast $\Delta T_B/\Delta T_s$ in the long-wave part of the range considered practically does not change. This peculiarity of emission transfer was used in complexing the space-borne experiments using support radiometers carried by either the flying laboratory or the RV.

The results of water temperature retrieval of Lakes Ladoga and Onega using microwave instruments of the centimeter and decimeter ranges, carried out between 1966 and 1998, enabled determination of the boundaries and the areas of TAA and TIA as climatic parameters, detection of a trend for TB shifting in the spring–summer–autumn period, and drawing of maps charting spatial variability in lake temperatures. Some differences in T_s values from IR data and microwave measurements are explained by different thicknesses of the layer which forms the emission in these ranges (effect of cold water-surface films).

Mass model calculations based on data of multi-year studies on thermal regimes of large lakes and data of aerological sounding by HMS at the cities of Sortavala, Petrozavodsk, Bologoye, Erevan enabled a series of algorithms to be suggested for temperature retrieval of fresh-water inland water bodies under different conditions of microwave survey (Kondratyev *et al.*,1992):

—1-channel method (aircraft), algorithm 1:

$$T_s = 1.299(T^{60}_{B,v,18.0} - 160.5)$$ (4.3)

—1-channel method (aircraft), algorithm 3:

$$T_s = 1.250\{(T^{60}_{B,v,18.0} - 161.0) - [12.8 - 1.110(\rho^{60}_{18.0} - 35.0)]\}$$ (4.4)

—1-channel method (aircraft), algorithm 4:

$$T_s = 0.9009\{(T^{60}_{B,v+h,18.0} - 219.1) - [38.8 - 2.980(\rho^{60}_{18.0} - 34.0)]\}$$ (4.5)

—2-channel method (aircraft), algorithm 7:
(a) clouds, precipitation, 3-4-point roughness

$$T_s = 1.923[(T^{0}_{B,18.0} - 97.5) - (0.65 - 0.06\rho^{40}_{0.8})]$$ (4.6)

(b) storm roughness, $\nu \geq 12\,\text{m/s}$

$$T_s = 1.923[(T^{0}_{B,18.0} - 97.5) - (28.6 - 2.252\rho^{40}_{0.8})]$$ (4.7)

The development of a great number of algorithms to retrieve the thermal regime of water bodies from the data of microwave survey has become necessary in connection with a relatively low accuracy of the correlation methods for the solution of inverse problems of satellite meteorology. Figure 4.15 presents the results of surface temperature retrieval of Lake Onega from satellite measurements data for 30 models of water and air mass condition. Best agreement with the Petrozavodsk HMS measurements was obtained using algorithm 3: $T_{s\,true} = 13.46°\text{C}$; $\sigma_{n-1} = 4.27°\text{C}$; $T_{s\,retr} = 13.64°\text{C}$; $\sigma_{n-1} = 4.28°\text{C}$.

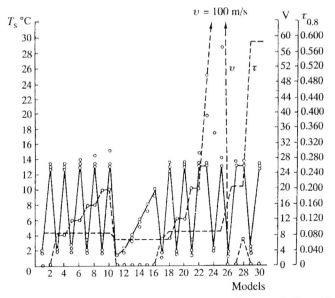

Fig. 4.15. Results of microwave retrieval of surface temperatures for water bodies with different water surface and atmospheric conditions (τ is optical depth of atmosphere).

Figure 4.15 demonstrates retrieval results by different models of the condition of the lake–atmosphere system covering practically the whole spectrum of possible meteorological situations on lakes of moderate climate zones using algorithms 3 and 4. Best agreement of numerical experiment data with true values of T_s were obtained with algorithm 4.

The results of the experimental microwave survey of water bodies and model experiment data show the prospects for using spectral-polarization methods for mapping the condition and quality of waters. The developed method is an all-weather one. In contrast to IR techniques, it enables information on the state of extended thick surface layers of water bodies to be obtained. Knowledge about parameters of the condition of this layer is very important for understanding numerous physical and biological processes taking place in lake waters.

The technological scheme of thematic processing of microwave data to estimate SST using the 2-channel sign-contrast method ($\lambda_1 = 0.8$ cm, $\lambda_2 = 5.0$ cm) was suggested for analysis of microwave data obtained from the "Almaz" satellite (Kondratyev et al., 1992). A modification of the method consists in using routine meteorological, hydrological data from ships located at sub-satellite points at the moment of measurement as well as support radiometer data carried by the ships to calibrate the space-borne measuring complex and to introduce atmospheric correction. It is expedient to equip ships with key radiometers which serve as stationary locations of WMO oceanographic observations; for example, point C in the Atlantic Ocean ($52°45'$N/$35°30'$W). The technological scheme of satellite information processing includes three types of control programs and, apart from the routine (orbit-to-orbit) database, the bank of background climatological data for test areas (the bank of seasonal masks), the bank of meteorological information from sub-satellite supporting ships, and the bank of data of the ship-borne key microwave radiometer.

The accuracy of retrieval of geophysical parameters from microwave measurement data substantially increases by the use of background information, the data on key regions of the World Ocean, large lakes and reservoirs, test sites of watershed territories, the condition of which is adequately described by complex meteorological and hydrological observations (Kondratyev and Melentyev, 1992). The contribution of background atmospheric emission for such key locations (sub- satellite points) is considered through calculations.

Areas of water surface of different length, located in different geographical regions of the globe (fresh-water lakes, parts of the sea and oceanic structures) as well as land surface areas selected by assumed decoding indications were taken as test areas. Spatial and temporal variability in basic meteorological elements for such test areas is well studied, but it is desirable to choose either homogeneous conservative emitting domains or, on the contrary, domains with considerably contrasting emission.

During sub-satellite measurements the following locations were selected as test areas: Lake Ladoga, several areas in the North Pacific and North Atlantic (Sakhalin Island, Kurill Islands, and WMO point C, correspondingly), as well as land surface areas in the Saratov region, in the Lake Balkhash region (Taukum Desert), lakes and taiga in the Russian sub-Arctic (Komy Autonomous Republic). The choice of regions was determined by differences in emissivity as well as high provision of standard meteorological, hydrological, aerological and other accompanying measurements. For the test areas calculations were made of diurnal, seasonal, interannual variability in water and land surface emission Σ on $\lambda = 0.8–18.0\,\text{cm}$, $\theta = 0–75°$. Calculations were made for radiometers mounted on satellite, flying laboratory, and RV.

Figures 4.16 and 4.17 show examples of background information and calculated values of brightness temperature (seasonal mask) for Lake Ladoga.

The data of Figs. 4.16–4.17 demonstrate the possibility of using microwave radiometry to monitor TB shifting, retrieval of diurnal, seasonal and interannual variability in lake temperature. Maps on microwave emissivity of water and land surface similar to that shown were used for external calibration of instruments mounted on aerospace vehicles and for development of schemes for retrieval of geophysical parameters.

4.2.3   Ice formation and its temporal/spatial evolution: combined *in situ* and remote-sensing studies

4.2.3.1   *Type of winter severity and climatically significant factors affecting the environment: assessment from satellite and in situ observations of lakes*

This section deals with problems of ice-cover parameter retrieval and the ice regime of large water bodies of the moderate climatic zone and developing techniques for remote monitoring of natural and anthropogenic impacts on inland water bodies by satellites, including ERS-1&2 SAR data. To more accurately retrieve ice parameters

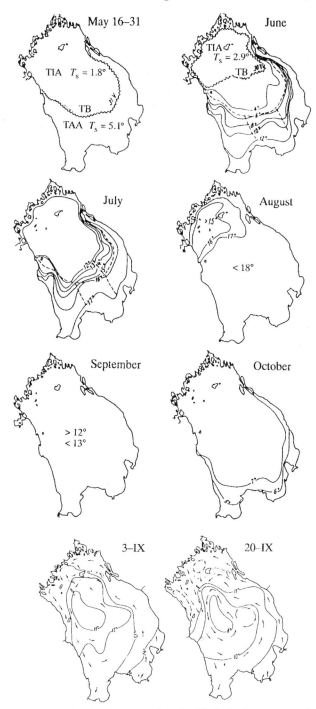

Fig. 4.16. Background information for the seasonal mask of Lake Ladoga: open water-surface monthly averaged temperature variability, open water period, 1957 (a), and currents 1962 (b).

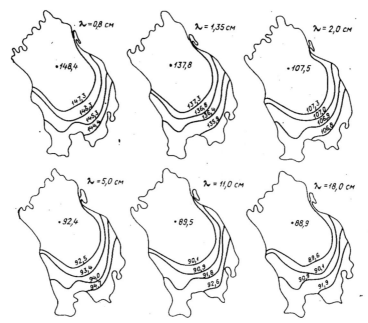

Fig. 4.17. Brightness temperature of Lake Ladoga, 7 June 1983, $\lambda = 0.8, \ldots, 18.0$ cm, $\theta = 30°$, horizontal polarization.

for large lakes of the St. Petersburg region (Ladoga, Onega, Chudsko-Pskovskoye, and others), their annual thermal cycle has been taken into account. The analysis of SAR, microwave and visual images was based on the results of multi-year studies of the thermal structures and peculiarities of their variability in the autumn–winter hydrological seasons, on calculational data of heat supply and temperature of lake waters, and on data of sub-satellite, ship and aircraft experiments (Kondratyev *et al.*, 1992, 1996b; Melentyev *et al.*, 1997; Tikhomirov, 1982).

The authors hold the archive of ERS-1&2 SAR images of the St. Petersburg region, including those of the ice cover of the largest lakes of north-western Russia for 1993–1998. Some results are given below of the use of LRI SAR survey to study environmental conditions and assess the contribution of climatically significant natural and anthropogenic factors of external forcing.

The ice regime of inner water bodies is determined by their geographic location and climatic conditions of regional experiments (Betin, 1957; Church, 1943; Hupfer, 1979; Medres, 1957; Tikhomirov, 1982). The moderate climatic zone is located approximately from 40° to 65°N and from 42° to 58°S. The winters in most of this zone are characterized by a long period of frosty weather and stable ice cover. In the northern and more continental eastern regions of the moderate zone, adjacent to the cold zone, ice conditions on lakes are relatively more severe.

The processes of ice formation and melting depend on climate. The traditionally used indicator for ice regime of the seas and inland water bodies is their ice-cover

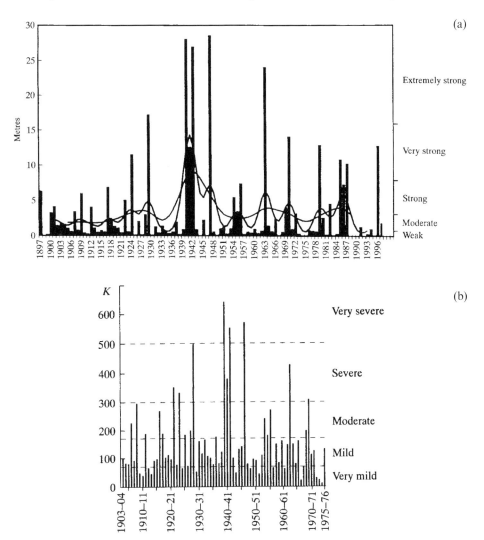

Fig. 4.18. (a) Area-related ice volume with classification of the Baltic Sea 1896/1897–1995/1996 winter season by ice-cover extent (Alpers, 1997). (b) Sums of the degree-days of frost for Rostok (from the winter 1963/64 for Varnemiunde) with the classification of all winters by ice-cover extent (Hupfer, 1979).

extent; that is, the ratio of the ice-covered area to the whole area of the water body considered—Fig. 4.18a (Alpers, 1997).

Typifying winters by their severity may be based on different principles, including navigational ones—by capability of ice routing of different classes of ships (Betin, 1957). However, for these purposes the sum of days with negative temperatures (degree-days) accumulated during the summer which is closely connected with the area of maximum propagation of ice is most often used (Betin, 1957; Medres, 1957; Tikhomirov, 1982; Hupfer, 1979; Melentyev *et al.*, 1997). According to Hupfer

(1979), the following gradations of ice cover due to winter severity are introduced: very mild, mild, moderate, severe and very severe—Fig. 4.18b. The amplitude of oscillations of each type of winter corresponds to one-fifth of that of the whole area of the ice cover.

There are few studies of the quantitative estimates of relationships between ice propagation over large inland water bodies (ice cover) and the sum of days with negative temperatures accumulated by a certain time due to special features of wintertime limnological work and necessity to take into account regional features of lakes. For example, our study (Tikhomirov, 1982; Melentyev *et al.*, 1997) in which an empirical formula was suggested to calculate the ice area of Lakes Ladoga and Onega: $S = K(\sum T_{air})$ (where S is in square kilometers, $K = 22$—the dimensionless coefficient) from data on diurnal negative air temperatures from coastal hydrometeorological stations of the town of Vidlitsa and the city of Petrozavodsk. For sea-water areas this problem has been much better studied. Studies of the relationships mentioned above are exemplified in Hupfer (1979) where the sums of degree-days are given for the Baltic Sea located in the same climatic zone. The generalized row of ice-cover extent for the observational period from 1903 to 1976 with the classifications of all the winters by their ice-cover extent has been drawn from the data of ships and island stations, aerial surveys, and recently from the data of TV-satellite imagery.

The satellite survey refers to the third phase of hydrological winter (Melentyev *et al.*, 1997) and fixes very high ice-cover extent in the winter 1986–1987 (very strong winter, according to W. Alpers' classification); all large lakes of the north-western region of Russia (including Ladoga and Onega) and all large lakes of Finland and Estonia have 100% ice cover. Calculations by the formula given above for the case when Lakes Ladoga and Onega are completely ice-covered showed that the sum of diurnal negative temperatures over the St. Petersburg region for the winter 1986–1987 exceeded $500°C$ (Onega turned out to be totally covered with ice) and constituted $\sim850°C$ (Ladoga was practically frozen, too). So, by satellite observations the severity of this winter should be referred to as very severe (very strong) winters. The conclusion drawn from the satellite survey data is also confirmed by the complex of meteorological and limnological observations carried out in this region at this time.

The known advantages of radar and microwave radiometry are connected with the possibility of obtaining information from relatively deep-penetrating microwaves into the sounded surface (Kondratyev *et al.*, 1992, 1996b; Johannessen *et al.*, 1994). As a result, it is possible to use ERS-1/2 SAR imagery not only to study the general severity of winter (from the quantitative estimates of total ice cover) but also to obtain a detailed picture of the development of frosty weather in the region and to assess the seasonal-spatial variability in surface wind. It has been shown (Kondratyev *et al.*, 1992, 1996; Melentyev *et al.*, 1997) that ice cover as a more inert medium compared with open water, capable of responding to several external forcings, can also keep and "remember" some of them, preserving, accumulating and conserving the external forcing, which is fixed on the fall/winter radar "portrait" of the lake. The proposed method of SAR diagnostics is based on study of the laws of the process of ice-cover formation on large water bodies and study of the dependence of fresh-water ice reflectivity on the water-surface condition at the moment of freezing and its variability during the hydrological winter and in the period of ice melting. The study of subsequent series of SAR images enables one to reveal the phase change of freezing and to systematize radar signatures of fresh-water ice in winter, to study their relationships with climatic factors: the type of lake climate, air temperature, variability in surface wind speed and direction (Melentyev *et al.*, 1997).

4.2.3.2 *Use of ERS-1&2 SAR wintertime data to assess seasonal/spatial forcing on large inland water bodies of the moderate climatic zone*

The main factors determining the process of formation of fresh-water-body ice cover are as follows: (a) intensity of heat release from the surface to the atmosphere; (b) wind activity; (c) heat supply in the lake; (d) morphological and morphometric features of the lake hollow (Medres, 1957; Tikhomirov, 1982).

According to the thermal classification (Melentyev *et al.*, 1997), the first phase of freezing on large lakes begins with the second half of the hydrological autumn, when the fall-time thermobar appears. At this time the thermal conditions of water masses are characterized by vertical isothermy with its spatial deep inhomogeneity. Ice formation on lakes starts after the air temperature becomes negative and only in regions where the water temperature in vertical constitutes $2°C$ and less. Under these conditions the intensity of heat release from the surface becomes such that a thin surface layer turns out to be cooled down to freezing point. The heat release quantity determined by air temperature and efficient radiance increases sharply in thin-clouds weather.

Light breeze at negative air temperatures favours formation of ice—smooth ice crust on water surface (looks like nilas). The radar signature of such a crystalline structure which has no visible inner boundaries and which usually lacks air bubbles and admixtures of mineral and organic particles is a deeply homogeneous dark tone of signal (Fig. 4.19a). New ice formed at negative temperatures and strong winds (fast and drift ice) is characterized by reduced density and small-grain structure. Shuga—white porous ice lumps—forms in windy, rough conditions from grease ice and inner ice emerging at the surface. The radar signatures of such ice are signals of bright white exposure (Fig. 4.19b).

In Figs. 4.20a, b–4.21a, b showing the ERS-1/2 SAR images and decoded schematic maps of Lake Onega on 13 January 1995 and 18 February 1996, the age types of fast ice formed during the first phase of freezing at weak and strong winds are marked as gradations 1a and 1b, correspondingly.

Parts of lake area covered with shuga formed of slush due to solid air deposits onto the open water are marked in Figs. 4.20–4.21 as 1c. These areas of bright white exposure differ from areas 1a and 1b by their lacking clearly expressed orientation of the optical axes of ice crystals. They are particularly clearly seen in the ERS-1&2 images of the period of lake ice melting—Fig. 4.22.

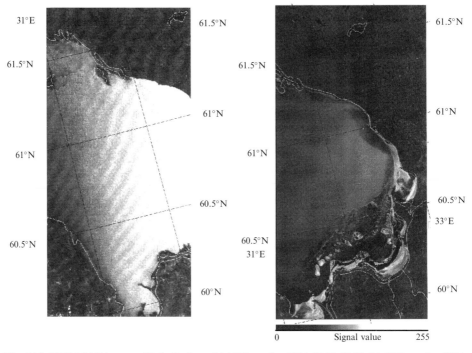

31°E

61.5°N

61.5°N

61.5°N

61°N

61°N

61°N

61°N

60.5°N

60.5°N

60.5°N

60.5°N

33°E

60.5°N

60.5°N

31°E

60°N

60°N

0   Signal value   255

Fig. 4.19. ERS-2 SAR images of Lake Ladoga, (a) 12 December 1996, 09.03 GMT; (b) 28 December 1996, 09.00 GMT: 1996/97 winter was moderate.

Comparison of subsequent ERS-1&2 images (e.g. comparison of images and decoding maps in Figs. 4.20–4.21) shows that measurements of areas covered with ice of gradations 1a, 1b, and 1c, assessments of their spatial distribution over the frame enables estimation of the ratio of the number of days with breezy and windy weather as well as the surface wind direction and forcing. It is also possible to estimate the parameters of solid air deposits onto the water surface up to the end of the first phase of freezing (for Lake Ladoga up to 14 December). Referencing of the ERS-1&2 SAR images to the bathymetric map (see Fig. 2.23) of the water body makes it possible to estimate not only the sum of degree-days but also the change of surface air temperature in the region during the first phase of freezing since, as calculations show, the sum $\sum T_B$ determining the motion of fast ice per each subsequent 10-m isobath is $-166°C$. That is, analysis of the parts of SAR images referring to the first phase reveals a possibility to carry out remote control of air temperature and wind speed in the surface layer and to assess their variability and the results of forcing.

According to Melentyev *et al.* (1997), the onset of the second phase of freezing should be related to the moment when the temperature of deep lake waters is lower than 4°C, and the wintertime dome of dense water tries to smooth and flatten. The second phase of freezing characterized by heat advection at the bottom with water temperature 3°C leads to slowing down (or even to temporary total cessation) of ice

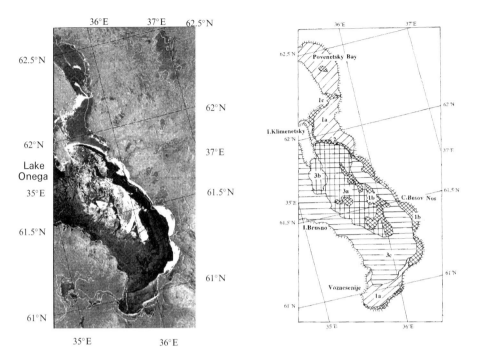

Fig. 4.20. ERS-1 SAR image (a) and composite ice map (b) of Lake Onega, 13 January 1995, 08.46 GMT: 1994/95 winter was weak.

formation on the lake. Only after the effect of advection ceases, does floating ice start forming again and fast ice accretion renews, the fast ice edge becoming the center of ice formation—Fig. 4.23, and 4.26.

In Figs. 4.20a, b–4.21a, b, mentioned above, and in Figs. 4.24a, b–4.25a, b showing the ERS-1&2 SAR images and decoding schematic maps of Lake Ladoga as of 21–24 January 1995 (mosaic) and 6 March 1996, the ice formed during the second phase of freezing is marked as gradation 2. Analysis of the ERS-1&2 SAR imagery archive shows that the radar signature of the ice of phase 2 can be a signal of both bright white and dark tones (depending on parameters of surface wind).

According to the proposed classification, the onset of phase 3 of freezing is connected to the moment when the temperature of deep waters of a large inland water body drops to 2°C and below—Figs. 4.24 and 4.26 illustrate the ice conditions of partial freezing of the lake. The main type of ice forming during phase 3 becomes the ice rind—the transparent crystalline ice of increased hardness (ice-cutter)—it is indicated as 3a, 3b, and 3c. In mild weather followed by strong winds sub-dome water moves to the shore and, being involved in the all-winter vertical cycle together with heat preserved during freezing, is taken to the center of the lake and keeps the deep-water part of lakes and water reservoirs open until spring.

The ice formed during phases 1 and 2 may be wind-driven over the entire water basin. Note should be taken that at this time (in contrast to early winter conditions)

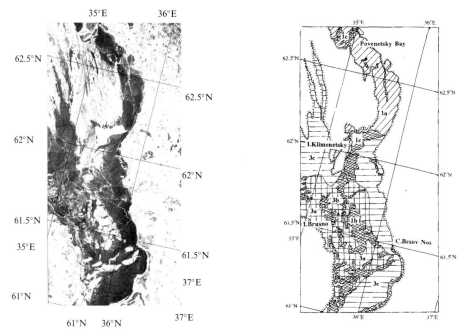

Fig. 4.21. ERS-1 SAR image (a) and composite ice map (b) of Lake Onega, 18 February 1996, 19.25 GMT: 1995/96 winter was very strong.

ice melts neither in the central nor in other parts of the lake. During phase 3 this ice floe becomes the center of newly formed ice—on schematic maps it is indicated as 1a, 1b, and 2 within area 3. Knowledge of signature features of the identified types of fresh-water ice makes it possible to estimate atmospheric forcing on the lake, their seasonal change, and to indicate direction and speed of surface wind blowing in the region during phase 3 of freezing as well.

In severe winters when a spell of calm frosty weather lasts for several days, the ice rind covers vast areas of large lakes. Ice rind capturing floating ice over the entire water basin, combining fast-ice edges from shore to shore, fixes various ice processes and phenomena taking (and having taken) place in the lake at the moment of time preceding its formation. Depending on the character of wind forcing, ice rind differs somewhat by its radar signatures: areas 3a, 3b, and 3c in Figs. 4.20–4.21 and Figs. 4.24–4.26, respectively. According to the authors' classification, the time of ice rind formation of type 3c is shifted to later terms.

4.2.3.3   Study of ice parameter and ice regime features of large lakes

Analysis of the data archive shows that the wintertime ERS-1&2 SAR survey of inland water bodies enables fixing, apart from parameters of open water condition, of all age types of ice corresponding to the identified three phases of freezing of inland water bodies. An analysis has been performed of archived TV, IR, passive

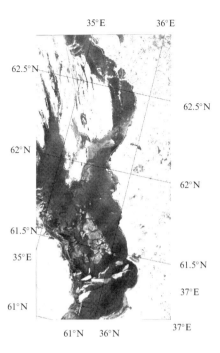

Fig. 4.22. ERS-1 SAR images of Lake Onega, 28 April 1996, 19.25 GMT: 1995/96 winter was very strong.

microwave, and SAR images which refer to both typical (mean-climatic) and abnormal ice conditions characteristic of north-western Russia for winters at both extremes: very mild and very severe. As regime-forming factors characteristic of the period of freezing, air temperature and atmospheric circulation were considered as well as surface wind characteristics (speed and direction), annual change, seasonal-spatial and interannual variability in winds over the water surface.

Using ERS-1&2 data to retrieve ice-cover parameters and to reconstruct the ice regime we will now demonstrate features based on analysis of the 13 January 1995 SAR image of Lake Onega (Fig. 4.20a, b).

The initial formation of Onega fast ice in the winter 1994–1995 (weak winter, according to Alpers' classification) started along the south-eastern (and southern) coastline—these are areas of bright white exposure 1b on the SAR image. Ice formation took place in conditions of strong westerlies and south-westerlies. Due to repeated bouts of warm weather the snow cover on the surface of fast ice re-froze and melted again. In mid-November there was frosty and low-wind weather in the region corresponding to areas of smooth fast ice 1a formed in Povenetsky Gulf. The subsequent period of deep cyclones was followed by stormy weather and heavy snowfall: at this time an area of shuga formed at the shallow-water cross-piece separating Povenetsky Gulf from Small Onega—area 1c on the SAR image. Then followed a sufficiently long spell of frosty low-wind weather when the ice froze-up Small Onego Gulf.

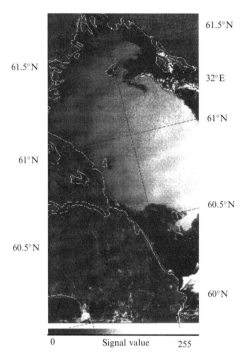

61.5°N

61.5°N

32°E

61°N

61°N

60.5°N

60.5°N

60.5°N

60°N

0   Signal value   255

Fig. 4.23. ERS-2 SAR image of Lake Onega, 31 December 1996, 09.06 GMT: 1996/97 winter was moderate.

Phase 2 was rather short: areas of bright white exposure 2 at the southern point of Small Onego Gulf and in the southern part of Svir' Bay were very small.

The onset of phase 3 was followed by heavy winds. The vast zone of bright exposure (1b) observed in the center of the frame had been part of the fast ice. In the first 10 days of January, under the influence of heavy southern and south-eastern winds it was torn off Svir' Bay and the eastern shore of the lake, drifting to the center of the lake. This event has been exactly timed as substantiated above in the deep waters of the lake showing that ice frozen in phase 3 does not melt until spring (see Fig. 4.22: 28 April 1996 is the beginning of melting processes).

So, phase 3 of Lake Onega freezing proceeded in frosty and low-wind weather—areas of smooth newly-formed fast ice along the 20-m isobath in Svir' Bay correspond to this phase as does ice rind freezing-up the fast ice breccia in the center of the lake and covering at the moment of survey (13 January 1995) the whole shore-to-shore water surface.

Due to the fast ice boundary closely following the 20-m isobath on the image, it can be stated that the sum of negative air temperatures in the air surface layer for the Onega region at the end of phase 1 (mid-December 1994) should constitute $-332°C$. Considering the mean-climatic data, the onset of winter 1994–1995 can be attributed to mild winters (Hupfer, 1979). The conclusion drawn from SAR data is confirmed

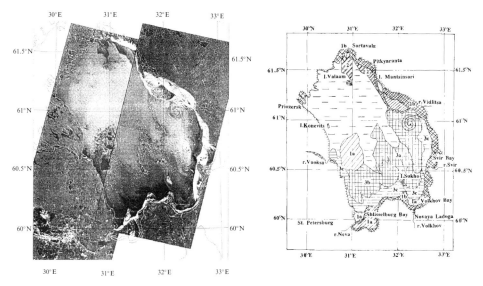

Fig. 4.24. ERS-1 SAR image (a) and composite ice map (b) of Lake Ladoga, 21–24 January 1995: 1994/95 winter was weak.

by the Petrozavodsk HMS (hydrometeorological station): the sum of air temperatures for November–December reached $-328°C$.

Independent comparison of the data of our analysis with synoptic maps has shown their good agreement. Note also that from SAR images one can identify trends of wind and temperature development typical of the St. Petersburg region as well as their interannual variability. Comparative analysis of Lake Onega SAR survey (Figs. 4.20–4.21) shows, for example, that the early winter 1995–1996 turned out to be more frosty than early winter 1994–1995 (considerably exceeding areas 1b in Fig. 4.21), but dynamic forcing on the ice cover was also greater: practically all the initially formed fast ice turned out to be fractured and under the influence of south-easterlies spread over the lake. SAR images confirm thereby the known climatic conclusion about the average south-western transfer of air masses prevailing in winter (January) in the north-western region of Russia.

4.2.3.4　*Use of ERS-1&2 SAR data for study of wintertime dynamic and circulation processes in large lakes*

The results of analysis of 21 January 1995 and 24 January 1995 ERS-1&2 SAR images of Ladoga (Figs. 4.24a, b) show that by the moment of survey the fast ice closely following the 20-m isobath had settled along the southern and eastern coasts of the lake. The ice is covered with snow, and under the snow layer it is rather inhomogeneous: the structural and textural changes in thickness are determined by

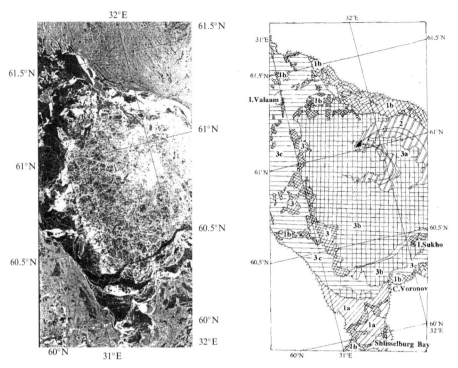

Fig. 4.25. ERS-1 SAR image (a) and composite ice map (b) of Lake Ladoga, 6 March 1996, 09.03 GMT: 1995/96 winter was very strong.

variability in meteorological conditions in the process of ice formation, by repeated melting and wet snowfalls.

In Schlisselburg Bay and farther eastward to Volkhov Bay, the ERS-1&2 image fixes spatially inhomogeneous fast ice. In the zone of fast ice there are numerous dry cracks both going through the whole ice thickness and not reaching the ice bottom as well as areas of impending and rafted ice—their signatures are bright white bands of different width and brightness. Spatial orientation of these deformations points to the fact that they formed under the influence of various thermal and dynamical factors, but mainly because of the southern winds prevailing in this part of the lake in November–December. Prevalence of these winds in this period has resulted in repeatedly fractured fast ice and off-shore ice drifting. It is interesting that even from rough assessments the areas on the image covered with rafted ice within the fast ice (zones of signal that are extra bright) are approximately equal to the areas of darker signal exposure corresponding to the thinner smooth fast ice.

In the authors' opinion, in the first half of January 1995 Lake Ladoga was repeatedly subjected to spells of heavy southern and south-eastern winds: fields of ice breccia ranging from vast down to large are clearly seen on the SAR image and stretching along the south-western coastline of Lake Ladoga right up to Konevits Island apparently originated from the fast ice zone near the southern shallow waters of the lake. Apparently, this drifting had taken place at the same time when in Lake Onega, due to stormy south-eastern winds, part of the fast ice was broken off and taken to the center of the main reach (see the § 4.2.3.3).

The ice rind connecting vast areas in the southern part of Lake Ladoga formed, in our conclusion, in a spell of calm weather: it is confirmed by its structural features, closely following the isobaths' spatial orientation.

The satellite image supports the conclusion that a vast zone of ice rind in the southern part of the lake broke up somewhat later and its northward shift was impeded. These winds were negligible (not stronger than 7–8 m/s) and their impact did not last long—this conclusion was drawn on the basis of estimates of the width of impeded (tucked-under) ice of this type and lacking deformations of ice rind "brackets" indicating the isobaths. Open water areas in this part of the lake were confined in the east to the sandbank of Sukho Island. The ice rind newly formed here on shallow water has originated by the action of small winds. It is almost snowless and thinner—its signature is the signal of darker and more homogeneous exposure.

Near the western point of Volkhov Bay is Ptinov Island, to the south of which lies the shallow sandbank of Sukho Island, where ice formed approximately at the same time as coastal fast ice. As seen on the SAR image, the island fast ice whose signature is the signal of inhomogeneous bright white exposure had also suffered repeated transformations of melting and accretion. The sickle-shaped spit of shuga north of Sukho Island marks the shallow waters in this part of the lake.

It turns out that the distribution of ice east and west of the "Ptinov Island–Sukho Sandbank" line is different. This reflects the features and the character of water circulation in the southern part of Ladoga. That is, based on SAR imagery data the conclusion can be drawn, extremely important for description of the ecological situation in the St. Petersburg region, that the "Ptinov Island–Sukho Sandbank" line turns out to be a watershed, limiting penetration of River Volkhov water (the most polluted in this region) directly to the River Neva. This circumstance is extremely important for understanding the problem of provision of St. Petersburg with pure water and its solution.

Some special features of ice-cover construction in this part of the lake on the ERS-1&2 SAR image (character of the spatial orientation of the arcs of newly formed ice rind and southward direction of these arcs) support the conclusion that Svir' and Volkhov waters in fall and winter seasons are involved in the cyclonic circulation, rushing in a narrow band along the fast-ice edge round Lake Ladoga's coastline. Some special manifestations of this circulation near the eastern shore of the lake will be demonstrated below.

The fast ice in Svir' Bay is different in its properties: on shallow waters within the bay area of the River Svir' the ice is smooth, without any roughness at the water–air interface (its signature is the dark tone of the signal exposure), but in most of Svir' Bay the ice-cover structure of fast ice due to external forcing is deformed. Beyond the coastal fast-ice line the ice rind binds the shuga, zones of ice breccia involving fast-ice fractures. Floes of large and small-broken ice are seen on the SAR image in the form of a narrow thin bank stretching along the boundary with the open water surface far to the north. A small part of this bank wedges southward to the center of the lake.

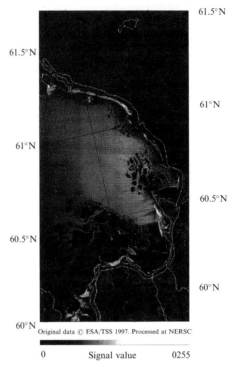

Fig. 4.26. ERS-2 SAR image of Lake Ladoga, 13 December 1997, 09.01 GMT: 1997/98 winter was moderate.

The 13 December ERS-2 SAR image of Lake Ladoga is shown in Fig. 4.26. As seen in the figure, ice cover at the southern part of the lake has contrasting signatures compared with the 1994/95 and 1995/96 winter seasons. This is connected to the fact that ice formation and ice dynamics on Lake Ladoga were affected in 1997/98 by another atmospheric circulation: beginning of the winter was calm and frosty, and only strong south-easterly winds that occurred over 11–12 December 1997 were able to strengthen the cyclonic circulation around the whole lake, lessening over the winter, and broke level fast ice along the southern shore.

4.2.3.5 Anthropogenic factors affecting the environment of the St.-Petersburg region: experience of ERS-1&2 SAR diagnostics of water pollution and processes of self-clearing in large lakes

To solve the problem, we shall consider the ice-cover conditions in the north-eastern part of Lake Ladoga (Fig. 4.24a, b); that is, in the part where, north of the mouth of the River Vidlitsa, the Kondushskaya Shoal projects into the lake and where, from this shoal towards the southern point of Mantinsari Island, stretches a chain of banks. The satellite SAR image makes it possible to retrieve the history of ice formation in this region.

So, we can reliably state that initially fast ice settled here in conformity with the bottom relief around the shoal mentioned above and the entire chain of banks, and later on (at the beginning of the second phase of freezing) the consolidated shuga ice joined the fast ice from the off-shore edge. In the first 10 days of January 1995 during a spell of stormy westerlies the fast ice broke along the shore and in the mouth of Vidlitsa River. The coastline configuration in the region considered is such that westerlies can strengthen the constant cyclonic circulation of dense waters. The presence of ice in the region of the Kundushskaya Shoal can prevent water masses from mixing and formation of the closed spiral of cyclonic circulation in the near-deep layer north of the Vidlitsa River. So, it turns out that the satellite SAR survey gives a unique opportunity to fix in winter the regulated structures forming on the surface of large lakes as a result of forcing by atmospheric meso-vortices: the data of the 6 March 1996 SAR survey (Fig. 4.25a, b) fixed a spiral of larger-scale cyclonic circulation appearing in the same part of Ladoga in the winter 1995–1996 (very strong winter, according Fig. 4.18a).

Let us dwell on analysis of this phenomenon from the viewpoint of its significance for ecological safety and self-clearing processes of the waters of large lakes.

Lake Ladoga is known to be the main source of fresh water for more than the 5-million population of St. Petersburg and its region. Water sampling for the city is now made in the region of Schlisselburg Bay near the source of Neva River. The existence in this gigantic fresh-water reservoir of the cyclonic cycle of waters considerably lengthening the path of polluted waters before they get to Neva favours, as mentioned above, self-clearing of the lake. But on the other hand, the presence in Ladoga at fall and at beginning of the winter season of water masses with drastically different properties (thermally active zone in the near-shore area and thermally inert zone in the center of the lake) has negative consequences for supplying the city with drinking water.

It is this circumstance that leads to the fact that only a narrow band of near-shore water is involved in the cycle which, heavily affected by habitable and economic territories and the lake watershed, turns out to be most polluted. Moreover, due to the mechanism of natural phenomena functioning in the fall-winter period, part of water involved at this time in the cycle happens to be particularly narrowed, so water taken for the city in the period of hydrological winter requires special control and purification. But meso-scale vortices of the type identified on the 21/24 January 1995 and 6 March 1996 ERS-1/2 SAR images favour self-clearing of lake waters because under their influence polluted waters along the coastline are thrown to the center of the lake and taken away from the cycle of water used by the city. It can be supposed that such vortices and regulated circulation also exist in other large water basins. They can form with certain combinations of wind forcing and currents and in the period of open water, and their effect on the life of inland water bodies may be quite different during the year. Up to the present time, they have not been observed simply because of lack of respective measuring instruments: only ERS-1&2 SAR diagnostics have given the unique possibility of monitoring such natural phenomena.

Consider ice conditions in the northern and north-eastern parts of Lake Ladoga. An assessment of the state and quality of the waters in this region is extremely important, since several of the largest pulp-and-paper industrial centers of north-western Russia are located here, which substantially affect the ecological condition of the water body.

The satellite image of the skerries part of the lake shows the fast ice (bright white tone of the signal exposure), and north and north-west of Mantinsari Island between depths of 40–60 m the drifting shuga ice—ice formation of the second period of freezing—is observed. Note the characteristic ice formation with the deep dark inhomogeneous tone of signal exposure located in the strait between Valaam Island and the skerries region of the lake. All this vast area is covered with even ice formed around small islands north of Valaam Island. Here ice appeared during the second phase of freezing due to circulation pushing coastal waters to the center of the lake. This ice formed by accretion of light nilas, and at the moment of satellite survey it preserves some areas of open water, the traces of rafting and hummocking are seen—these characteristics of the gray-ice surface (thickness 10–15 cm) create the inhomogeneous radar signatures mentioned above. The gray-ice area with the polluted waters of Sortavala Gulf as its source, in our opinion marks the boundaries of propagation of polluted waters from the pulp-and-paper plants of Sortavala and Pitkyaranta. The heavily polluted waters of this region lead to a change in surface viscosity of lake waters favouring smoothing out of capillary waves and formation of a vast slick zone—this feature results in a deep dark tone of the signal signature.

4.2.3.6 Using ERS-1&2 SAR data for study of winter seiche currents in large lakes

The 13 March 1995 ERS-1 SAR image of the Chudskoye, Teploye and part of the Pskovskoye lakes is shown in Fig. 4.27a. According to Alpers (1997), the 1994/1995 winter season was classified as a weak winter.

Our conclusion is that freezing-up of Chudskoye Lake took place in three phases: phase 1 was characterized by formation of fast ice near the eastern coastline, a small part of which after breaking was driven by south-westerly winds to the center of the lake. During phase 2 drifting shuga ice formed near the fast ice of the eastern coastline. The lake became frozen-up with the formation of ice breccia fields and ice rind (phase 3).

As seen in the figure, ice in Lakes Teploye and Chudskoye have contrasting signatures. Bright white tone of the signal in Lake Teploye is explained by structural features of rotten ice which can form on this lake. This is connected to the fact that ice formation on the Lake Teploye follows quite different laws (Melentyev *et al.*, 1997). Here total freezing is systematically observed. However, the ice of Lake Teploye forms much later compared with Lakes Chudskoye and Pskovskoye (which is indirectly reflected in the name of the lake—Teploye (Warm)). The point is that in Lake Teploye, which is in fact a narrow strait connecting Lakes Chudskoye and Pskovskoye, seiche currents are constantly active, catching warm bottom-water masses of both larger water bodies. It is this circumstance that strongly prolongs the process of ice formation here. Figure 2.35 shows a picture of the polynya in

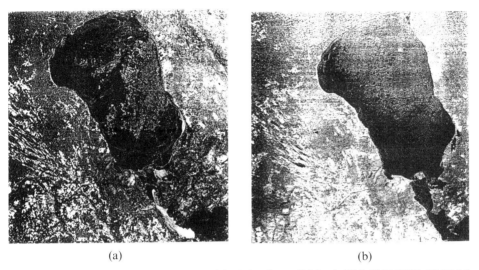

(a)   (b)

Fig. 4.27. (a) ERS-1 SAR image of Lake Chudsko-Pskovskoye, 13 March 1995, 19.25 GMT. (b) ERS-1 SAR image of Lake Chudsko-Pskovskoye, 18 March 1994, 19.05 GMT.

Yakimvarsky Gulf of Lake Ladoga not freezing at all with the same seiche origin as in Lake Teploye. Due to the long period of ice formation in the straits where seiche currents are manifested, ice is always porous, unstable and easily broken under load (rotten ice). It constantly threatens both the people and transport using ice roads.

Results of thematic decoding of 13 March 1995 ERS-1 SAR image have been used by authors in an unexpected way: to reconstruct and explain a paradoxical event which vitally affected Russian history (Melentyev et al., 1998). As is known, it is here that the battle between Russian troops under Saint Grand Duke Alexander Nevsky and shock troops of the Teutonic Order had taken place (April 1241) (the battle known in history as "Ledovoye Poboishche"). Historical chronicles testify to the fact that the decisive crucial moment which led to the victory of the Russians was when heavily armed German soldiers started falling under the ice of Lake Chudskoye. In our conclusion, it is connected to the likelihood that troops of the Order did not know the local ice conditions: the troops happened to be in that part of the lake where ice was of the seiche origin. The present SAR images help archaeologists, to locate at the places of seiche manifestations the most probable location of archaeological rarities at the bottom of the lake.

The 18 March 1994 ERS-1 SAR image of the region is shown in Fig. 4.27b. According to Alpers (1997), 1993/1994 winter season was classified as moderate winter and, as seen in the figure, manifestation of the seiche was respectively weak.

Results of long-periodical wintertime seiche forcing were also revealed from the 6 March 1996 SAR image of Lake Ladoga on which they are seen as a vast ice-rind-covered polynya stretching in the latitudinal direction in the southern part of the lake. An appearance of this type of circulation in this part of the lake is explained by differences in the temperature regime of total and partial freezing of the lake (Fig.

4.25a, b), when warm waters remaining at depth in the totally frozen lake under the influence of atmospheric forcing can push out from there to the regions of shallow waters. This type of circulation is the basis for the mechanism of lake self- clearing in conditions of severe and very severe winters.

On the whole, the revealed possibility to fix wintertime hydrological, ecological and dangerous ice phenomena in fresh-water basins using the ERS-1&2 SAR imagery is advantageous for this type of information.

Analysis shows that the ice cover of inland water bodies is a tracer of various natural processes and phenomena, including climatically and ecologically important ones. Our efforts have been concentrated on revealing the main laws of ice formation, enabling identification of three phases of freezing and the main different types of lake ice characteristic of these phases: fast ice, shuga ice including shuga, slush, broken drifting ice and finally the ice rind as a major structure forming fields of ice breccia. The radar signatures of these structures have been traced in detail.

Experience of complex decoding of ERS-1&2 SAR images of the ice cover of the largest lakes in north-west Russia has shown the prospects for using SAR information not only to assess the parameters of ice-cover conditions and diagnostics of ice conditions in lakes at the moment of satellite survey but also to reconstruct and retrieve the history of lake-ice formation. Moreover, the possibility has arisen to describe from satellite data the various transformations of thermal structures in lakes, to carry out quantitative estimates of climatically significant meteorological parameters as well as to reference and to perform temporal and spatial indexing of the distribution of ice, variations and transformation of the sums of negative temperatures in the region over the given period of hydrological winter. Revelation on SAR images of anatural phenomenon unknown before—the winter spiral-shaped cyclonic circulation of water masses and ice cover—is of considerable scientific and practical interest.

4.3   HYDROBIOLOGICAL STUDIES

4.3.1   Spatial and temporal variations in lake phytoplankton and dissolved organics distribution and macrophyte stands

Until recently, the most important achievements in remote sensing of phytoplankton community status and dynamics stemmed from utilization of water-color characteristics, that is, from the analysis of amplitude and spectral parameters of radiance signals of ERTS-1, Landsat, and Nimbus-7 satellite data as well as many airborne sensors (e.g. ROSIS, CASI, etc.).

Passive remote-sensing techniques in the visible spectrum have some clear advantages. However, they are subject to operational restraints inherent in water-surface illumination conditions. Difficulties also are related to hydro-optical complexity of coastal and inland waters arising from the influence of discharging rivers, run-off and anthropogenic admissions, etc.

Analyses of past (see, e.g., Bukata *et al.*, 1995) and current publications (Fisher *et al.*, 1991; Dekker, 1993; Yacobi *et al.*, 1995; Arenz *et al.*, 1996; George, 1997; see also references in § 3.1.1.3) dedicated to passive optical remote determinations of case-II (lake and/or marine coastal) water-quality parameters indicates that the major work in this area has been done so far by establishing area-specific or employing literature regression algorithms relating either to water-leaving radiance or volume-reflectance values at various wavelengths to phytoplankton chlorophyll concentration. In § 3.1.1.3, we have already discussed in detail the limitations and shortcomings of such an approach.

Gower and Borstad (1990), Fisher and Kronfeld (1990) and some others explored phytoplankton natural fluorescence as a tool to remotely determine phytoplankton chlorophyll abundance in top-water layers of natural water bodies. These investigations revealed (see also § 3.1.1.2) that the signal remotely recorded at wavelength 685 nm could be practically used as a measure of chlorophyll concentration. However, the intensity of chlorophyll natural fluorescent emission is variable and generally fairly low.

As discussed in § 3.1.2, active (lidar) remote sensing relaxes some of the limitations characteristic of passive remote sensing in the visible spectrum and can be considered as an alternative or rather complementary approach to studying *in situ* phytoplankton spatial/temporal distributions. Indeed, this technique largely facilitates remote assessment of concentrations of fluorescent components of natural waters, such as phytoplankton chlorophyll (*chl*) and dissolved organic matter (*doc*).

Pozdnyakov and Kondratyev (1997) applied this approach in studying lake time-dependent patterns of spatial distributions of *chl* and *doc* in a number of large lakes and water-storage reservoirs of the former USSR. In experiments, helicopter-borne lidar was used. Developed at the state University of Yerevan (Arst *et al.*, 1990), the lidar complex consisted of an Nd : YAG laser, calibration light source, transceiver, optical system, receiving optical telescope, polychromator, a system of sensors and amplifiers, delay chains, analog-to-digital converter, and a computerized data acquisition, processing and readout system.

As discussed in § 3.1.2, the optical response signal encompasses a component caused by water Raman backscattering as well as components arising from phytoplankton and dissolved organic matter fluorescence. The response signal also contains background noise resulting from reflection of direct and diffuse solar radiation at the water surface. Identification of these components and subtraction of background noise were performed automatically. Background noise-signal recording was several microseconds delayed in the regime of legitimate response signal reception, but without laser operation. With such a delay (the carrier's flight velocity being about 100 m/s), the background signal is generated by the target (water volume) sounded previously by laser beam.

In studying 2-D distributions of *chl* and *doc* concentrations in inland waters, the sensing impulses were lased at 532 and 337 nm, respectively. When sensing chlorophyll, the confusing component caused by *doc* fluorescence was extracted from the total response signal using a "pedestal" extraction technique: assuming $P_R = 0$ at $\lambda = 600$ nm and $P_{doc} = 0$ at $\lambda = 700$ nm, the P_{doc} signal can be subtracted provided the *doc* fluorescence band contour is known (Abroskin *et al.* 1986; Fadeev

Table 4.1. Parameters of a regression relationship relating C_{chl} to ratio $n = F/R$ (for notation see § 3.1.2) for different types of water basins: $C_{chl} = a + bn$. N is the number of stations used for establishing the retrieval algorithms.

Water basin	a	b	r	N
Baltic Sea	−0.38	2.114	0.69	13
Lake Ladoga	−0.48	4.015	0.67	12
Lake Onega	−0.60	3.330	0.63	17
Rybinsk Reservoir	−0.62	5.375	0.85	12
Lake Sevan	−0.68	3.480	0.65	10
Samara Reservoir	−2.40	10.000	0.66	6

1978). When sensing *doc*, the Bristow *et al.* (1981) technique of filtering up the legitimate response signal was used.

Table 4.1 gives numerical values of regression coefficients for C_{chl} retrieval obtained from analysis of experimental data from several water basins. For the water bodies examined, the correlation coefficients of regressions ranged from 0.63 to 0.85 (Table 4.1.) for stations where concurrent remote and *in situ* measurements of C_{chl} were carried out. The number of stations in different water bodies subjected to surveillance ranged from 6 to 17.

During the 1990 field campaign, helicopter flights with lidar on board were conducted over Lake Onega along a transect (straight line in Fig. 4.28) running from Petrozavodsk to Besov Nos Cape. A high resolution 2-D distribution of C_{chl} (Fig. 4.28) was retrieved (Kondratyev, Pozdnyakov, 1987a, b).

Ground truth data were concurrently collected from a research vessel. As shown in Figure 1a, a patchy distribution of C_{chl} in the open and deep part of the lake is apparent (the test area is centered at 61°45′N, 35°E). Analyses of the thermal maps of this part of the lake for pre- and post-survey periods (Figure 4.28b, c) indicate that the thermal field was considerably heterogeneous in the region of sensing. On August 11, a warm spot was discovered in the center of an anticyclonic gyre generated in this part of the lake. Apparently, by the time of sensing, further enhancement of this gyre had taken place and there was a drift of the warm spot (core) to the south-west (with concurrent spreading of its area). A subsequent storm (it started at the moment of sensing) led to intensive turbulent mixing of the lake upper layer and, as a result, its surface temperature became homogeneous (Figure 4.28c). It should be emphasized that the *en route* temperature profile obtained by a ship-towed electrical sensor justifies the application of thermal remote-sensing data to interpret phytoplankton distribution. This temperature profile showed that C_{chl} faithfully followed lake-surface temperature variations. In addition, according to model simulations (Filatov, Zaytsev 1993) and shipborne sensing data, the observed anticyclonic gyre only involved the upper 5-m layer, below which a cyclonic circulation took place due to the effects of baroclinicity and bottom relief. This explains the stability of the area

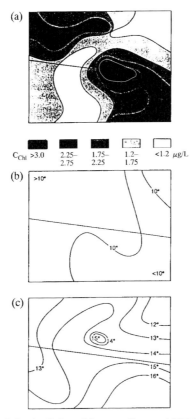

Fig. 4.28. Spatial variation (a) of C_{chl} (µg/l) from lidar soundings and SST (°C) registered several days (b) before, and (c) after lidar soundings.

with a heightened *chl* concentration: water layers involved in cyclonic circulation, as if supported from below the "eye" of the anticyclonic water gyre, prevented phytoplankton from submersion into deeper layers of the lake. The actual occurrence of such situations in Lake Onega was previously substantiated by data obtained from multispectral space imageries of Lake Onega (Kondratyev *et al.* 1988).

As Karabyshev (1987) points out, for practical purposes and understanding of spatial and temporal variability of biological and associated oceanological characteristics (*chl* fluorescence, in particular), it is important to reveal the regularities in temporal variability of hydrodynamic parameters with periods ranging from tens of minutes to several days, and spatial variability from hundreds of meters to tens of kilometers (i.e. the periods and spatial scales starting at the upper limit of small-scale phenomena and covering the whole mesoscale range). Lorenzen (1967), Platt (1972), Denman (1976) and Denman *et al.* (1977) were among the first workers who studied 1-D distributions (both in time and along the ship route) of the C_{chl} fluorescence in marine and oceanic environments.

However, 1-D distributions recorded at a fixed horizon are not easily interpretable since it is pretty difficult (sometimes even impossible) to determine from such data the actual configuration of spatial inhomogeneities in the distribution of the parameter under investigation.

Steel (1978) was among those pioneers who obtained 2-D mesoscale distributions of *chl-a* fluorescence from on board a research vessel (at a depth of 3 m) along meridional transects within an area of 24×24 km. It should be noted that in studies conducted by Steel (1978), Karabyshev (1987) and Marra (1982), the distribution of *chl-a* fluorescence was not similar to the relevant water-surface temperature field, and local enhancements in C_{chl} were observed several kilometers off the front.

Karabyshev (1987) hypothesized that the phytoplankton community can respond to variations in external conditions with a delay determined by the rates of growth and decay of aquatic plants, i.e. approximately 24 hours.

Given a high spatial and temporal resolution of C_{chl} distribution throughout Lake Onega estimates of the characteristic scales of the inhomogeneities were done using correlation matrices, autocorrelation functions, and C_{chl} variance spectra in the horizontal plane in the upper layers of the lake. The field of *chl* spatial distribution (Fig. 4.28) was subdivided (one pixel size corresponds to the spatial resolution of C_{chl} data along the flight path, i.e. about 300 m), with subsequent assessment of the coefficients of correlation between the lines and columns of the resulting triangular matrix. Figure 4.29 shows the results of such statistical calculations. Interestingly, the autocorrelation function intersects the abscissa at $\tau = 12$ km.

As regards the spectrum of C_{chl} variability, it is characterized by a maximum followed by a decrease with a slope proportional to $k \sim -5/3$. Below, we shall consider these results and their implications to Lake Sevan (Figs. 4.30 and 4.31).

With regard to the problem of deterioration of Lake Sevan's ecological status caused by a dramatic 18-m drawdown due to excessive withdrawals during the previous decades, a series of comprehensive limnological studies were conducted. In conjunction with the Sevan Hydrobiological Station, Pozdnyakov and Kondratyev (1997) mapped the spatial distribution of C_{chl} for four consecutive days in early September 1990 from on board an MI-8 helicopter carrying the lidar complex. The airborne data was supported by concurrent ground truthing.

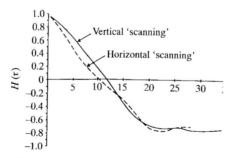

Fig. 4.29. Averaged autocorrelation function $K(\tau)$ calculated from *chl* distribution shown in Fig. 4.28a.

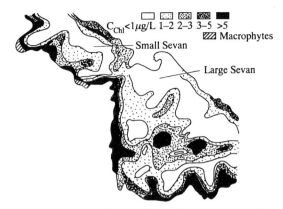

Fig. 4.30. Spatial distribution of C_{chl} in upper layers of Lake Sevan as retrieved from lidar soundings.

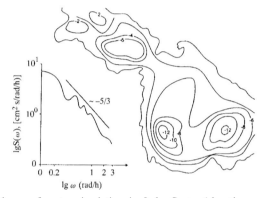

Fig. 4.31. A scheme of water circulation in Lake Sevan (elevation contour in mm) inferred from simulation runs with the use of diagnostic models (Filatov, 1983). The current power spectrum is shown in the inset (Filatov, 1983).

As in the case of Lake Onega, a map of C_{chl} 2-D distribution at a similar spatial resolution has been obtained for Lake Sevan (Fig. 4.30). The patterns correspond with bathymetry and with results of multi-annual surface-layer observations of C_{chl} (Parparov, 1979). In Small Sevan (the northern part of the lake), the average *chl-a* concentration is less than in Large Sevan (the southern part of the lake). The C_{chl} distribution over Large Sevan is characterized by the formation of a large area of enhanced C_{chl} located in the central, pelagic (i.e. the deepest) part of Large Sevan.

Simulations of the water dynamics performed for the same season (Fig. 4.31) explain this somewhat unexpected effect (Filatov 1983).

As numerical experiments have shown, a cyclonic water circulation exists in Large Sevan in this season (with counterclockwise motion of water). Due to adjustment of water velocity and density fields at the center of the lake, an upwelling of isopycnic layers towards the lake surface is bound to be formed. The vertical velocity component reaches about 10 cm/s. As a result, in this part of the lake, the upper

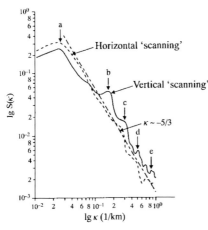

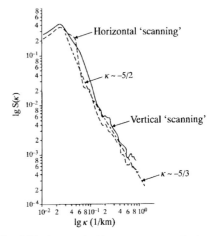

Fig. 4.32. Averaged *chl* power spectra calculated from *chl* distribution.

Fig. 4.33. Averaged *chl* power spectra calculated from *chl* distribution.

quasi-homogeneous layer turns out to be propped up to the surface. Within the center of the cyclonic circulation, this layer becomes thinner than in the coastal zone and the concentration of suspended particles in it (including indigenous phytoplankton) becomes enhanced. Spatial equalization of chlorophyll concentration is hampered by the fronts. These were revealed during sensing experiments by observing foam stripes. Moreover, water temperature gradients of about 1°C/km were recorded by the airborne IR spectrometer which operated concurrently with the lidar sensings. The fronts show up (in the form of quasi-concentric circles) on the lake surface when the cyclonic circulation of water periodically becomes intensified due to atmospheric forcing.

In Small Sevan, in pelagic waters, C_{chl} concentration proves to be uniformly distributed. From lidar sensings and results of numerical experiments, there is no persistent cyclonic circulation in Small Sevan. Thus, there are no conditions for the formation of the patchy structure of the *chl*-field, like that observed in Large Sevan.

The power spectrum $S_{chl}(k)$ for chlorophyll in Large Sevan is characterized by a uniform slope (Fig. 4.32) with a wave number dependence of $k \sim -5/3$. In this range of spatial frequencies, the distribution of the *chl*-field is affected by such thermohydrodynamic processes as regular cyclonic circulation of water, internal motions and waves (see the inset in Fig. 4.30). Similar results have been also reported in other publications (see, e.g, Denman 1976).

Analysis of *chl* spatial variance spectra lg $S(k)$ (Figs. 4.32 and 4.33) and the behavior of the autocorrelation function (Figs. 4.34) (i.e. the variance spectra and autocorrelation functions averaged over the horizontal lines and the vertical columns of the matrices, whose elements are C_{chl} in a pixel calculated for the Lake Onega test area (Fig. 4.28) and that of Large Sevan (Fig. 4.30), suggests several interesting conclusions.

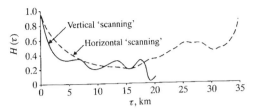

Fig. 4.34. Averaged autocorrelation functions $K(\tau)$ calculated from *chl* distribution.

First of all, for Lake Onega, the autocorrelation function transects the abscissa at $\tau = 11$–12 km, whereas in the case of Large Sevan, it never becomes zero. One possibility is that in Large Sevan, the characteristic length scale of the largest inhomogeneities is comparable with the size of the lake itself. As for mesoscale inhomogeneities, it is seen in Fig. 4.34 that they extend about 5 to 6 km and are determined by traces of a series of frontal zones of concentric shape related to quasi-periodic intensifications and subsidings of cyclonic vorticity of the field of water motion velocities in Large Sevan (see above). These conclusions are further substantiated by specific features of $S_{chl}(k)$ spectra (Fig. 4.34).

In the case of the Lake Onega test area, the prevailing maximum in the $S_{chl}(k)$ spectrum corresponds to characteristic length scales of about 30 km, which are comparable with the actual size of the test area. In a shorter wavelength region (i.e. in the region of large wave numbers), there is a sudden change in the slope. For the interval corresponding to characteristic spatial scales of about 6 km, the slope is proportional to $k^{-5/3}$; in the longer wavelength interval (about 10 to 15 km), this slope is steeper ($\sim -5/2$). As shown by numerical modelling (see, e.g., Filatov, 1983), the slope in question, when proportional to $k^{-5/3}$, is characteristic of conditions where the hydrodynamic situation is determined by locally isotropic turbulence. It is clear that the presence of an interval with a wave-number dependence of $-5/3$ in the $S_{chl}(k)$ spectrum points to the existence of mesoscale quasi-regular inhomogeneities in the field of *chl* distribution, probably dominated by such mechanisms as internal motions, zones of upwelling and multi-nodal seiches. Such quasi-regular inhomogeneities are clearly seen in Fig. 4.28 as the *chl* field vorticity.

In the case of Large Sevan, the $S_{chl}(k)$ spectrum also exhibits a maximum (Fig. 4.33), corresponding to inhomogeneities in the distribution of the C_{chl} field of about 40 km, which is comparable with the size of this part of Lake Sevan. In the interval of larger wave numbers, the *chl* variance spectrum decreases less sharply and is proportional to $k^{-5/3}$, which is typical of *chl* distributions dictated by turbulent motions.

On the whole, it can be stated that both Lake Onega and Large Sevan C_{chl} distributions (being subject to general regularities of basin-water circulation) do not show any pronounced regular pattern in the region of mesoscale length scales. Being caused by numerous thermohydrodynamic factors, the spectra of spatial variability of C_{chl} typical of water-eddy motions correspond to these fields.

As pointed out above, the patchiness in spatial distribution of C_{chl} is inherent in a number of specific mechanisms of physical and/or biological nature. Within the range of spatial scales 0.001 to 1 m, molecular diffusion and viscosity are controlling mechanisms which form an isotropic flow of the fluid. In the spatial ranges exceeding several meters, the density stratification prevents development of vertical motions, which results in rendering eddy as well as periodic motions progressively horizontal. Within the range of spatial coastal waters, the rotation of the earth becomes a controlling factor.

A phytoplankton patch persists if the growth rate $b(t^{-1})$ is sufficient to compensate for the influence of eddy diffusion. Hence, there should be a characteristic length scale at which b starts prevailing over eddy diffusion. This transition in the power spectrum $S_{chl}(k)$ reveals itself through a change in slope. According to Platt and Denman (1975), slopes with wave-number dependence of k^{-1} and $k^{-5/3}$ are characteristic, respectively, of the first and second mechanisms mentioned above. For the region of small frequency numbers, geostrophic motions and the spectrum proportional to k^{-3} are most appropriate. It is precisely in this range where slopes of *chl* power spectra and water-surface temperature are likely to coincide (parallel) since they result from the action of the same physical mechanisms.

The above data provided by Pozdnyakov and Kondratyev (1997) seem to substantiate the scheme outlined by Platt and Denman (1975). Indeed, in Fig. 4.34, the region of the biological "window" (k^{-1}) as well as the region with slope proportional to $k^{-5/3}$ are readily discernible, thus strongly supporting the interpretation of *chl* 2-D distributions suggested hereinabove.

It is noteworthy that the results reported by Pozdnyakov and Kondratyev (1997) for large and deep temperate lakes in many ways are very close to those reported by Smith *et al.* (1988) for oceanic waters. We believe that this correspondence is indicative of similarities (with some appropriate limitations) in limnological and oceanographic processes and is supportive of the idea of using large deep lakes as physical models of oceans (see also § 1.5).

Lidar studies of *chl* mid-summer distributions in the upper layers of Lake Ladoga (Pozdnyakov and Kondratyev, 1997) revealed (Fig. 4.35) for central and southeastern parts of Lake Ladoga a highly inhomogeneous character of the C_{chl} 2-D distribution in the zone confined by the 10°C isotherm, covering the inlets of the Svir' and Volkhov Rivers. Within the zone limited by 10°C and 6°C isotherms, the field of C_{chl} becomes more homogeneous. The homogeneity becomes further enhanced beyond the 6°C isotherm. The field of C_{chl} distribution is particularly interesting in the inlet of the Volkhov River. Here, one can observe a large horseshoe-shaped intrusion of pelagic lake water with comparatively small content of *chl*. Closer to the shoreline, this water mass gets "broken" into small fragments, regularly alternating in space with coastal waters more heavily laden with *chl*. This pattern of C_{chl} distribution in the south-eastern part of the lake is not permanent. In fact, it varies considerably in time, apparently due to wind increasing or subsiding as well as wind direction changes which actually occurred during our flights over the lake.

Figure 4.35 gives an example of C_{doc} spatial distribution inferred by us from lidar sensings of Lake Ladoga in July 1989. Retrieval was carried out making use of a

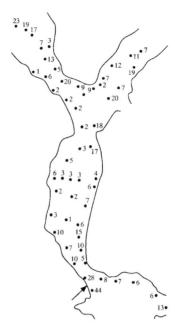

$C_{Chl}(1\mu g/L)$ >6   4–6   2–1   <1   lacunae   isotherm

Fig. 4.35. A fragment of the *chl* distribution in the upper layers of Lake Ladoga as retrieved from lidar soundings.

Fig. 4.36. Spatial distribution of C_{doc} (mg C/l) in Samara Water Storage Reservoir as retrieved from lidar soundings.

regression relationship ascertained in the course of preliminary flights synchronized with samplings for the eventual laboratory determination of C_{doc}. It can easily be seen from Fig. 4.36, the heterogeneity is inherent not only in spatial distributions of C_{chl} but also in C_{doc} fields.

This heterogeneity is initially provoked by non-uniform influx of organic matter with runoff. With the onset of summer, as the lake starts warming up and the thermobar begins its retractive movement to the central region of deep water, organic matter both recently introduced (allochthonous) and newly formed (autochthonous) finds itself involved in the steady motion of water masses. One of the temporal phases of this evolution is shown in Fig. 4.36.

Between 1985 and 1987, a series of extensive studies were carried out on Samara water storage reservoir (Pozdnyakov and Kondratyev, 1997). The *chl* concentration was measured on several test sites using the airborne lidar technique. The data obtained from lidar sensings illustrated the basic regularities of formation of horizontal heterogeneity in *chl* spatial distribution in a follow-through-type of water storage reservoir with areas of different stream flows. In the region close to the dam, very high concentrations of *chl* were invariably recorded. It is worth noting that in flow-through basins, coastal areas will differ in *chl* concentration depending on the distance from the mainstream. This effect is clearly discernible in retrieved C_{chl} spatial distributions over Samara Water Storage Reservoir.

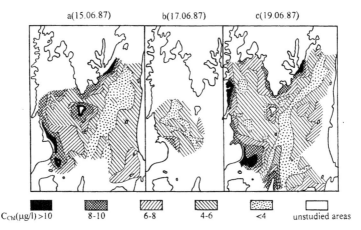

a(15.06.87)   b(17.06.87)   c(19.06.87)

$C_{Chl}(\mu g/l) > 10$ 8–10   6–8   4–6   <4   unstudied areas

Fig. 4.37. The *chl* distribution at Rybinsk Water Storage Reservoir based on three-day lidar soundings.

Within the international experiment, "Inland Waters-87", lidar sensings C_{chl} 2-D distributions in surface waters within a specially chosen test site at Rybinsk Water Storage Reservoir (Fig. 4.37) was conducted over three days June 15(a), 17(b), 19(c) 1987 (Pozdnyakov and Kondratyev, 1997).

The C_{chl} distribution for June 15 is characterized by increased content of *chl* in coastal waters of the south-western coastline of the reservoir. As ground-truth sensings of water temperature and currents showed, upwelling formed in this region on June 15 (apparently caused by persistent easterlies). Two large areas of uniform C_{chl} distribution were observed with *chl* concentration averaging 4–6 and 6–8 µg/l, respectively.

On June 18, rainy and stormy weather persisted and continued to worsen. By the morning of June 19, the winds ceased and sunny weather set in. The retrieved C_{chl} distribution for this day revealed only residual patches of the June 17 pattern. The absolute values of C_{chl} proved to be significantly less than those recorded on June 17. In the coastal zone, an area of enhanced C_{chl} reappeared. With prevailing southern winds, the occurrence of the area enriched with *chl* could be caused by renewed upwelling. Among other salient features of C_{chl} distribution at Rybinsk Reservoir was the presence of a stable area with low *chl* concentrations, stretching from Cherepovets (upper part of Fig. 4.34) to Rybinsk (bottom of Fig. 4.37). The origin of this area can be arguably explained by rapidly flowing water from the Sheksna River taken up by the reservoir.

The above lidar data reported by Pozdnyakov and Kondratyev (1997) provide convincing evidence that, under certain conditions, remote measurements in the visible spectrum can confidently reveal not only patterns of *chl/doc* spatial/temporal distributions, but also the occurrence of such lake hydrodynamic events as upwelling, gyres, fronts, currents, and stagnant zones as well as specific features of bottom topography.

To recap then, it could be underscored that the practical application of remote-sensing techniques has good potential as an efficient instrument for assessing not only trophic status of inland water basins, but also for revealing and characterizing thermodynamic processes taking place in them. This makes it possible to move on from recording purely surficial expressions of various limnological phenomena to an analysis of quantified specific features of in-water dynamics, which has been demonstrated above with examples from Lakes Onega and Sevan.

An important result of our analysis of *chl* 2-D distributions, made using hydrodynamic models, is that the efficiency of the remote-sensing approach as applied to the problem of eutrophication of inland waters can be sufficiently improved, provided the planning of remote survey is based on thermodynamic (and ecological—in future) simulations of the water body under study (Pozdnyakov *et al*, 1998). It promises important perspectives for studies in this direction. However, in order to meet these objectives, combined efforts by many experts in interdisciplinary areas will be required.

4.3.2   Remote assessment of primary production and advective carbon fluxes in aquatic environments

In Chapter 3, methodological aspects of assessing phytoplankton primary production P from remotely sensed data have been discussed in detail. The practical realizations of various approaches to quantify P generally relate to marine/oceanic (case-I) waters (for ref. see Bukata *et al.*, 1995), and relatively few of them concern case-II waters.

Hoepffner (1993) reported on using spaceborne ocean color data for assessing phytoplankton primary production in the northwest coast of Africa. The protocol suggested by Hoepffner is based on the relationship derived by Platt and Sathyendranath (1988) (see References to Chapter 3):

$$P_{(z,t,\lambda)} = C_{chl}(z)P'_{max}(1 - \exp[-(\alpha'I_0 \sin(\pi t/D)e^{-Kz})/P'_{max}]) \qquad (4.8)$$

where t is the local time, D is the Gregorian date, and all the ramining notation is given in § 3.1.1.4.

The chlorophyll concentration at each depth in a vertically homogeneous water column was determined from a satellite-weighted surface estimate following Gordon and Clark (1980). In the case of a subsurface chlorophyll maximum, inaccessible to satellite sensors, a theoretical profile was assumed and defined by three parameters of a Gaussian distribution superimposed on a background chlorophyll value C_0:

$$C_{chl}(z) = C_0 + \frac{h}{\sigma\sqrt{2\pi}}\left[\frac{-(z - z_m)^2}{2\sigma^2}\right] \qquad (4.9)$$

where parameters h, z_m, σ were identified in § 3.1.1.4, and the sum $C_0 + h/\sigma\sqrt{2\pi}$ can be empirically related to remotely determined chlorophyll concentration. Therefore, this approach requires knowledge of five biological parameters: three of them are associated with the biomass profile (the position of chlorophyll maximum z_m, width of the chlorophyll maximum σ, and ratio of the peak height to the background chlorophyll concentration) and two photosynthetic parameters P'_{max}, α', discussed in § 3.1.1.4.

In order to remotely determine chlorophyll concentrations in surface waters, Hoepffner used the following retrieval algorithms:

for case-I waters:

$$\ln C_{chl} = 0.334 - 1.68X_1 \quad \text{for } R_s(\lambda_3) < R_s(\lambda_3)_{lim}$$
$$\ln C_{chl} = 1.31 - 3.674X_2 \quad \text{for } R_s(\lambda_3) < R'_s(\lambda_3)_{lim}$$

(4.10)

and for case-II waters:

$$\ln C_{chl} = 0.768 - 2.61X_1 + 0/791X_1^3 - 0.338X_1^3 \quad \text{for } C < 2.0 \text{ mg/m}^3 \quad (4.11)$$

with

$$R_s(\lambda_3)_{lim} = 0.01 \exp[1.05 - 0.02X_1 - 0.429X_1^2 + 0.094X_1^3]$$
$$R'_s(\lambda_3)_{lim} = 0.01 \exp[-3.58 + 0.678X_2 - 3.77X_2^2 - 3.43X_2^3]$$

(4.12)

where $X_1 = \ln(R(\lambda_1)/R(\lambda_3))$, $X_2 = \ln(R(\lambda_2)/R(\lambda_3))$, $\lambda_1, \lambda_2, \lambda_3$ are the CZCS spectral bands (443, 520, and 550 nm).

Following Platt *et al.* (1992), Hoepffner assumed the existence of a relationship between α' and P'_{max} if P'_{max} is under 12 mg C (mg *chl-a*)$^{-1}$ h^{-1}. Based on shipborne data on nutrient availability within the five delineated areas and retrieved concentrations of *chl-a*, Hoepffner established appropriate values of α' and P'_{max} which proved to be in the ranges $(0.03-0.07)$ mg C mg *chl-a*$^{-1}$ (W m^{-2})h^{-1} and $(1.5-7.3)$ mg C mg *chl-a*$^{-1}$ h^{-1}, respectively.

For each pixel of chlorophyll data from CZCS, the vertical profile of biomass was estimated through available *in situ* data and the primary production was assessed from eq. (4.8) every half an hour at a vertical resolution of 1 m, and at 61 wavelengths. The results were then integrated over the entire daylight period (assuming clear-sky conditions), over the euphotic zone and over the PAR spectral region.

The computed rates of phytoplankton primary production within the delineated five provinces varied between 0.75 and 1.1 g C m^{-2} day^{-1} which proved to be in good agreement with the relevant data for these locations reported in the literature. Comparison of these estimations with the estimations made assuming a homogeneous case (i.e. when the targeted area was not divided into regions with different trophic properties but was taken as a uniform water mass) indicates that the above differentiation is conducive to perhaps more adequate assessments, although differences in both kinds of assessments were not really significant.

The protocol suggested by Pozdnyakov *et al.* (1998b) and described in § 3.1.1.4 has been used to process field data and evaluate primary production in Lake Ladoga.

It will be recalled that this protocol suggests a series of operations which can be outlined as follows:

1. Obtain an upwelling radiance spectrum of high spectral resolution above the water surface being studied or monitored. Also directly measure or otherwise obtain /estimate global radiation at the air–water interface.
2. Convert the atmospherically adjusted upwelling radiance spectrum into the subsurface volume reflectance spectrum just beneath the air–water interface.
3. Use the multivariate optimization technique in conjunction with *a priori* or directly determined pertinent scattering and absorption cross-section spectra to deconvolve the subsurface volume reflectance spectrum into co-existing concentrations of near-surface chlorophyll, suspended sediments, and dissolved organic carbon (or additional components if the bio-optical model discussed throughout this book can be expanded).
4. From global radiation and Fresnel reflectivities, determine the downwelling irradiance just beneath the air–water interface.
5. Determine the attenuation coefficients for the direct, diffuse, and total subsurface PAR irradiance; i.e. determine $K_{d\,sun}$, $K_{d\,sky}$, and $K_{d\,PAR}$. The subsurface diffuse fraction η_w is required for assessing $K_{d\,PAR}$.
6. From *a priori* knowledge of the area, from direct measurements in the area, from model outputs, or from mathematical simulations, obtain reliable estimates of chlorophyll, suspended mineral, and dissolved organic carbon dimensionless depth ($\zeta = x/z_{eu}$) profiles ($C_{chl}(\zeta)/\langle C_{chl\,eu}\rangle$, etc.).
7. Use Platt *et al.*'s (1991) model or any comparable primary production model (e.g. the Vollenweider eq. 3.115) to determine phytoplankton primary production.

Although no remote sensing *per se* was performed during this campaign, optical measurements conducted above and below the air–water interface are pertinent to support the suggested procedure.

Daily spectral measurements of the incident global PAR were made aboard a small research vessel in the northern coastal zone of Lake Ladoga (Fig. 4.38).

Concurrent with these observations were *in situ* measurements of subsurface upwelling and downwelling irradiances and the vertical distribution of phytoplankton *chl* concentration throughout the euphotic zone. The field data were collected under conditions of a cloudless sky and an almost perfectly calm air–water interface at local noon ($\theta = 35°$). The diurnal value of the incident global PAR was calculated from continuously recorded data to be $27\,\text{Ein}\,\text{m}^{-2}$. The subsurface volume reflectance spectrum displayed a peak value of 0.040 at $\lambda = 580\,\text{nm}$.

The use of absorption and backscattering optical cross-section spectra for the principal constituents of Lake Ladoga water as well as the Livenberg–Marquardt multivariate optimization technique (Kondratyev *et al.*, 1990c) resulted in retrieval from the volume reflectance spectrum of $7.1\,\text{mg}\,\text{m}^{-3}$ of *chl-a*, $0.8\,\text{g}\,\text{m}^{-3}$ of *sm*, and $8\,\text{g}\,\text{C}\,\text{m}^{-3}$ of *doc* as estimates of co-existing near-surface aquatic concentrations. The value of $K_{PAR}(z = -0)$ deduced on the basis of the bio-optical model and retrieved concentration values of *chl*, *sm*, and *doc* was $1.8\,\text{m}^{-1}$. This $K_{PAR}(z = -0)$ corresponds to a penetration depth of *ca.* 0.6 m.

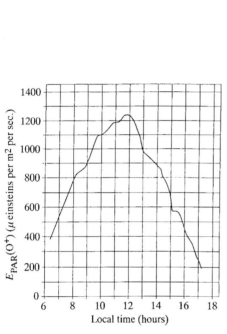

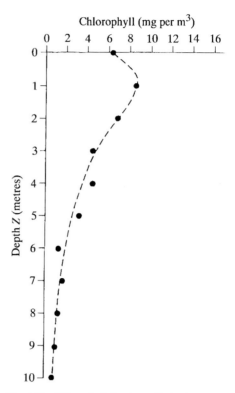

Fig. 4.38. Daily profile of E_{PAR} over Lake Ladoga, June 1989.

Fig. 4.39. Chlorophyll depth profile in the north-western coastal zone of Lake Ladoga, June 1989.

In our computations we employed the phytoplankton primary production model suggested by Platt *et al.* (1991). It was assumed that depth profiles of *sm* and *doc* remained invariant from their estimated near-surface values throughout the euphotic zone. The chlorophyll depth profile (Fig. 4.39), as stated above, was taken from *in situ* measurements.

Integration of the Platt *et al.* equation yielded a daily phytoplankton photosynthesis value of $352\,\mathrm{mg\,C\,m^{-2}\,day^{-1}}$. This estimate was in good agreement with direct measurements of primary production in Lake Ladoga performed during the same vegetation season (relevant values in the range $320\text{–}380\,\mathrm{mg\,C\,m^{-2}\,day^{-1}}$ were reported by Petrova (1990)). The primary production value determined from volume reflectance estimates was not corrected for scalar irradiance since direct measurements reported in Petrova were values for downwelling vector irradiance.

Remote determinations of phytoplankton primary production present crucially important means of virtually assessing carbon exports through aquatic environment systems such as, e.g. rivers, estuaries, lake/marine coastal zones, open lake/marine/ oceanic areas. It is believed that these intimately relate to climate change on scales ranging from local to regional and global. Bakun (1990) reported, for instance, climate-change-related enhancement of coastal upwelling in several locations throughout the global ocean which, by increasing primary production and carbon export, may set up an important negative feedback mechanism for increasing anthropogenic CO_2 in the atmosphere. In the context of the global carbon cycle it is mandatory not only to identify but adequately quantify regional fluxes and sinks of biological carbon, and through such an inventory to evaluate the present status and dynamics of the atmosphere-to-ocean carbon flux on a global basis.

Gabric *et al.* (1993) used CZCS satellite imagery to measure the phytoplankton biomass that has been advected off the continental shelf and examined the carbon flux temporal evolution during the upwelling event in the vicinity of Cape Blank (Mauritania). Since the targeted marine area qualifies as case-I waters, Gabric *et al.* used pigment retrieval algorithms given by Andre and Morel (1991):

$$\ln C_{sat} = 0.347 - 2.73 \ln r_{13} + 2.14[\ln r_{13}]^2 \qquad C_{sat} < 2\,\mathrm{mg\,m^{-3}} \qquad (4.13)$$

$$\ln C_{sat} = 0.661 - 8.48 \ln r_{23} + 11.52[\ln r_{23}]^2 \qquad C_{sat} > 2\,\mathrm{mg\,m^{-3}} \qquad (4.14)$$

where r_{ij} is the ratio of reflectance in CZCS bands i and j (443, 520, 550 nm) and C_{sat} is the pigment concentration (in milligrams chlorophyll per cubic meter) which is an exponentially weighted value vertically averaged over one optical penetration depth Z_{opt}:

$$C_{sat} = \int_0^{Z_{opt}} C(z) \exp(-K_d z)\, dz \Big/ \int_0^{Z_{opt}} \exp(-K_d z)\, dz \qquad (4.15)$$

$C(z)$ is the actual phytoplankton profile, K_d is the downwelling irradiance attenuation coefficient, Z_{opt} is typically $\sim$20–30% of the euphotic zone Z_e in clear waters and $\sim$50% in eutrophic waters (Smith, 1981). As cited in Chapter 3, for oceanic waters Morel and Berthon (1989) suggested an empirical relationship relating the euphotic zone pigment concentration C_e to C_{sat}:

$$C_e = 40.6(C_{sat})^{-0.459} \qquad (4.16)$$

Retrieved euphotic zone phytoplankton chlorophyll concentrations were then used in hydrodynamic numerical modelling to determine the desired carbon fluxes.

The authors used an empirical relationship between particulate organic carbon (POC, $\mathrm{mg\,C\,m^{-3}}$) and chlorophyll concentration C ($\mathrm{mg\,m^{-3}}$):

$$POC = 91 + 23C \qquad (4.17)$$

The wind stress vector τ ($\mathrm{N\,m^{-2}}$) at the sea surface was derived using the conventional quadratic expression:

$$\tau = \rho_a c_d |\mathbf{U}_{10}| \mathbf{U}_{10} \qquad (4.18)$$

where ρ_a is the density of air (1.25 $\mathrm{kg\,m^{-3}}$), and c_d is the drag coefficient (0.0014).

The Ekman transport vector $\mathbf{M}_c$ ($\mathrm{kg\,m^{-3}\,s^{-1}}$) was estimated from the surface wind stress by

$$\mathbf{M}_c = (\tau \times k)/f \qquad (4.19)$$

where k is the unit vector in the vertical direction (positive upwards), and f is the Coriolis parameter (which is in the range $[4.7\text{–}5.9] \times 10^{-5}\,\mathrm{s^{-1}}$ for latitudes of

Mauritania). The Ekman volume transport vector $\mathbf{Q}_e$ $(m^2 s^{-1})$ per unit width cross-slope of the shelf was computed from:

$$\mathbf{Q}_e = \mathbf{M}_e/\rho_w \tag{4.20}$$

with ρ_w the density of water in the offshore flowing Ekman layer.

To take into account the effects of wind periodicity, varying shelf topography and water-column stratification, Gabric *et al.* used a 3-D baroclinic hydrodynamic model (Eifler *et al.*, 1991) to compute the cross-shelf velocity profile $\nu(z)$ normal to the 200-m isobath. The model formulation is based on hydrostatic approximation, and the Boussinesq assumption with respect to the effect of variable water density. Baroclinic effects have been computed using separate transport equations for heat and salinity with the turbulent kinetic energy profile predicted from momentum equations and a depth-dependent mixing length. Model predictions for $\nu(z)$ were integrated over the assessed mixed layer depth Z_m to obtain the net cross-slope volume transport Q_n $(m^2 s^{-1})$:

$$Q_n = \int_0^{Z_m} \nu(z)\, dz \tag{4.21}$$

The net horizontal advective flux of POC into the oceanic part of the targeted area was then calculated from:

$$A = \int Q_n POC\, dL \tag{4.22}$$

where L is the path length.

The carbon net flux $[(g\,C\,s^{-1}) \times 10^5]$ values thus obtained were in the range 0.9 to 1.1 which proved to be comparable with relevant values reported in the literature for this marine area. Net primary production was estimated at about $2.5 \times 10^{13}\,g\,C\,yr^{-1}$.

Although strictly applicable to a definite location, this study is of considerable interest in terms of the general approach employed in it (i.e. a combination of remotely obtained *chl* concentration spatial distributions and numerical hydrodynamic modelling of water-advective movements) and as such it could be a model for assessing carbon fluxes in other areas of the global ocean.

REFERENCES

Abroskin, A. 1986. Field experiments on laser sensings using the Tchayka airborne instrumentation. *Proceedings of the IOFAN*. Moscow: IOFAN, Vol. 1, pp. 23–29 (in Russian).

Alpers W. 1997. Private communication.

Andre J.M., Morel A. 1991. Atmospheric corrections and interpretation of marine radiances in CZCS imagery revisited. *Oceanol Acta*, **14**, 3–22.

Arenz R.F., Lewis W.M., and Saunders J.F. 1996. Determination of chlorophyll and dissolved organic carbon from reflectance data for Colorado reservoirs. *Int. J. Remote Sensing*, **18**, No. 8, 1547–1566.

Bakun A. 1990. Global climate change and intensification of coastal upwelling. *Science*, **247**, 198–201.

Betin V.V. 1957. Ice conditions in the region of the Baltic Sea and its multiyear variations. *Trudy GOIN*, **41**, 54–125 (in Russian).

Bolgrien D.W., Granin N.G., Levin L. 1995. Surface temperature dynamics of lake Baikal observed from AVHRR images. *Photogammet.Engng Remote Sensing*, **61**, No. 2, 211–216.

Bristow, M. 1981. Use of water Raman emission to correct airborne laser fluorescensor data for effect of water optical attenuation. *Appl. Opt.*, **20**, 2889–2906.

Bukata R.P., Bennett, J.E. Bruton. 1978. Landsat-1 observation of an internal standing wave pattern in Lake Superior. *Canad. J. Spectros.*, **23**, No. 2, 57–65.

Bukata R.P., Haras W.S., and Bruton J.E. 1975. The application of ERTS-1 digital data to water transport phenomena in the Point Pelee—Rondeau area. *Ver. Internat. Verein. Limnol.*, **19**, 168–178.

Bukata R.P., Jerome J.H., Bruton J.E. 1988. Particulate Concentrations in Lake St. Clair as Recorded by a shipborne Multispectral Optical Monitoring System. *Remote Sensing Environ.*, **25**, 201–229.

Bukata R.P., Jerome J.H., Kondratyev K.Ya., Pozdnyakov D.V. 1995. *Optical Properties and Remote Sensing of Inland and Coastal Waters.* Boca Raton: CRC Press, 350 pp.

Bukata R.P., McColl W.D. 1973. The utilization of sun-glint in a study of lake dynamics. In: *Remote Sensing and Water Resources Management. American Water Res. Assoc. Proc*, No. 17, June 1973, 351–367.

Church P.E. 1943. The annual temperature cycle of Lake Michigan. 1. Cooling from late autumn to the terminal point, 1941/42. Inst. Meteorol. University Chicago, Misc. Rep. No. 4, 48 pp.

Dekker A.G. 1993. Detection of optical water quality parameters for eutrophic waters by high resolution remote sensing. Doctoral thesis. Amsterdam: University of Amsterdam, 222 pp.

Denman, K. 1976. Covariability of chlorophyll and temperature in the sea. *Deap-Sea Res.*, **91**, 83–90.

Denman, K., Okubo, A. Platt, T. 1977. The chlorophyll fluctuation spectrum in the sea. *Limnology, Oceanography*, **22**, 1033–1038.

Eifler W., Kupusovic T., Schrimpf W. 1991. Numerical verification exercises with different computer models for simulating sea circulation pattern: the verical diffusion of momentum in a forced baroclinic sea. *Environ. Qual. Life Rep. EUR 13370 EN*, Commission of the European Communities, 59 pp.

Fadeev, V. 1978. Laser remote sensing of photosynthesizing organisms. *J. Quantum Electron.*, **5**, 2221–2226.

Fedorov N.P., Kuz'mina N.P. 1977. *Meso-scale Variations of the Ocean Temperature.* Moscow: Nauka Press, 201 pp. (in Russian).

Filatov, N. 1983. *Dynamics of Lakes.* Gidrometeoizdat.: Leningrad, 345 pp. (in Russian).

Filatov N.N., Beletsky D.V., Zaitsev L.V. 1990. *Variability of Hydrophysical Parameters in Lake Onega.* Petrozavodsk: Nauka Press, 115 pp. (in Russian).

Filatov, N. and Zaytsev, L. 1993. On the thermodynamic processes in Lake Onega. *Water Resources*, **2**, 175–181 (in Russian).

Fisher J., Doerffer R., and Grassl H. 1991. Remote sensing of water substances in rivers, estuarine and coastal waters. In: Dergens E.T., Kempe S., Richey J.E. (eds.), *Biogeochemistry of Major World Rivers.* London: John Wiley & Sons Ltd, pp. 25–55.

Fisher J., Kronfeld U. 1990. Sun-stimulated chlorophyll fluorescence. 1: Influence of oceanic properties. *Int. J. Remote Sensing*, **11**, No. 12, 2125–2147.

Forel F.A. 1895. *Le Leman: Monographie Limnologique.* Tome 2, *Mécanique, Chemie, Thermique, Optique, Acoustique.* Lausanne: F. Rouge, 651 pp.

Gabric A.J., Garcia L., Camp L.V., Nykjaer L.,Eifler W., Schrimpf W. 1993. Offshore export of shelf production in the Cape Blanc (Mauritania) Giant Filament as derived from Coastal Zone Color Scanner Imagery. *J. Geophys. Res.*, **98**, No. C3, 4697–4712.

George D.G. 1997. The airborne remote sensing of phytoplankton chlorophyll in the lakes and tarns of the English Lake District. *Int. J. Remote Sensing*, **18**, No. 9, 1961–1975.

Gordon H.R., Clark D.K. 1980. Remote sensing of optical properties of a stratified ocean: an improved interpretation. *Appl. Opt.*, **19**, 3428–3432.

Gower J.F.R., Borstad G.A. 1990. Detection of phytoplankton in marine waters due to chlorophyll-a fluorescence. *Int. J. Remote Sensing*, **11**, No. 2, 313–320.

Hoepffner N. 1993. An analytical model to estimate the primary production from satellite ocean color data in a coastal upwelling environment. Technical Note No. 1.93.143. Ispra: Institute Remote Sensing Apll., 40 pp.

Hupfer P. 1979. *Die Ostsee—Kleines Meer mit Grossen Problemen.* Leipzig: BSB B.G. Teubner Verlagsgesellschaft, 136 pp.

Johannessen O.M., Pettersson L.H., Sandven S., Melentyev V.V. 1994. Real-time sea ice monitoring of the Northern Sea Route using ERS-1 satellite radar images. NERSC Technical Report No. 2, Bergen, Norway, July 1994, 33 pp.

Karabyshev, G. 1987. *Fluorescence in the Ocean.* Leningrad: Gidrometeoizdat., 275 pp. (in Russian).

Khomskis V.P. 1969. *Dynamics and Thermics of Small Lakes.* Vilnius: Nauka Press, 204 pp. (in Russian).

Kondratyev K.Ya. 1987. *Global Climate and its Changes.* Leningrad: Nauka Press, 231 pp. (in Russian).

Kondratyev K.Ya. *et al.* 1996a. Combined application of remote sensing and in situ measurements in monitoring environmental process. *Hydrologia*, **322**, 227–232.

Kondratyev, K.Ya., Filatov, N., Zaytsev, L. 1988. Specific features of Lake Ladoga dynamics inferred from remote sensing data. *Dokl. Acad. SSSR*, **293**, 1224–1227 (in Russian).

Kondratyev, K.Ya., B. Henderson-Sellers, D.Pozdnyakov. 1990a. Using large lakes as analogues for oceanographic studies. In: Henderson-Sellers B. (ed.), *Modelling Marine Systems.* Boca Raton: CRC Press, Vol. 2, pp. 249–344.

Kondratyev K.Ya., Johannessen O.M., Melentyev V.V. 1996b. *High Latitude Climate and Remote Sensing.* Chichester, UK: Wiley-Praxis, 200 pp.

Kondratyev K.Ya., Melentyev V.V. 1991. *Modern Situation on the Small Rivers of USSR and Ways for its Using, Saving and Re-construction.* Leningrad: Gidrometeoizdat. Proceedings GGI, Issue 2, 76 pp. (in Russian).

Kondratyev K.Ya., Melentyev V.V. 1992. Retrieving the surface temperatures of the key regions of the World ocean supplied with background hydrometeorological information (microwave technique). In: *Proceedings of Joint Conference 4th Symposium of the International Society of Acoustic Remote Sensing and 2nd Australian Conference the Physics of Remote Sensing of Atmos. and Ocean, Australia Defence Force Academy, Canberra,* 2-80, pp. 1–3.

Kondratyev K.Ya., Melentyev, V.V., Mishev D.N. 1990b. Comprehensive study of watershed, wetlands and shallow waters with using microwave (Rybinsk Reservoir as example). *Earth Obs. Remote Sensing,* No. 3, 136–138 (in Russian).

Kondratyev K.Ya., Melentyev V.V., Nazarkin V.A. 1992. *Remote Sensing of Water Areas and Watersheds (Microwave Methods).* St. Petersburg: Gidrometeoizdat., 248 pp. (in Russian).

Kondratyev, K.Ya., Pozdnyakov D.V. 1987a. Lidar fluorescence sensing of chlorophyll in Lake Onega, In: Kondratyev K.Ya. (ed.), *Comprehensive Remote Sensing of Lakes.* Leningrad: Nauka Press, pp. 156–163 (in Russian).

Kondratyev, K.Ya. Pozdnyakov, D. 1987b. *Optical Properties of Natural Waters and Remote Sensing of Phytoplankton.* Leningrad: Nauka Press, 165 pp. (in Russian).

Kondratyev K.Ya,. Pozdnyakov D.V., Isakov V.Yu. 1990c. *Radiation and Hydrooptical Experiments on Lakes.* Leningrad: Nauka Press, 115 pp. (in Russian).

Kondratyev K.Ya., Vasil'yev L.N., Vedeshin L.A. 1988. International experiment "INTERKOSMOS"—"Inland waterbodies". *Earth Obs. Remote Sensing,* No. 3, 119–122 (in Russian).

Lorenzen, C. 1967. Determination of chlorophyll and pheopigments: spectrometric equations. *Limnology, Oceanography,* **12**, 343–346.

Marra, J. 1982. Variability in surface chlorophyll-a at a shelf break front. *J. Mar. Res.,* **40**, 575–592.

Medres P.L. 1957. Ice regime of Lake Ladoga from materials of aircraft survey. *Trudy GGI,* **66**, 98–139 (in Russian).

Melentyev V.V., Mel'nichenko I.G. 1995. Landscape and ecological microwave diagnostyc of the system water body-watershed accomodation. *Water Resources,* **22**, No. 1, 110–114 (in Russian).

Melentyev V.V., Tikhomirov A.I., Kondratyev K.Ya., Johannessen O.M., Sandven S., Pettersson L.H. 1997. Study of the freeze-up phase change on large temperate-zone inland water bodies and possibilities of their microwave diagnostics. *Earth Obs. Remote Sensing,* No. 3, 61–72 (in Russian).

Melentyev V.V., Johannessen O.M., Kondratyev K.Ya., Bobylev L.P., Tikhomirov A.I. 1998. Experience of the ERS 1&2 SAR using ice cover for lake diagnosis: ecology and history. *Earth Obs. Remote Sensing,* No. 2, 91–101 (in Russian).

Morel A., Berthon J.-F. 1989. Surface pigments, algal biomass profiles, and potential production of the euphotic layer: Relationships reinvestigated in view of remote sensing applications. *Limnology, Oceanography*, **34**, No. 8, 1545–1562.

Naumenko M.A., Karetnikov S.G. 1993. Application of IR spaceborne data for assessing the Lake Ladoga thermal status. *Earth Obs. Remote Sensing*, No. 4, 69–78 (in Russian).

Naumenko M.A., Karetnikov S.G., Furmanchik K., and Prais E. 1991. A comprehensive study of frontal zones in large lakes utilizing spaceborne observations. In: *Proceedings International Geoscience Remote Sensing Symposium (IGARSS '96)*. Helsinki: IGARSS, Vol. 4, pp. 1199–1202.

Oppenheimer C. 1997. Remote sensing of the colour and temperature of volcanic lakes. *Int. J. Remote Sensing*, **18**, No. 1, 5–37.

Panin G.N., Karetnikov S.G. 1991. On the thermal regime of a fresh water body upper- top layer. *Meteorol. Hydrol.*, **6**, 79–83 (in Russian).

Parparov, A. 1979. Primary production and chlorophyll-a content in the phytoplankton of Lake Sevan. In: Abramyan A. (ed.), *Ecology of Hydrobiontes in Lake Sevan*. Yerevan: Armenian Academy of Sciences, pp. 83–100 (in Russian).

Petrova N.A. 1990. *Successions of Phytoplankton with the Anthropogenic Eutrophication of Large Lakes*. Leningrad: Nauka Press, 200 pp. (in Russian).

Platt T., Caverhill C., and Sathyendranath S. 1991. Basin-scale estimates of oceanic primary production by remote sensing: The North Atlantic. *J. Geophys. Res.*, **96**, 15 147–15 159.

Platt, T. 1972. Local phytoplankton abundance and turbulence. *Deep-Sea Res.*, **19**, 183-187.

Platt, T. and Denman, K. 1975. Spectral analysis in ecology. *Annu. Rev. Ecol. Syst.*, **6**, 189–210.

Platt T., Sathyendranath S., Ulloa O., Harrison W.G., Hoepffner N., and Goes J. 1992. Nutrient control of phytoplankton photosynthesis in the western Atlantic. *Nature*, **356**, 229–231.

Pozdnyakov D.V., Kondratyev K.Ya. 1997. Lidar fluorescence technique application for studying hydrodynamic processes. *Mar. Technol. Soc. J.*, **30**. No. 4, 46–53.

Pozdnyakov D.V., Kondratyev K.Ya, Kotov S.V. 1998a. Optimization of coastal zone airborne monitoring through thermohydrodynamic modelling anf interpolation techniques. In: *Proceedings of the 27th International Symposium on Remote Sensing of Environment, 8–12 June 1998, Tromso, Norway*, pp. 209–212.

Pozdnyakov D.V., Kondratyev K.Ya., and Pettersson L.H. 1998b. Remotely assessment of algae primary production in clear and turbid waters. *Earth Obs. Remote Sensing*, **4** (in press).

Roumyantsev V.B. 1987. Hydrooptical inhomogeneities in Lake Ladoga. In: *A Comprehensive Remotely Monitoring of Lakes*. Leningrad: Nauka Press, pp. 19–35 (in Russian).

Selegey V.V., Selegey T.S. 1978. Lake Teletskoye. In: *Hydrometeorological Regime of the USSR Lakes*. Moscow: Nauka Press, pp. 1–140 (in Russian).

Sherstyankin P.P. 1993. Optical structures and fronts of oceanic type in Lake Baikal. Summary of the Dissertation for a Doctor's Degree. Shirshov Institute of Oceanology, Russian Academy of Sciences, 37 pp. (in Russian).

Smith R.C. 1981. Remote sensing and the depth distribution of ocean chlorophyll. *Mar. Ecol.*, **5**, 359–361.

Smith, R., Zhang, X., Michaelson, J. 1988. Variability of pigment biomass in the California current system as determined by satellite imagery spatial variability. *J. Geophys. Res.*, **93**, No. D9, pp. 10 863–10 882.

Steel, J. 1978. Some comments on plankton patches. In: Steel J.S. (ed.), *Spatial Patterns in Plankton Communities*. New York: Plenum Press, pp. 1–20.

Tikhomirov A.I. 1982. *Thermal Regime of Large Lakes*. Leningrad: Nauka Press, 232 pp. (in Russian).

Victorov S. 1996. *Regional Satellite Oceanography*. New York and London: Taylor & Francis, 300 pp.

Vinogradov V.V. 1987. Private communication.

Voronin V.A., Melentyev V.V., Shannikov D.V. 1986. Antenna system for underlying surfaces research. Patent No. 1246197 (3886815/24-09. 15.02.1985). *Patent Bull.*, No. 27 (in Russian)

Witting R. 1929. Beobachtungen in Ladoga See in den jahren 1898–1903. Inst. Meeresforschung. Sch. No. 60. Helsinki, 34 pp.

Yakobi Y.Z., Gitelson A., and Mayo M. 1995. Remote sensing of chlorophyll in Lake Kinneret using high-spectral-resolution radiometer and Landsat TM: spectral features of reflectance and algorithm development. *J. Plankton Res.*, **17**, No. 11, 2155–2173.

Zaykov B.D. 1955. *Limnology*, Part 1. Leningrad. Gidrometeoizdat., 271 pp. (in Russian).

5

Geographical Information Systems in limnology

5.1 GEOGRAPHICAL INFORMATION SYSTEMS: DEFINITIONS, TYPES, COMPONENTS

The statement that limnology has a complex character will hardly have any objection. Any lake is a complex system itself due to the fact that processes of different nature interact there; we can mention hydrological, physical, chemical, biological and geological processes. Moreover, there are strong connections between a lake and its surroundings, which link a lake and an adjoined area to a unique system. Very often the spatial distribution and linkages of environmental parameters is of key importance in a particular science. Interactions among air, land, water and biota most frequently occur when they are adjoining in space, so information about their distribution is essential to understanding their interconnectedness. Geographical information systems (GISs) are used to store information with a geographical component to investigate interactions among different parts of a system, which is described by collected data and information, and to manage complex and delicate environmental structures.

Maps have long been used to represent the location of Earth-system components and their superposition in space, but maps are static and qualitative. GISs grew out of the need to derive quantitative information about the distribution of and interactions among the Earth's resources. GIS technology integrates geographic analysis benefits offered by maps and database operations such as query and statistical analysis and it is probably the best technology currently available for spatial and temporal analysis of data.

A GIS differs from a map in several ways. A map is an analog depiction of the Earth's surface, while a GIS records spatially distributed features in digital form. A map simultaneously depicts a variety of features of landscape or waterbody (e.g. topography, land cover type, and bathymetry), while a GIS usually stores each feature as a separate data layer, identifying relationships between different layers. A map is static and difficult to update, while a GIS data layer can be easily revised. Although maps can be a form of GIS input or output, a GIS greatly increases the versatility of mapped data because of its wealth of techniques for quantitative analysis of geodata. The end product of a GIS analysis may be a map or data.

A geographic information system is a computer-based system for the manipulation and analysis of spatial information in which there is an automated link between the data and their spatial location (*Encyclopedia...*, 1992). A GIS consists of computer hardware and software for entering, storing, transforming, measuring, combining, retrieving, displaying and performing mathematical operations on digitized thematic data that have been registered to a common spatial coordinate system (*Encyclopedia...*, 1992). If we add to the given definition a qualified personnel and powerful hardware we will come to the ESRI's (Environmental Systems Research Institute) view on GIS contents, which determine, that GIS integrates five key components: hardware, software, data, people and methods (http://www.esri.com/base/gis/abtgis/comp_gis.html).

Possibly the most important component of a GIS is the data. There are two basic types of GISs, which differ in the way they store data. Raster-based GISs (grid- or pixel- based systems) present the surface as a matrix of grid cells, each with an individual data value. Vector-based GISs portray surface as an aggregation of points, lines and polygons, which are encoded and stored as a collection of x, y coordinates. Each of these data structures has advantages and disadvantages, depending on the type of GIS application. Modern GISs are able to handle both data structures, but still they are more efficient in handling one type of data.

Raster-based GISs are in widespread use for spatial analysis because of their computational simplicity and compatibility with remotely sensed data. Scanned maps can also be used in raster format or as an image in another terminology. If the image is a remotely sensed satellite image, each pixel represents the electromagnetic radiation reflected, emitted or scattered back from a portion of the Earth's surface corresponding to an element of spatial resolution of the particular satellite. However, if the image is a scanned document (traditional map, for example), each pixel represents a brightness value associated with a particular point on the document. A raster image could also be called rasterized database, in which each cell can be assigned only one numeric value, so different geographic attributes must be stored as separate data layers, even when they refer to the same object. In a raster image numeric values given to each cell may represent either a feature identifier, a qualitative attribute code or a quantitative attribute data.

The individual raster is the minimum unit of spatial resolution. Therefore, features that are dimensionless or basically smaller than the minimum raster dimension are difficult to depict in a raster-based system. At coarse levels of resolution (large rasters) data appear blocky and lines appear stair-stepped. At finer levels of resolution, a raster representation looks more like a map, but data-storage requirements increase exponentially. The appropriate raster size should be comparable with the scale at which limnological analysis is going to be provided.

Quadtrees are a type of raster data-storage structure, which take into account the inherent resolution of the data. The quadtree is a hierarchical raster data structure based on the successive division of a $2^n \times 2^n$ array into quadrants, in which the data for four quadrants having the same value are aggregated into the next largest block in the hierarchy. As a result, coarse-grained features are stored in large cells and fine-grained features are stored in small cells.

Raster data can be organized in a number of ways depending upon the particular image format. Typically, the image data file contains a header record that stores information about the image such as the number of rows and columns in the image, the number of bits per pixel, the color requirements and the georeferencing information. Following the image header is the actual pixel data for the image. Sometimes, as in GIS IDRISI, raster data is represented simply by a matrix of pixel values, the header information being kept in the separate file.

Raster systems are superior to vector systems for depicting continuously varying features, such as reflectance values recorded by satellite sensors. Whereas conventional maps usually group data categorically and spatially, raster-based systems can maintain the spatial diversity of the input data. This capability is particularly important for applications involving the analysis of spatial heterogeneity. Images are often used as the background to a combined GIS product, with vector data being drawn on top of them.

Vector-based GISs represent surface features as points, lines, or polygons, stored as a series of coordinates. Vector data can accurately record the actual ground location of features. It is very well suited to record the location of discrete geographic features with precise locations like roads, rivers, water bodies, boundaries. However, vector data are highly dependent on the number of x, y coordinates points that are chosen to represent features, especially natural objects like rivers or coastlines. The size of point-objects and the width of lines do not depend on scale (resolution) of the data. Homogeneous patches (or, strictly speaking, areas described as homogeneous) are bounded by lines, as they would appear on a customary map.

Most vector-based GISs use an arc-node data structure, in which arcs are the individual line segments defined by each pair of x, y coordinates and nodes occur at the intersections and ends of lines. Polygons are defined by the arcs that bound them, topologic relationships being stored in associated tables. Some vector-based GISs use their own specific elements. For example, ArcView GIS puts to use such elements as route and region, the first is a linear feature composed of one or more arcs, sets of routes with similar attributes being stored as a subclass; the second is a polygon feature composed of one or more polygons. The more distinct features you describe in the beginning, easier it will be to conduct spatial analysis later.

Vector data in a map can have a whole range of attribute values, specifying their nature. Attribute data may be stored with spatial data, but usually it is kept in a separate table as a relational database. Attribute data tables describe each point, line and polygon in the data layer—each class of features stored in a layer has its own attribute table. In GISs with relational databases, all of these items are assigned a unique identifier, which is used to link spatial and attribute information.

Vector-based GISs are preferred to raster-based systems for describing discrete features. The use of vector structure is essential in applications where boundary locations play the most important role in geographical analysis, the description of coordinates must be precise, and the mentioned locations well known.

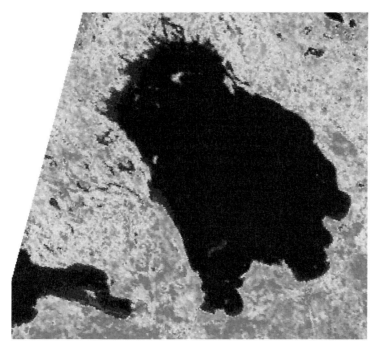

Fig. 5.1. The AVHRR (second channel) image of the area around Lake Ladoga, received on 18 August 1995. The image has been placed into GIS IDRISI.

To clearly show the difference between raster and vector representation of geographical data we display two pictures of Lake Ladoga, which is located in the north-western part of Russia. Figure 5.1 represents the image of the area, taken by the second channel of the Advanced Very High Resolution Radiometer (AVHRR), carried by NOAA satellite. The image has been placed into raster GIS IDRISI and registered according to Mercator's projection (because of registering the image has a shape of irregular rectangle). The special palette has been designed to make the image look like a customary map. Figure 5.2 shows the vector map of the same area, displayed by means of ArcView GIS on the base of Digital Chart of the World (DCW) data, which was tested and updated using ARC/INFO GIS. The map is composed of two layers—Political/Oceans (PONET) and Drainage (DNNET) of DCW for the region around Lake Ladoga.

The spatial resolution of the satellite image shown in Fig. 5.1 is only 1.1 km. It complicates mapping linear objects (e.g. rivers) and small objects (e.g. small lakes), which are clearly seen on the vector map, shown in Fig. 5.2. On the other hand, the satellite image shows the actual complexity of land-cover characteristics in the area. To tell the truth researchers seldom realize the full potential of remote-sensing data. The development of retrieval algorithms for evaluating environmental characteristics on the base of satellite measurements is a key problem in utilizing remote-sensing data.

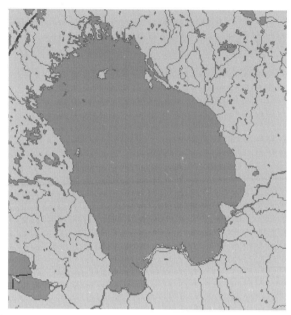

Fig. 5.2. Map of the area around Lake Ladoga, Russia, produced using ArcView GIS. Geographical coordinates: 60°N–62°N; 29.5°E–33°E. Components of the ArcView project: PONET coverage; DNNET coverage.

Before geographic data can be used in a GIS, the data must be converted into a suitable digital format. The process of converting data from paper maps into computer files is called digitizing. The result will be a vector file(s). If remotely sensed images are used as input data, GIS should supply the possibility of transforming satellite images into the particular GIS format, or, if it does not have such an option (probably these data are rarely entered into GIS) a special application should be developed.

Modern GIS technology can automate this process fully for large projects using scanning technology; smaller jobs may require some manual digitizing (using a digitizing table). Today many types of geographic data already exist in GIS-compatible formats and it is one more reason to use GIS technology. These data can be obtained from data suppliers and loaded directly into a GIS. Nowadays, all information regarding available geographical data can be found on the Internet. Some data is readily obtainable on-line. As an example of information on available digital data we place the view of Internet Bartholomew Digital Data Home Page in Fig. 5.3 (http://www.harpercollins.co.uk/maps). At that home page one can find links to demos, price lists and other useful information on available data.

Most likely, data types required for a particular GIS project will need to be transformed or manipulated in some way to make them compatible with the system. Very often, required geographic information is available at different scales: for example, for studying wetland pollution and its influence on water quality in the lake

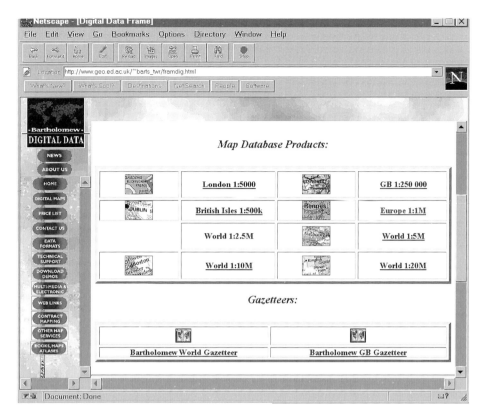

Fig. 5.3. Example of the home page of a data supplier. Vector maps available from Bartholomew Digital Data.

we have detailed wetland descriptions; less detailed water basin boundaries; and pollution sources at a regional scale. Data can also be presented in different projections. Before this information can be integrated, it must be transformed to the same scale (degree of detail or accuracy) and projection. This could be a temporary transformation for display purposes or a permanent one required for analysis. GIS technology offers many tools for manipulating spatial data and for weeding out unnecessary data.

The management of geodata collected within GIS becomes an important issue if the project needs a lot of data to be processed. For small GIS projects it may be sufficient to store geographic information as simple files. However, when data volumes become large and the number of data users becomes more than a few, it is often best to structure data as a database and to use a database management system (DBMS) to help store, organize and manage data. DBMS is computer software for managing a database. It has been in use much longer in the computer-using community than has GIS. It does not have the analytic and visualization tools common to GIS.

There are many different schemes of DBMSs to be constructed, but in GIS the relational structure has been the most useful. In the relational design, data are stored conceptually as a collection of tables. Common fields in different tables are used to link them together. This surprisingly simple scheme has been so widely used primarily because of its flexibility and widespread use in applications both within and outside GIS. By creating a shared database, one team can benefit from the work of another—data can be collected once and used many times.

5.2   GIS FUNCTIONS

5.2.1   Visualization

Probably, the human eye and brain are the most powerful tools in assessing spatial data. Automated analytical techniques cannot always take into account all information associated with map/image features which can be essential for their explication. It is especially true in the interpretation of satellite imagery, which is quite often best performed by an expert who is familiar with the geographical area under study and who has got basic knowledge on the significance of image features in terms of physical, biological or other characteristics of the environment.

Moreover, visualized GIS output (the end product of performed analysis, map or graph data), being very important for research process, can also be used in support of the decision-making practice. It can assist in tasks such as presenting information for environmental impact assessment, executing sustainable development planning, helping resolve fishery problems, etc.

Maps are very efficient at storing and communicating geographic information. While cartographers have created maps for millennia, GIS provides new tools to extend the art and science of cartography. Map displays can be integrated with reports, making them more clear and succinct, with three-dimensional views, photographic images and other output such as multimedia.

5.2.2   Query and analysis

GIS provides both simple point-and-click query capabilities and sophisticated analysis tools to provide timely information to researchers and analysts. GIS technology really comes into its own when used to analyze geographical data to look for patterns and trends and to undertake "what if" scenarios.

Once you have a functioning GIS containing your geographic information, you can begin to ask simple questions such as

- What is the area of lake's basin?
- What is the length of shoreline?
- Where is the biggest nitrogen/phosphorus concentration in the lake?
- What types of soils are located in the drainage basin?
- How many towns with the population more that 500 000 people are located in the lake's basin?

And analytical questions such as

- How is the distribution of water temperature related to the bathymetry in the lake?
- How has the distribution of chlorophyll *a* concentration changed over time?
- What is the dominant soil type in the basin?
- If we build a new plant in the location (x, y) how will water quality be affected?
- What changes are supposed to occur in water parameters if we place a purification factory/hydropower plant in the location (x, y)?

The important part of GIS intelligence are "AND", "OR", and "NOT" logical operators. GIS uses them to answer questions of the type "Does the studied object meet the selected condition?" Spatial analysis is an operation that examines geographical data with the intent to extract or create new data that fulfills some required condition or conditions. It includes such GIS functions as *polygon overlay* and *proximity analysis* as well as the concepts of *contains*, *intersects*, *within* or *adjacent*.

- How many populated places lie within 500 m distance from the shoreline of the lake?
- What is average zooplankton biomass in the lake at 0–10 m depth in July?

To answer such questions, GIS technology uses a process called buffering to determine the proximity relationship between features. To carry out this operation an area or zone of a given distance is generated around geographical objects. Buffers are user- defined or can be generated for a set of features based on those features' attribute values. The resulting buffer zone forms the field representing the area that is within the specified buffer distance from the object.

Overlay analysis is a spatial operation that merges overlapping polygons from two different thematic layers to analyze those intersected areas or to create a third layer of new polygons. For example, overlay of a map of iron-rich soil type and a watershed area will show the area of a river/lake basin, which supplies iron to the water.

Geographic operators are used to select objects on the basis of their spatial relationship to some other object. MapInfo GIS includes, for example, the following geographic operators:

Contains	Object A contains object B if B's centroid is anywhere within A's boundary.
Contains Entire	Object A contains entire object B if B's boundary is entirely within A's boundary.
Within	Object A is within object B if its centroid is inside B's boundary.
Entirely Within	Object A is entirely within object B if A's boundary is entirely within B's boundary.
Intersects	Object A intersects object B if they have at least one point in common.

5.2.3   Classification

Classification is a technique, which is usually applied to remote-sensing images, i.e. raster data, with the aim of generalizing data. Strictly speaking, it is the component of the Image Processing module of a GIS. It is a procedure whereby data cells are assigned to one of a broad group of surface cover classes according to the values of the specific reflectances found at that place. Classification attempts to imitate the activity of the human interpreter, who subdivides imagery into a series of classes, based on experience and on the requirements of the project. The human interpreter uses color, texture and context to identify surface cover types. In spite of the fact that computer classifiers have a strong mathematical base, they face the problem that the rules they use cannot take into account the complexity of nature in all details.

Unsupervised classification is a technique for computer interpretation of remotely sensed imagery, which distinguishes typical patterns in reflectance data without any prior knowledge of the image. The digital values of each pixel are examined and the image is sub-divided into a pre-set number of classes. Some image classification procedures allow the user to select the number of classes, or the amount of difference in reflectances there should be between classes.

Supervised classification is a technique where the operator interactively trains the computer, informing it that a particular group of pixels in the image represents a specific cover type of the surface, and the computer searches for pixels of similar spectral characteristics. Of course, the training area should be as homogeneous as possible and it should be representative of the class which the system is required to identify. Most image processing systems allow at least two types of supervised classification. The simplest classifier defines the mean and range of digital values for each class, based on training areas, and then classifies all pixels lying within these ranges as belonging to the class. A more sophisticated method of classification uses statistical techniques to assign each pixel to a class. For example, the maximum likelihood classifier is based on the probability density function associated with a particular training area signature. It also has the ability to incorporate prior statistical knowledge using Bayes' theorem. This knowledge is expressed as a prior probability that each class exists. It can be specified as a single value applicable to all pixels, or as an image expressing different prior probabilities for each pixel.

All described classifiers are known as hard classifiers, while modern GISs suggest *soft classifiers* to process imagery. Unlike hard classifiers, soft classifiers defer making a definitive judgement about the class membership of any pixel in favor of producing a group of statements about the degree of membership of that pixel in each of the possible classes. Like traditional supervised classification procedures, soft classification uses training area information for the purpose of classifying each image pixel. However, unlike traditional hard classifiers, the output is not a single classified surface cover map, but rather, a set of images (one per class) that express (according to your choice) either the belief or plausibility that each pixel belongs to each class.

Accuracy of classification is higher for spectrally more homogeneous images displaying low intra-class variability. Classification can be improved by using multidate imagery. Unsupervised clustering is more appropriate in areas of complex distribution of surface parameters, but good-quality ground data is needed to interpret the results. Accuracy can be significantly improved by pre-classification segmentation of imagery in order to exclude unwanted areas from the classification. The image can be segmented by digitized contour mask, or by use of intrinsic properties of the satellite data (Legg, 1994).

5.2.4   Time-series analysis

The detection of change of limnology-related parameters over time can be performed fairly effectively with remotely sensed imagery, if parameters of interest can be retrieved from satellite data with substantial accuracy. The task may be formulated to monitor changes in chlorophyll concentration in a lake, or to study water-temperature changes. It should be taken into account that a change in digital values between two dates of imagery of the same area can be caused by solar elevation or atmospheric differences. That means that images must be corrected for the mentioned effects to guarantee precise comparison.

5.3   DATA

As already stated, data is the main component of any GIS. Fortunately, volumes of existent geographic vector data are readily available. Main GIS producers have established a partnership with leading commercial data vendors to provide information in a format compatible with their products. For example, ESRI's GIS Store and ArcData Online both offer a convenient way to get the most popular geographic data (http://www.esri.com). Recently issued by Bartholomew Digital Data, UK, CD ROM with complete up-to-date vector data set for the whole of Europe contains data at a scale of 1 : 1 000 000 (see Fig. 5.3). It covers 40 different themes and provides data in two formats—compatible with ESRI's ArcView and with MapInfo. Moreover, a variety of digital geodata comes bundled with GISs to help users get started quickly.

Unfortunately, the problem of vector data quality and accuracy arise after the computer-using community have gained experience in using available digital maps. Most geographic data we use at present is derived not from direct measurements in the real world, but from representations of that world in existing maps. These maps can be considered, for many reasons, as artistic interpretations of reality rather than accurate representations of the Earth's surface (Barr, 1997). Additional mistakes arise during the process of digitizing existing maps. A piece of Digital Chart of the World, available at very low price, is shown in Fig. 5.4. This map represents the area between the Baltic Sea and Ladoga Lake, Russia and it shows some mistakes which were made during the digitizing process (they are indicated by pointers and numbers). Mistakes marked by the numbers 2, 4, 5 and 6 are obvious. Number 3 indicates an irrelevant feature, which does not correspond to any object on the

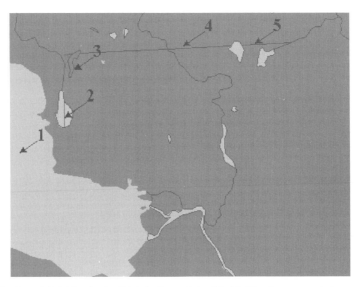

Fig. 5.4. Mistakes of digitizing in the Digital Chart of the World. The shown area is between the Baltic Sea and Lake Ladoga.

surface. Number 1 indicates the mistake caused by human activities in the area in the last decade. Now there is a dam between the island and the land.

Usually, making geographical analysis or other investigations, people presume the geodata they use is correct. However, defining correctness' is not an easy matter. It has become clear that there cannot be any absolute standard of correctness, and that geographic data quality can only be judged in relation to the applications to which that data will be put. "Fitness for purpose" becomes the fundamental criterion for determining data quality; no absolute completeness or precision is taken into consideration (Barr, 1997).

Data suppliers face a conflict of interest. In the absence of wide competition and external audits of data quality, they are seeking to supply data of a quality high enough to ensure that their customers are satisfied and are ready to buy it. They have little motivation to develop costly quality assurance procedures that may not produce additional sales. That means we need clear and applicable test routines and methodologies to check digital maps available from data suppliers. Organizations providing independent quality assurance services could act as arbitrators between data suppliers and their customers, providing an objective assessment of data quality. Remote-sensing images can be of use in correcting and updating digital maps.

Scientists carrying out environmental research have another problem with available digital data, that the data in the market which are announced to be environmental can be used in the best case for environmental control or management and there is too little data for scientific studies. In many cases (the limnology case among them) scientists have to make thematic digital maps at home, based on

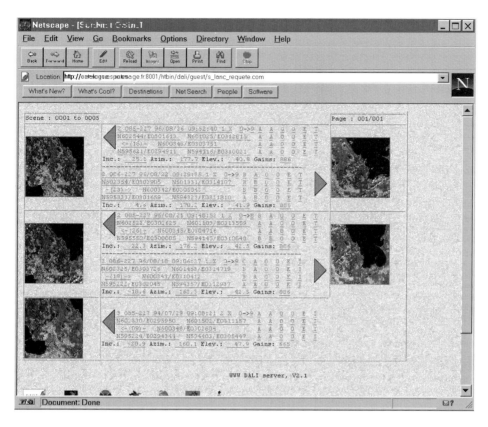

Fig. 5.5. SPOT IMAGE Home Page: quick look at the results of the query.

collected field data, on aerial photography, or on satellite images. Obviously, it is an expensive and time-consuming process.

As to remote-sensing imagery, a wide range of satellite data is available at the moment. The main advantages of satellite measurements are their wide spatial coverage (it is obviously connected with spatial resolution) and good temporal resolution (since satellites regularly return to the same position above the Earth, we can monitor time changes of environment features). It is very convenient that all main remote-sensing data suppliers have their home pages on the Internet, where there exist on-line services that let you access their catalogues of satellite data, allowing you to browse thumbnails and metadata of the scenes covering your geographic area and time range of interest, and download "quick looks". As an example, the results of the query to the SPOT IMAGE on-line catalogue are shown in Fig. 5.5. The program has found five scenes covering theuser's request.

The choice of an appropriate sensor for a given task is not always easy and indeed it is exceptional for a single sensor to have characteristics which are ideal for a specific problem. Sensor design is usually a compromise between the conflicting requirements of diverse user communities, constrained by spacecraft power and payload, data transmission rates, and by sensor/antenna possibilities and atmospheric physics. The main factors to be considered in the choice are the spatial and spectral resolution, available algorithms for environment parameters retrieval, and the cost. In some cases, particularly in areas often covered by clouds, the choice of imagery may be dictated more by availability of suitable data than by any other factor.

The problem of accuracy of remote-sensing data, which are used in environmental studies, is different from one we have discussed for vector digital maps. Here the user has a wide choice of data of different spatial resolution so he or she can order the data which covers all important details of the region under study. The main problem here is interpretation of remote-sensing data its intricacy is probably of the same level as nature's complexity. That means we need advanced algorithms for retrieving environmental parameters (including the problem of atmospheric correction) and methods for raster data classification in GIS adequate to data complexity.

5.4   INTERNET AND GIS

5.4.1   Internet Map Servers

The Internet has made a revolution in connection possibilities and data availability. Until recently, it was mostly used to send and receive mail messages, exchange files and access data on remote computers. This was made possible by a number of protocols, such as TCP (Transmission Control Protocol), Telnet, FTP (File Transfer Protocol) and others. Generally speaking, these were text-based and unsuitable for spatial query and analysis. Advent of the World Wide Web and Hypertext Markup Language (HTML) has created a graphical browsing environment for the Internet and Internet Map Servers (IMSs) have been developed to provide a means of sharing spatial information over the Net. Different GIS vendors implement the new technology developing IMSs which are compatible with their GIS products (Toon, 1997b). The Web links of the most popular IMSs are as follows:

ESRI ArcView Internet Map Server   —http://www.esri.com/;
MapInfo ProServer   —http://www.mapinfo.com/;
Autodesk Map Guide   —http://www.mapguide.com/.

An IMS works in a client/server framework, the Web browser being the client. Browsers (such as Netscape Navigator or Microsoft Internet Explorer) work by connecting to a remote server via the Internet, requesting information and then formatting that information appropriately for viewing on a local machine. This functioning is due to hypertext transfer protocol (HTTP). HTTP (Web) servers wait for requests from HTTP clients (browsers) and then return the specified information, offering mapping services in IMS case. If the client performs processing, the machine hosting a browser must be powerful enough.

Modern Web browsers support functions that allow users to process data, either on the client machine or on the server. For additional functionality HTML is customized using a variety of browser scripting languages (JavaScript, Jscript or VBScript). Mostly Java applets and plug-ins are used, Java being an object-oriented programming language with built-in support for the Web. The power of Java applets (programs that execute within a browser) and applications (programs that run on a machine without a host browser) is that they are written independently of hardware and operating systems—any machine that supports Java can run a compiled Java program and we can have the same functionality on a wide range of machines. Many spatial Java applications are being developed to use with IMS. Contrarily, plug-ins are applications that are compiled for a particular platform to extend the functionality of the Web browser, allowing it to handle new data types. They are tightly integrated with the functionality of the browser (Toon, 1997a).

5.4.2   GIS on-line for the Great Lakes

The Great Lakes Information Network (GLIN) has been organized on the Internet, its Web link being http://www.great-lakes.net. It is a partnership that provides one place on-line for people to find information relating to the binational Great Lakes region. GLIN has become a necessary component of informed decision making, and a trusted and reliable source of information for those who live, work or have an interest in the Great Lakes region. Appreciated across the Great Lakes region and around the world, GLIN offers a wealth of data and information about the region's environment and economy, tourism, education, and more.

Governments in the Great Lakes region have built a legacy of cooperation since the Boundary Waters Treaty of 1909. They collaborate through meetings, telephone conferences, newsletters and joint projects. At the same time, the shelf life of information has decreased, while the volume, diversity and need for quick access have increased dramatically. Conventional communication methods alone are no longer sufficient. So, the need to use the World Wide Web arises and GLIN has been established.

GLIN went on-line in July 1993, with leadership from the Great Lakes Commission and technical support from CICNet. It is supported by many organizations; details can be found by visiting their Web site. GLIN has an Advisory Board which oversees network content development, assists in training new GLIN partners, promotes GLIN membership and use at its meetings and conferences, and recommends regional goals for the project.

The GLIN Index of available topics offers quick access to subject matters through an alphabetical subject listing of all GLIN documents. Most of the topics and terms are hyperlinks to corresponding documents. In addition, the index also includes references to lead users to relevant information. For example, the following themes can be found there

- Weather and Climate
- Coastal Zone

- Economic Development
- Pollution, including
    ⇒ Contaminated Sediments
    ⇒ Oil and Hazardous Materials
    ⇒ Toxic Contaminants
    ⇒ Wet Weather.

The Great Lakes GIS Online Project, governed by the Great Lakes Commission, is GLIN's first step toward providing quick and efficient access to accurate Great Lakes spatial data. This project is aimed at providing on-line mapping capability for a variety of consistent spatial data layers covering the Great Lakes Basin and to establish a solid foundation for inter-agency spatial data sharing and collaboration across the binational region. A lot of work has to be done to gain this aim.

Nowadays (in 1998) Great Lakes GIS on line (http://www.great-lakes.net/gis/glgis.html) has two pilot areas: Ohio Lake Erie Commission pilot area and Minneapolis/St. Paul pilot area (see Fig. 5.6). The first pilot area merges spatial data from the Ohio Department of Natural Resources and the Ohio Environmental Protection Agency. This pilot area currently covers the Ohio counties of Lucas, Fulton, Henry, Wood, Ottawa, and Sandusky. The second pilot area covers the Minneapolis/St. Paul metropolitan counties of Anoka, Carver, Dakota, Hennepin, Ramsey, Scott and Washington. Data are intended for use by community planners and spill responders with information on the potential resources at risk during an oil spill.

The interface of Ohio Lake Erie Commission on-line GIS is shown in Fig. 5.7. The two GISs allow you to select data layers to be mapped. You may select more than one layer.

In the Ohio Lake Erie Commission on-line GIS the following layers are available:

Fig. 5.6. Great Lakes region with two pilot areas.

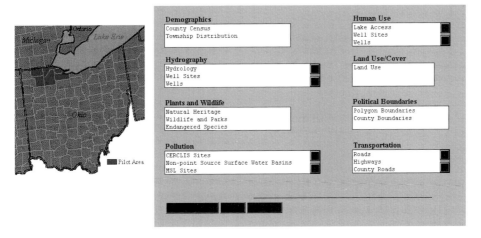

Fig. 5.7. The menu for choosing data layers of Ohio Lake Erie Commission on line GIS. The pilot area is shown to the left of the menu.

- Demographics
    - ⇒ County Census
    - ⇒ Township Distribution
- Human Use
    - ⇒ Lake Access
    - ⇒ Well Sites Wells
    - ⇒ Water Supply (1)
    - ⇒ Water Supply (2)
    - ⇒ Water Impoundments
    - ⇒ Non-point Source Surface Water Basins
- Hydrography
    - ⇒ Hydrology Well
    - ⇒ Sites Wells
    - ⇒ Flood Hazard
    - ⇒ Water Impoundments
    - ⇒ Surface Water Basins
    - ⇒ Wetland Distribution
- Land Use/Cover
    - ⇒ Land Use
- Plants and Wildlife
    - ⇒ Natural Heritage
    - ⇒ Wildlife and Parks
    - ⇒ Endangered Species
- Political Boundaries
    - ⇒ Polygon Boundaries
    - ⇒ County Boundaries

- Pollution
    - ⇒ CERCLIS Sites
    - ⇒ Non-point Source Surface Water Basins
    - ⇒ MSL Sites
    - ⇒ Spills 1994-95
    - ⇒ National Priority List
- Transportation
    - ⇒ Roads Highways
    - ⇒ County Roads
    - ⇒ Municipal Roads

The result of processing the query to this GIS is shown in Fig. 5.8 as it appears on the monitor of a client-computer. Two layers, which were selected to be mapped on-line, are "Land Use" and "Wetland Distribution". The picture, shown in the background, represents the map composed of two selected layers. The interface of Great Lakes GIS on-line is typical IMS interface, written with the use of Java applet Map Cafe. The user is able to zoom the map in and out, to use the find option and to

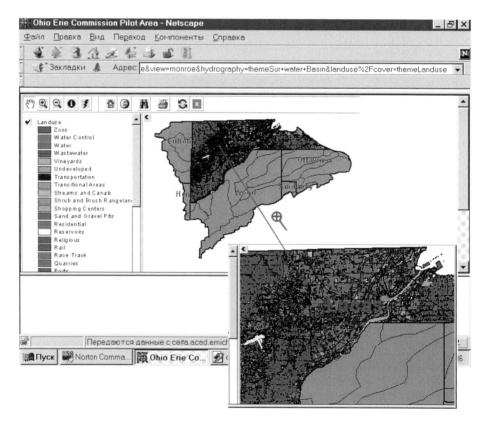

Fig. 5.8. Two layers of Lake Erie GIS "Land Use" and "Wetland Distribution" have been chosen and displayed together. A part of the map is zoomed and placed in the foreground.

do some other actions. To the left of the map the legend is placed. In the foreground of Fig. 5.8 you can see a part of the initial map, which has been zoomed in.

The other partner of the Great Lakes GIS Project, the Minneapolis/St. Paul Area Contingency Planning Subarea, has a slightly different view on information content and principles of presenting data in Great Lakes GIS on-line. Their GIS includes the following coverages:

- Human Use
  ⇒ Boat
  ⇒ Access/Marinas
  ⇒ Water Intakes
  ⇒ Navigational Locks and Dams
  ⇒ Fixed Petroleum Facilities
  ⇒ Pipelines
  ⇒ Pipeline Crossings
- Hydrography
  ⇒ Water Intakes
  ⇒ Hydrography (Line)
  ⇒ Hydrography (Polygon)
  ⇒ 1 : 100 000-Scale Hydrography
  ⇒ Rivermiles
  ⇒ Shorelines
  ⇒ Navigational Locks and Dams
- Plants and Wildlife
  ⇒ Managed Areas (Polygons)
  ⇒ Managed Areas (Lines)
  ⇒ Managed Areas (Points)
- Boundaries
  ⇒ Managed Areas (Polygons)
  ⇒ Managed Areas (Lines)
  ⇒ Tribal Lands
  ⇒ Shorelines
- Pollution
  ⇒ Fixed Petroleum Facilities
  ⇒ Pipelines
  ⇒ Pipeline Crossings
  ⇒ Railroad Crossings
  ⇒ Highway Crossings
- Transportation
  ⇒ Boat Access/Marinas
  ⇒ Navigational Locks and Dams
  ⇒ Rivermiles
- Miscellaneous
  ⇒ Index to Tiles
  ⇒ 1 : 100 000 Scale Map Index

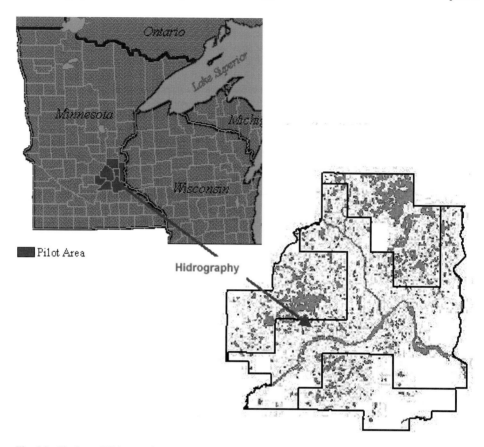

Fig. 5.9. The layer "Hidrography 1 : 100 000" of Mineapolis/St. Paul GIS. The location of the pilot area is shown in the upper left corner.

  ⇒ 1 : 24 000 Scale Map Index
  ⇒ Shoreline
 • Digital Raster Graphics
  ⇒ DRG 1, 2, . . . , 16 = 16 images

As an example of collected data, the layer "1 : 100 000 Scale Hydrography" has been displayed and is shown in Fig. 5.9.

5.5   CONCLUSION

It is of great importance to the limnology community to have an adequate instrument to study the complex system of a lake and its surrounding. Without any doubt GIS can be such an instrument.

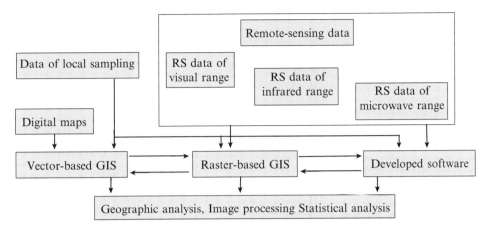

Fig. 5.10. General scheme of limnological GIS.

The list of tasks to be fulfilled before starting system development includes the following steps:

- to analyze what geographical data are available for the project;
- to understand what additional data need to be obtained;
- to find possible sources of the necessary data.

The general scheme of integrated GIS, which can be suggested for limnological studies, is shown in Fig. 5.10.

It can be said in advance that the most wealthy source of spatial distribution of environmental parameters is satellite data. From images in the visual range of the spectrum one can derive land-cover or land-use maps, vegetation maps (SPOT, Landsat/TM, Resurs/MSU-E) and maps of chlorophyll distribution in the water (SeaWiFS). Measurements in this range are also very important in testing and updating existing digital maps. Infrared images (NOAA/AVHRR, ERS/ATSR) are usually used for temperature calculations. SAR data can be utilized for spill detection, for creating digital elevation models and for estimating the level of the pollution of the environment.

Development and application of adequate algorithms for remote-sensing data processing is of key importance here. This demands good understanding of physical processes taking place in the natural system, which is impossible without data of local sampling. Successful environmental parameter retrieval or image classification can be conducted only if sufficient field data is available to develop, calibrate and tune algorithms of remote-sensing data processing and to validate the results.

REFERENCES

Barr R. 1997. Never mind the quality—feel the width! *GIS Europe*, **6**, Issue 9, 14–15.

Encyclopedia of Earth System Science, Nierenberg W.A. (ed.), Vol. 2. 1994. San Diego, CA: Academic Press, 620 pp.

Legg C.A., 1994. *Remote Sensing and Geographic Information Systems.* Wiley-Praxis series in remote sensing. Chichester, UK: J. Wiley & Sons/Praxis Publishing, 166 pp.

Toon M. 1997a. The roundabout route to convergence. *GIS Europe*, **6**, Issue 11, 18–21.

Toon M. 1997b. The world by your window. *GIS Europe*, **6**, Issue 12, 38-41.

Afterword

Tracing back the developments in limnology and environmental remote sensing over the last decades, it becomes evident that both sciences have matured spectacularly since the early 1950s. Based on the incontestable assets of traditional discovery-descriptive limnology (which, in the first place, are extensive field observations, laboratory analyses and long time series of data on the principle limnological characteristics and processes), contemporary limnology is steadily evolving into a science whose principal objectives are no longer focused almost exclusively on studying separate limnological elements or processes and revealing appropriate alterations occurring to them, but rather on investigating an inland water body as a semi-closed natural environment including the water body *per se*, the catchment and the atmosphere above them. This systematic approach, supplemented with an extensive development of mathematical/numerical modeling of processes going on in the catchment as well as in lake biota, hydrochemistry, thermohydrodynamics, and, finally, in atmospheric transboundary transport, opened a way of forecasting lake ecology status changes that could be expected under different scenarios. This, in turn, enables modern limnologists to come up with appropriate recommendations for decision makers at local and federal levels responsible for sustainable development of natural environments.

Environmental remote sensing has also experienced important progress since the early 1960s. This progress relates both to methodology and instrumentation capabilities. Additionally, powerful impetus has been received from recent fundamental studies of electromagnetic radiation transfer in the atmosphere and natural waters as well as its interaction with the Earth's surface. Based on these accomplishments, remote sensing, both air- and spaceborne, has become an indispensable tool for local, regional and global environmental monitoring.

This monograph seems to be a faithful reflection of the above achievements in both sciences. But it also was supposed to show that an organic fusion of purely limnological and remote-sensing methods furnishes a qualitatively new type of integrated information enabling scientists and, ultimately, environmentalists to considerably extend the limits of traditional limnology. At the same time, this fusion provides unprecedented means to investigate limnological environments as elements of a complex hierarchical system whose functioning is regulated through a multitude of direct and feedback interactions on various temporal/spatial scales. Climatological effects on lakes and *vice versa* (i.e. of lakes on the ambient environments) are extensively illustrated in this book.

At the same time, the authors tended to explicitly indicate throughout the book the limitations and shortcomings inherent in both sciences and their associated technologies as well as the urgent methodological, technological and logistical problems awaiting their solution.

As far as environmental remote sensing is concerned, the following needs can be listed without attempting to prioritize them. It seems appropriate to start with the necessity of further improvement of physical models simulating the interaction of natural media prone to remote sensing (i.e. natural water bodies, the atmosphere, the water/air and land/air interfaces, etc.) with the electromagnetic radiation in optical and microwave spectral regions. This is particularly mandatory for hydro-optical models of inland waters which due to their exceptional complexity present a real challenge for remote sensing in the visible spectrum. It seems highly recommendable to launch a program on a national or international level aimed at compiling an atlas of specific inherent properties of major optically active components indigenous to lakes and water storage reservoirs prone to remote surveillance.

On the basis of such models, a considerable amelioration of retrieval algorithms might be expected in the future. These algorithms could then be less area-specific and more readily translatable to wider regions of surveillance; nevertheless, it is hard to expect them to be universally applicable in view of the dramatic variability in biotic community compositions, suspended mineral nature and microphysics in various latitudinal zones and landscapes.

The same requirement holds equally for the microwave spectral region: being considerably less vulnerable to the influence of the intervening atmosphere, passive and active microwave sensors become an indispensable element of many environmental observational programs. At the same time, the presently available models of brightness and scattering properties of natural surfaces in the microwave spectral region still leave much to be desired.

Efficient removal of the optical impact of the atmosphere on airborne and satellite data in the visible and infrared portions of the spectrum is another pressing problem. Being generally highly absorptive in the visible spectrum, inland waters appear dark in space imagery. It implies a low level of the legitimate signals that should be detected by a remote sensor against a background signal composed mainly of atmospheric (path) radiance. Hence, requirements for precision of atmospheric correction techniques in the case of surface waters are very stringent. At the same time, atmosphere removal methods designed so far have been intended for open ocean areas where reference pixels from clear water could invariably be found, a prerequisite which obviously could not be easily met in many, if not almost all, surface waters. Unfortunately, at the present time this problem appears rather formidable. Before any principally new approaches appear, it seems to us inevitable to complement space optical data with concurrent ground-based measurements of atmospheric spectral transparency. Moreover, because many of the advanced retrieval techniques use remote-sensing reflectance as input value, it necessitates concurrent ground-based spectral measurements of incident radiation.

In this context it seems appropriate to mention the following logistical problem. The space data presently provided by national and international space agencies are not always an end-product; in fact they often require additional processing by the user. In the future it is very important to integrate the entire processing chain of a given sensor, *viz.* navigation related corrections, radiometric corrections, classification of land, cloud and water pixels, atmospheric corrections, inference of water-quality parameters, etc. The final stage of this processing should be creation of a video-server frame allowing a wide spectrum of users to exploit this information for scientifically, economically and socially important applications.

Analysis of world-wide activities in the area of remote studying of inland basins indicates that up to now the major efforts in each individual research/application case were mainly confined to concurrently obtaining information on rather limited limnological parameters. However, with some of the presently operating satellite sensors, and particularly with such forthcoming sensor packages as Envisat, a truly synergistic approach can be attainable. It certainly should be thoroughly exploited to furnish simultaneous data on a wealth of limnological parameters which conjointly will provide a more exhaustive characterization of a basin under investigation. This is further important in terms of a truly comprehensive analysis of remotely sensed data: first, the spatial and temporal distributions of different limnological parameters should be mutually consistent (and it could be used as a criterion of adequacy of retrievals), and second, they could also be used as inputs for relevant limnological models usually requiring for their performance an array of input parameters. Nowcasts and forecasts generated by modeling, in turn, would be a further test for the correctness of remotely derived data. However, this semi-closed approach poses some stringent prerequisites concerning, e.g. the compatibility of temporal and spatial scales of data provided by remote sensing and ground-based measurements.

In our opinion, serious attention should also be given to establishing reliable relationships between remotely sensed surficial or top-layer parameters (water temperature, phytoplankton/chlorophyll concentration, etc.) with bulk water counterparts. Such relationships should be basin- and season specific, and, apart from purely statistical data, numerical modeling can greatly facilitate this task. Understandably, establishment of such relationships would be crucially important for thermohydrodynamic modeling, modeling of primary productivity, and many other applications.

Limnology, in turn, is nowadays strongly challenged with furthering the art of mathematical modeling not only at the level of an individual process (physical, chemical, biological) restricted to a particular lake zone or level (littoral, pelagic, surficial, epilimnetic, hypolimnetic, air/water and bottom/water interfaces, etc.), but also at a level of interacting processes or biotic interrelationships. This applies equally to the lake watershed. A truly adequate simulation of a semi-closed lake system should be coupled with atmospheric transport and local/regional climate models.

Notwithstanding the limitations inherent in remote-sensing feasibilities, contemporary and future limnological modeling should be tailored with a much more serious consideration of widely using remotely derivable synergistic parameters as input and validation data. Realization of such models will inevitably require adjustment/harmonization of the time and space scales of the models themselves on the one hand, and, as mentioned above, integrated air-/spaceborne and ground based data on the other hand.

Most certainly, it will also require not only a further betterment of satellite but also ground/ship-based instrumentation which for many applications should be designed (and not "borrowed" from oceanographers, as often happens at present) to meet the specific limnological requirements discussed throughout this book.

As was explicitly, and hopefully convincingly, shown in Chapter 1, large and deep lakes can serve as physical models of oceans, and as such can open new horizons in oceanography. In fact, it is an appeal to oceanographers to seriously consider this option. It is obvious to us that the exploration of this new field would be mutually beneficial and cross-fertilizing both for oceanography and limnology.

Finally, while being perfectly aware of the futility of attempts to exhaustively list the requirements/recommendations for limnology of the 21st Century, we nevertheless believe it is worthwhile mentioning the issue of concerted activities on national and international scales. Unlike oceanic and more recently coastal zone studies, limnological research remains by and large considerably less integrated/coordinated by large-scale and amply funded national/international projects. Apart from a purely financial point of view, such projects prove to greatly stimulate rapid development of advanced laboratory, ground-, air- and spaceborne instrumental facilities, raw and value-added data stardardization/harmonization, establishment of efficient data flows and creation of meta-databases. Although achieving these goals implies the necessity of overcoming inevitable logistical and in many cases national/international legislative problems, this is the only way that limnological science can successfully develop in the next century.

Subject index

Page numbers refer to the first page on which the entry is mentioned/discussed. Some biological organisms are indexed as are the higher taxonomic categories of those organisms that are not indexed.